NOVEL APPROACHES AND THEIR APPLICATIONS IN RISK ASSESSMENT

NOVEL APPROACHES AND THEIR APPLICATIONS IN RISK ASSESSMENT

Editor

Keith Marcell

scitus academics

Novel Approaches and Their Applications in Risk Assessment
Edited by **Keith Marcell**

ISBN: 978-1-68117-618-5
Library of Congress Control Number: 2016946476

SCITUS Academics LLC,
616, Corporate Way, Suite 2, 4766,
Valley Cottage, NY 10989

www.scitusacademics.com

Notice

Printed in United States of America on Acid Free Paper

PREFACE

Risk assessment is to determine qualitative or quantitative risk estimates associated with a well-defined situation and a threat. An acceptable risk is a threat that is generally understood and tolerated because the cost or possibly difficulty of a good countermeasure for the associated vulnerability exceeds the expectation of losses. Risk assessment involves some important steps, such as problem identification, risk assessment and assessment review. Variations such as risk as type and severity of response, with or without a probabilistic context, are included in the assessment of health risks. Risk assessment consists of an objective risk analysis that clearly takes into account and presents assumptions and uncertainties. Part of the problem in risk management would be that it is difficult to measure both the number of risk assessments-potential loss and probability of occurrence.

Novel Approaches and their Applications in Risk Assessment are compiled to provide the latest information on risk assessment approaches and their efficiency. The materials presented also cover topics from environmental quality to the protection of human health.

TABLE OF CONTENTS

1

A QUANTITATIVE MODEL FOR ESTIMATING RISK FROM MULTIPLE INTERACTING NATURAL HAZARDS: AN APPLICATION TO NORTHEAST ZHEJIANG, CHINA

Baoyin Liu[1] , Yim Ling Siu[1] , Gordon Mitchell[2]

[1]School of Earth and Environment, University of Leeds, Leeds LS2 9JT, UK

[2] School of Geography and Water@leeds, University of Leeds, Leeds LS2 9JT, UK

ABSTRACT

Multi-hazard risk assessment is a major concern in risk analysis, but most approaches do not consider all hazard interactions when calculating possible losses. We address this problem by developing an improved quantitative model—Model for multi-hazard Risk assessment with a consideration of Hazard Interaction (MmhRisk-HI). This model calculates the possible loss caused by multiple hazards, with an explicit consideration of interaction between those hazards. There are two main components to the model. In the first, based on the hazard-forming environment, relationships among hazards are classified into four types for calculation of the exceedance probability of multiple hazards occurrence. In the second, a Bayesian network is used to calculate possible loss caused by multiple hazards with different exceedance probabilities. A multi-hazard risk map can then be drawn addressing the probability of multi-hazard occurrence and corresponding loss. This model was applied in northeast Zhejiang, China and validated by comparison against an observed multi-hazard sequence. The validation results show that the model can more effectively represent the real world, and that the modelled outputs, possible loss caused by multiple hazards, are reliable. The outputs can additionally help to identify areas at greatest risk, and allows a determination of the factors that contribute to that risk, and hence the model can provide useful further information for planners and decision-makers concerned with risk mitigation.

KEYWORDS

Multi-hazard risk modelling, Hazard interaction, Hazard-forming environment, Bayesian network, Zhejiang

1. INTRODUCTION

Multi-hazard risk assessment (MHRA) is a major concern in risk analysis and management (Carpignano et al. 2009; Frigerio et al. 2012; Marulanda et al. 2013; Ming et al. 2015). MHRA is a relatively new field, with little MHRA research conducted before 2000, although three recognizable phases in MHRA development can be identified. Initially, research focused on multiple hazards affecting a given area through the development of a synthetic indicator. Examples included the Australian Geological Survey Organisation Cities project (Granger and Trevor 2000), the Natural Hazard Index for Mega-cities (Munich Re 2003), the World Bank's global Natural Disaster Hotspot analysis (Dilley et al. 2005), European Spatial Planning and Observation Network (ESPON's) multi-hazard approach for the enlarged European Union (Schmidt-Thomé 2006), and the Calculation of the Total Place Vulnerability Index for the State of South Carolina, USA (SCEMDOAG 2009). These synthetic indicators are effective in comparing the relative danger experienced by different areas, but give no representation of the real risk situation in those areas—that is, the approach is useful in assessing relative but not absolute losses. To address this problem, research moved from synthetic indicators to assessing integrated losses caused by multiple natural hazards in a given region and time period. This included HAZUS-MH software for Risk Assessment in the USA (FEMA 2004), the Regional

RiskScape project for New Zealand (Schmidt et al. 2011), and the Central American Probabilistic Risk Assessment Program for Latin America and the Caribbean Region (Linares-Rivas 2012). However, this research still neglected hazard interaction, hence recently, more comprehensive MHRA research was proposed to assess possible loss due to multiple hazards with a consideration of domino effects (Marzocchi et al. 2012, Eshrati et al. 2015). Nevertheless, these studies, which typify a rather small body of MHRA work, are incomplete. The interaction between different natural hazards is complex and dynamic, and only addressing domino effect hazard interaction is not enough to cover all situations. For example, hazards can occur independently without evident common cause, yet in close proximity, spatially, temporally, or both, and thus interact to elevate risk. Thus the incomplete consideration of possible hazard interactions represents a significant gap in MHRA. This paper therefore aims to develop an improved MHRA model, Model for multi-hazard Risk assessment with a consideration of Hazard Interaction (MmhRisk-HI). The model calculates the possible loss caused by multiple hazards, with an explicit consideration of interaction between different hazards.

2. MULTI-HAZARD RISK ASSESSMENT

The MHRA process is based on that for single-hazard risk assessment, with its main advance being that it puts different hazard types into a single system for joint evaluation (Armonia 2006; Di Mauro et al. 2006; Marzocchi et al. 2009; Carpignano et al. 2009). MHRA is a relatively new field, with no clear definition, but in principle, it takes into account the characteristics (frequency, magnitude) of each hazardous event, and their mutual interactions and interrelations (e.g. a hazard may occur repeatedly in time; different hazards may occur independently in the same place; different hazards may occur dependently in the same place) (Kappes et al. 2012; Marzocchi et al. 2012). The aim of MHRA is to have a holistic view of the total effects or impacts by assessing and mapping the expected social, economic and/or environmental loss due to the occurrence of all natural hazards in a given area (Dilley et al. 2005; Armonia 2006; Kappes et al. 2012; Komendantova et al. 2014).

A conceptual model for multi-hazard risk assessment is as shown in Fig. 1, based on the disaster formation process (Burton et al. 1993; Shi 1996; Wisner et al. 2004). Natural hazards are natural events that arise from a specific geophysical environment, accompanied by concentrations of energy released to produce major threats to human life or economic assets (McGuire et al. 2002; ISDR 2004). Space, time, magnitude and frequency are the basic characteristics of natural hazards (Alexander 1993; Smith 2000). The hazard forming environment is the specific geophysical environment that natural hazards arise from, including environmental factors in the atmosphere, hydrosphere, biosphere and lithosphere. These factors are the basic conditions for the occurrence of hazards (Park 1994; Shi 1996; McGuire et al. 2002). Natural hazards are then caused by the substantial departure of one or more environmental factor from their mean values, either in a positive or negative direction. Thus flood can be induced when precipitation is above the normal level, and drought when it is below. Hence, the time, space, magnitude and frequency of hazards are determined by the hazard-forming environment. Some hazards can occur in close proximity, spatially, temporally, or both, in a specific hazard-forming environment. Exposure is defined as the number, type and/or monetary value of elements that are under threat from natural hazards (Alexander 2000; Blanchard 2005; Tong et al. 2009). Vulnerability is broadly defined as the conditions, determined by physical, social, economic, and environmental factors, which determine the potential damage following exposure to hazard (Pelling 2003; Brooks 2003; Ge et al. 2013). The hazards' interaction with vulnerability and exposures can result in losses and impacts of a human, material, economic and/or environmental nature, losses that can be increased if one hazardous event interacts with another (Gill and Malamud 2014). The induced consequence also influences the stability of the hazard-forming environment and the distribution of exposures, whilst the influenced hazard-forming environment has a chance of producing new hazards. This conceptual model provides a theoretical basis for the improved MHRA model (MmhRisk-HI) construction.

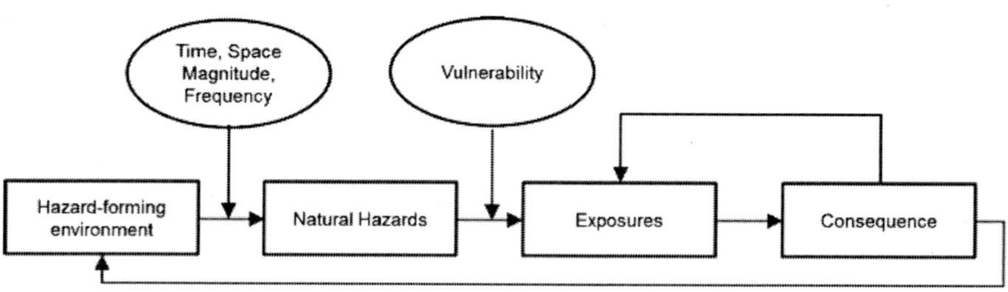

Figure 1. Conceptual model for multi-hazard risk

3. MODEL CONSTRUCTION

3.1 Framework

The structure of the MmhRisk-HI model, with its' explicit consideration of interaction between different hazards, is shown in Fig. 2. There are two main components. In the first, the hazard-forming environment is used in hazard identification and hazard interaction analysis to analyse the relationship between hazards and to calculate the exceedance probability of multiple hazards occurrence. The second component focuses on the calculation of the possible loss from multiple hazards with different exceedance probabilities. The methods used for the exposure analysis depends on the scale of the region addressed and the assessment units. A Bayesian network (BN) is used to measure the possible loss ratio (the ratio of total losses to the total value of the exposure) caused by multiples hazards with different exceedance probabilities, considering the relevant vulnerability indicators in the loss ratio assessment. Finally, a multi-hazard risk map can be drawn addressing the probability of multi-hazard occurrence and corresponding loss.

3.2 Calculation of the exceedance probability of multiple hazards

Natural hazards arise from specific hazard-forming environments, with environmental factors in each environment determining the preconditions for hazard occurrence. The hazard-forming environment can therefore be used to identify which natural hazards influence a given area. For example, proximity to a crustal plate boundary is a precondition for an earthquake. Substantial change in one or more environmental factor in a hazard-forming environment is the main reason that hazards are induced, thus these factors can be taken as triggers for natural hazards and used to analyse hazard interaction. Hazards are thus classified, into four types, according to their trigger factors (Liu et al. 2016), as follows.

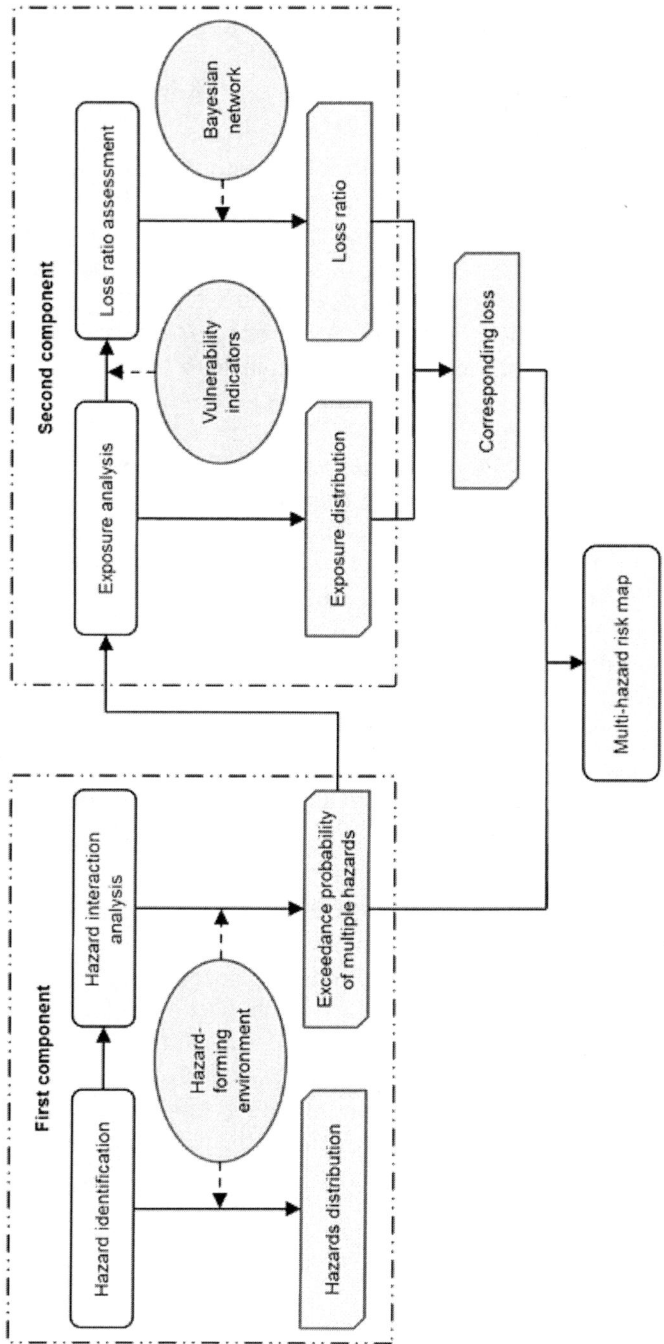

Figure 2. Framework of model for multi-hazard Risk assessment with a consideration of Hazard Interaction (MmhRisk-HI)

3.2.1 Independent relationship

The Independent relationship is where there is no evident common cause between two different hazards. This means that changes in trigger factors which induce hazard A are independent of those which induce hazard B. The relationship between these trigger factors and hazards can be expressed by Eqs. (1) and (2), whereby changes in trigger factors $t_1, t_2 \ldots t_i \ldots t_n$ are independent of changes in trigger factors $t_{n+1}, t_{n+2} \ldots t_s \ldots t_t$. Therefore, the probability of these two hazards occurring together can be calculated as Eq. (3).

$$f(p_{t_1}, p_{t_2} \cdots p_{t_i} \cdots p_{t_n}) = p(h_A) \tag{1}$$

$$f(p_{t_{n+1}}, p_{t_{n+2}} \cdots p_{t_s} \cdots p_{t_t}) = p(h_B) \tag{2}$$

$$\begin{aligned} P(A \cap B) &= p(h_A) \times p(h_B) \\ &= f(p_{t_1}, p_{t_2} \cdots p_{t_i} \cdots p_{t_n}) \times f(p_{t_{n+1}}, p_{t_{n+2}} \cdots p_{t_s} \cdots p_{t_t}) \end{aligned} \tag{3}$$

where f is the function used to calculate the probability of the change in trigger factors within the brackets occurring together, t_i is the trigger factor i, $i = 1, 2, \ldots n$, t_s is the trigger factor s, $s = n + 1, n + 2, \ldots t$, h_j is the hazard j, $j = A, B$, ptipti is the probability of the change in trigger factor i, ptspts is the probability of the change in trigger factor s, and $p(h_j)$ *is* the probability of hazard j occurrence.

3.2.2 Mutex relationship

A Mutex relationship is where the changes in trigger factors which induce hazards A and B are mutually exclusive, and so these hazards cannot occur together. The relationship between the trigger factors and hazards can be expressed as:

$$f(p_{t_{i+}}) = p(h_A) \tag{4}$$

$$f(p_{t_{i-}}) = p(h_B) \tag{5}$$

where f is the function, which is used to calculate the probability of the change in trigger factors within the brackets occurring together, t_i is the trigger factor i, $i = 1, 2, \ldots n$, t_{i+} represents the trigger factor i departure in positive direction from its mean values, t_{i-} represents the trigger factor *i* departure in the negative direction from its mean values, h_j is the hazard *j*, *j* = *A*, *B*, ptipti is the probability of the change in trigger factor *i*, and $p(h_j)$ *is* the probability of hazard *j* occurrence.

One trigger factor cannot move in two directions simultaneously, hence, the probability of these two hazards occurring together can be expressed as:

$$P(A \cap B) = 0 \tag{6}$$

3.2.3 Parallel relationship

Changes in one or more trigger factor may induce more than one hazard $A_1, A_2 \ldots A_m$ at the same time, that is, the relationship of hazard $A_1, A_2 \ldots A_m$ is parallel. This relationship between trigger factors and hazards can be expressed by Eq. (7), where hazards $A_1, A_2 \ldots A_m$ constitute a hazard group, with all hazards induced by the same trigger factor(s). Hence, the frequency and magnitude of this hazard group are determined by changes in the trigger factors, with the probability of the hazard group (hazard $A_1, A_2 \ldots\ldots A_m$) occurring expressed by Eq. (8).

$$\begin{aligned} f(p_{t_1}, p_{t_2} \cdots p_{t_i} \cdots p_{t_n}) &= p(h_{A_1}) \\ f(p_{t_1}, p_{t_2} \cdots p_{t_i} \cdots p_{t_n}) &= p(h_{A_2}) \end{aligned} \tag{7}$$

$$\begin{aligned} f(p_{t_1}, p_{t_2} \cdots p_{t_i} \cdots p_{t_n}) &= p(h_{A_m}) \\ P(A_1 \cap A_2 \cap \ldots \cap A_m) &= f(p_{t_1}, p_{t_2} \cdots p_{t_i} \cdots p_{t_n}) \end{aligned} \tag{8}$$

where *f* is the function, which is used to calculate the probability of the change in trigger factors within the brackets occurring together, t_i is the trigger factor *i*, *i* = 1, 2,...*n*, h_j is the hazard *j*, *j* = $A_1, A_2, \ldots A_m$, ptipti is the probability of the change in trigger factor *i*, and $p(h_j)$ *is* the probability of hazard *j* occurrence.

3.2.4 Series relationship

In the series relationship, hazard A induces changes in trigger factors, then the changes in these trigger factors induce hazard B. Hazards A and B are in a series relationship, which can be expressed by Eq. (9). Changes in trigger factors $t_1, t_2 \ldots t_i \ldots t_n$ induce hazard A, then hazard A changes the trigger factors $t_{n+1}, t_{n+2} \ldots t_s \ldots t_t$, which then induce hazard B. The probability of hazards A and B occurring together can thus be expressed by Eq. (10).

$$f(p_{t_1}, p_{t_2} \cdots p_{t_i} \cdots p_{t_n}) = p(h_A) \rightarrow f(p_{t_{n+1}}, p_{t_{n+2}} \cdots p_{t_s} \cdots p_{t_t}) = p(h_B) \quad (9)$$

$$P(A \cap B) = p(h_A) \times p(h_B) = f(p_{t_1}, p_{t_2} \cdots p_{t_i} \cdots p_{t_n}) \times f(p_{t_{n+1}}, p_{t_{n+2}} \cdots p_{t_s} \cdots p_{t_t} | p_{t_1}, p_{t_2} \cdots p_{t_i} \cdots p_{t_n}) \quad (10)$$

where *f* is the function, which is used to calculate the probability of the change in trigger factors within the brackets occurring together, t_i is the trigger factor *i*, *i* = 1, 2,...*n*, t_s is the trigger factor *s*, *s* = *n* + 1, *n* + 2,...*t*, h_j is the hazard *j*, *j* = *A*, *B*, ptipti is the probability of the change in trigger factor *i*, ptspts is the probability of the change in trigger factor *s*, and *p*(h_j) *is* the probability of hazard *j* occurrence.

This classification of trigger factors is useful as it helps ensure all possible relationships among hazards are considered. It fills a gap in current multi-hazard methods which to date only consider domino effects, just one type of possible hazard interaction. Based on the four part hazard interaction classification above, the probability of multiple hazards of different magnitudes occurring together can be calculated based on changes in trigger factors, with the degree of change representing the magnitude of hazards, and the probability of change representing the probability of hazards. This can be achieved using a mathematical statistics approach to define a function that determines event magnitude and frequency. For example, Grünthal et al. (2006) calculated exceedance probability-mean wind speed curves for windstorm magnitude assessment using Schmidt and Gumbel distributions (Gumbel 1958).

3.3 Calculation of the possible loss caused by multiple hazards

The second main component of the MmhRisk-HI model focuses on the calculation of possible loss caused by multiple hazards with different exceedance probabilities. This requires exposure analysis, and loss ratio assessment.

3.3.1 Exposure analysis

Exposure analysis is used to determine the spatial distribution of elements at risk (e.g. people, infrastructure). This is usually achieved using analysis of data contained in official statistical reports, or that obtained via on-site survey or remote sensing image. These data sources vary considerably in their characteristics: on-site survey data may be very detailed, but generally only exist on a very local scale, as it is time and resource intensive to collect. Conversely remotely sensed images provide wide area coverage, but the raster format means that the information conveyed is more limited in scope. Official statistics data, based on government administrative units, represent a common intermediate point, in terms of functional resolution. The exposure analysis module selects from these sources following a consideration of the study area size, and the data available for that area.

3.3.2 Loss ratio assessment

The loss ratio assessment module is used to measure the possible loss ratio for a given exposure, under conditions caused by multiple hazards with different exceedance probabilities, and then to determine how vulnerability-related indicators (physical, social, economic and/or environmental factors) influence the possible loss ratio. A BN, a probabilistic graphical model that encodes probabilistic interdependencies among a set of random variables (Jensen and Nielsen 2007), is used in this process. A BN is a good method for modelling uncertainties and interactions between related factors (Maldonado et al. 2015), and has previously been applied in risk assessment of earthquake (Bayraktarli et al. 2006), landslide (Straub 2005) and flooding (Li et al. 2010). The two key steps in the BN, discussed further below, are the construction of the BN structure and estimating the conditional probabilities of relationships within the network.

(1) Bayesian network structure

A BN is a complete model of the system of interest, comprising component variables and the probabilistic relationships between them. In this module, the BN modelling framework is constructed according to domain knowledge (e.g. Alexander 2000; Cutter et al. 2003; Villagran 2006; Schmidt-Thomé 2006). Figure 3 shows that the loss ratio, which is assumed to be a parent node of vulnerability and hazard related indicators, is the root node. Trigger factors are then used to construct the set of hazard-related indicators which represent the magnitude of multiple hazards. Indicators in economic, social, physical and environmental domains represent the sets of vulnerability-related indicators (Table 1).

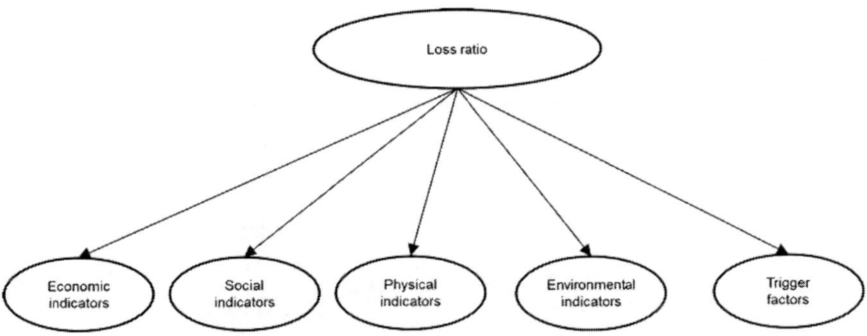

Figure 3. Generic Bayesian network framework for loss ratio assessment

Table 1 identifies some possible vulnerability-related indicators, with further details given below:

- GDP/capita: high GDP/capita indicates more economic activity is under threat of hazard events (Blaikie et al. 1994; Schmidt-Thomé 2006).

- Income of residents: high income indicates residents have more personal resources to absorb losses and speed up recovery after a disaster (Hewitt 1997; Cutter et al. 2003; SCEMDOAG 2009).

- Population density: high population density indicates more population is under threat of hazard events (Puente 1999; Pelling 2003).

- Gender ratio: females are often more vulnerable than males as they tend to have more limited education, lower income and family care responsibilities (Cutter et al. 2003; SCEMDOAG 2009).

- Age structure: children and the elderly are more vulnerable to hazard than young adults due to their limited physical strength (Cutter et al. 2000; Ngo 2001).

Table 1. Some possible vulnerability-related indicators

Domain	Indicator
Economic	GDP/capita
	Income of residents
Social	Population density
	Gender ratio
	Age structure
	Telecommunication
	Transport route
	Medical condition
	Social dependency
	Risk perception
	Warning systems
	Institutional preparedness
	Educational achievement
Physical	Technical infrastructure
Environmental	Significant natural areas
	Fragmented natural areas

- Telecommunication: high telecommunication capacity supports fast and precise hazard information transmission, thus targeted measures can be adopted quickly (Blaikie et al. 1994; Puente 1999).

- Transport route: good transport infrastructure makes it easier to evacuate people and to distribute emergency rescue and relief materials (Platt 1991; Villagran 2006).

- Medical condition: good medical services ensure wounded people get fast and effective treatment after a disaster, thus the recovery period can be shortened (Morrow 1999; Cutter et al. 2003).

- Social dependency: people who are wholly dependent on social services have very limited personal resources to absorb losses, and require more support in the post-disaster period, thus they are more vulnerable than the employed (Cutter et al. 2003; SCEMDOAG 2009).

- Risk perception: this measures an individuals' ability to discern and understand the characteristics and severity of risk from hazard events (Slovic 2000). Understanding risk promotes taking effective measures to cope with disaster, thus people with low risk perception are inherently more vulnerable (Armas 2006; Smith 2013).

- Warning system: a disaster warning system gives early warning to those at risk and promotes preparedness, mitigating against disaster (Zschau and Küppers 2003).

- Institutional preparedness: this comprises regulations or procedures (e.g. emergency response plan) developed to mitigate against potential disasters (Haque 2000; Schmidt-Thomé 2006).

- Educational achievement: higher education levels indicate people can better understand information about hazard events and take more effective measures to cope with disaster (Cutter et al. 2003; SCEMDOAG 2009).

- Technical infrastructure: this indicates presence of facilities needed to respond to hazard events (e.g. fire trucks, steamboats, helicopters). Good technical infrastructure makes it possible to evacuate people and exert more control during a disaster (Schmidt-Thomé 2006).

- Significant natural areas: these areas (e.g. national parks) are considered more vulnerable as they are unique and hard to recover (Schmidt-Thomé 2006).

- Fragmented natural areas: these are vulnerable because nature in larger undisturbed areas recovers faster than that in smaller areas (Schmidt-Thomé 2006).

In this framework, the indicators used to construct vulnerability-related indicators should be independent. To check for this, factor analysis, a classical statistical method to detect structure in the relationships between variables or indicators, is applied (Russell 2002). Using SPSS statistics software, principal component analysis (PCA) (Jolliffe 2002) is adopted to first make distinct the principal component, and then the varimax rotation strategy (Osborne 2008) is used to calculate the factor loading in each principal component. The factors (vulnerability-related indicators) having the highest loading in each principal component are then selected to construct the BN.

Next, trigger factors are chosen to construct the set of hazard-related indicators which represent the magnitude of multiple hazards. Hazard-related indicators for multiple hazards with different relationships are shown in Fig. 4.

In Fig. 4a, hazards A and B have an independent relationship. The changes in trigger factors t_1, $t_2 ... t_i ... t_n$ which induce hazard A are independent of the changes in trigger factors t_{n+1}, $t_{n+2} ... t_s ... t_t$ which induce hazard B. The two trigger factor groups (t_1, $t_2 ... t_i ... t_n$) and (t_{n+1}, $t_{n+2} ... t_s ... t_t$) can be used to measure the frequency and magnitude of hazard A and B respectively. Hence, the trigger factor group (t_1, $t_2 ... t_i ... t_n$) is chosen as the hazard-related indicator to represent the magnitude of hazard A, and the trigger factor group (t_{n+1}, $t_{n+2} ... t_s ... t_t$) is chosen as the hazard-related indicator to represent the magnitude of hazard B.

In Fig. 4b, hazards A_1, $A_2 ... A_m$ represent a parallel relationship. Hazards A_1, $A_2 ... A_m$ are all induced by the changes in the same trigger factors t_1, $t_2 ... t_i$. The frequency and magnitude of this hazard group (A_1, $A_2 ... A_m$) are determined by the changes in these trigger factors. Hence, the trigger factor group (t_1, $t_2 ... t_i ... t_n$) is chosen as the hazard-related indicator to represent the magnitude of group (A_1, $A_2 ... A_m$).

In Fig. 4c, hazards A and B are in a series relationship. The changes in trigger t_1, t_2 ... $t_i ... t_n$ induce the hazard A, then hazard A cause the changes in trigger factors t_{n+1}, $t_{n+2} \ldots t_s \ldots t_t$. The changes in trigger factors t_{n+1}, $t_{n+2} \ldots t_s \ldots t_t$ induce hazard B. Hence, the trigger factor group (t_1, $t_2 \ldots t_i \ldots t_n$) is chosen as the hazard-related indicator to represent the magnitude of hazard A, and the trigger factor group (t_{n+1}, $t_{n+2} \ldots t_s \ldots t_t$) is chosen as the hazard-related indicator to represent the magnitude of hazard B. The probability and degree of the changes in the trigger factor group (t_{n+1}, $t_{n+2} \ldots t_s \ldots t_t$) are decided by the magnitude of hazard A, that is, the changes in the trigger factor group (t_1, $t_2 \ldots t_i \ldots t_n$).

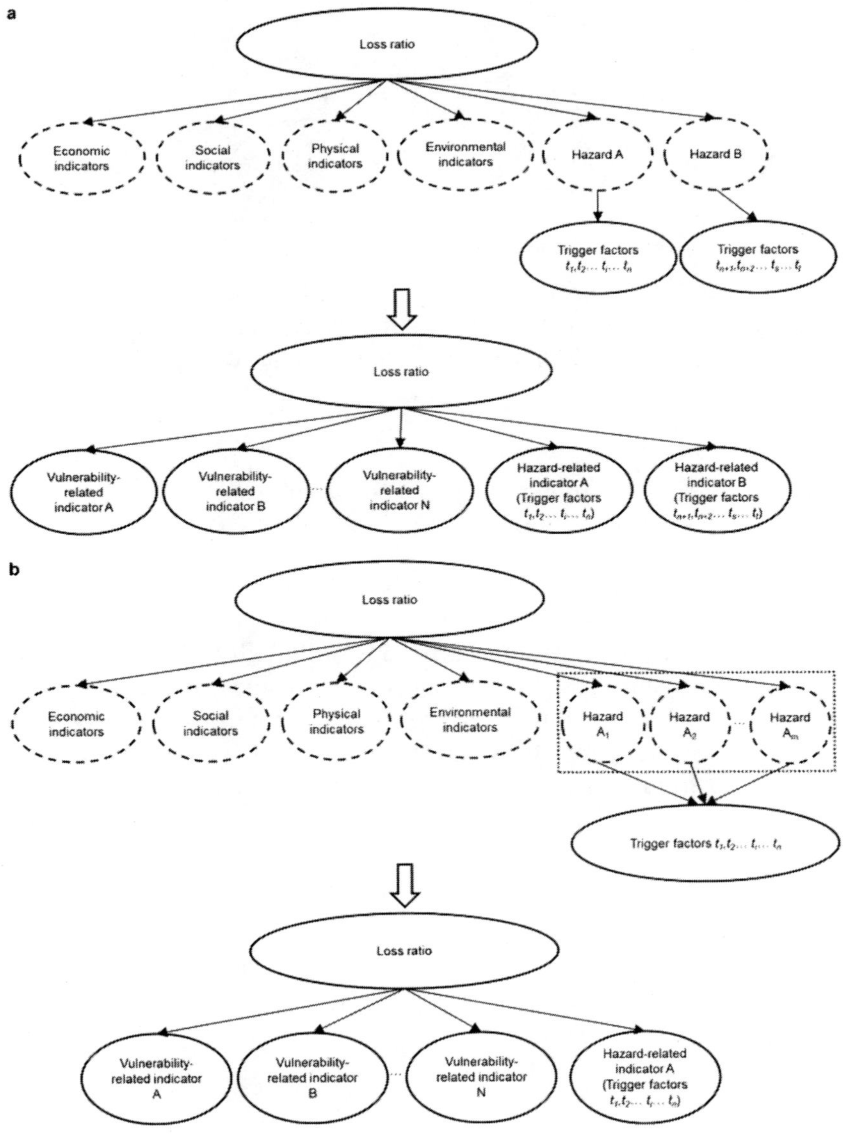

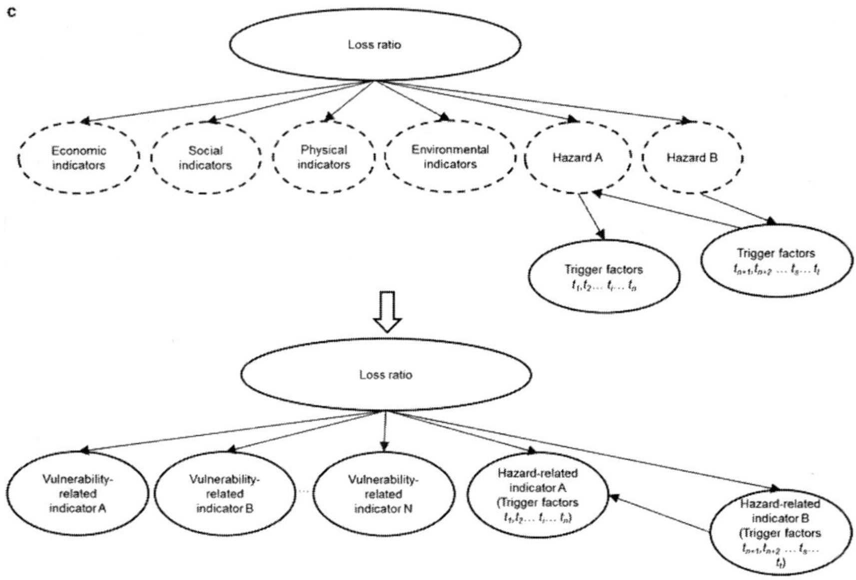

Figure 4. Bayesian network frameworks for loss ratio assessment of multiple hazards with different relationships **a**Independent relationship, **b** Parallel relationship, **c** Series relationship

Hazards in a mutex relationship cannot occur together, so the mutex relationship is not mentioned further.

(2) Determining the conditional probability

A conditional probability measures the probability of an event given that another event has occurred. Once a BN framework is constructed, the conditional probability of a node given their parent nodes should be determined. The conditional probability of a vulnerability-related indicator or hazard-related indicator given a loss ratio is determined in this module as:

$$p\left(v_{kj} \middle| l_i\right) \tag{11}$$

where l_i represent the i state of loss ratio l, $i = 1, 2, \ldots m$, and v_{kj} represents the j state of a vulnerability-related indicator or hazard-related indicator k, $k = 1,2,\ldots s$, $j = 1, 2,\ldots n$.

There are three commonly used methods for estimation of the conditional probability. When applied to a complete observed data set, maximum-likelihood estimation (MLE), a well-known statistics method is used to provide estimates for the model's parameters (the conditional probabilities) (Redner and Walker 1984; Grossman and Domingos 2004; Liu et al. 2015). If the model relies on incomplete observed data, an expectation–maximization (EM) algorithm, an iterative method for finding maximum likelihood estimates of parameters (conditional probabilities) in statistical models, can be used (Lauritzen 1995). When there is no observed data for the model, the model's parameters (conditional probabilities) can be estimated according to domain knowledge (Heckerman et al. 1995; Liao and Ji 2009). Hence, in this module, the methods used for estimation of conditional probabilities are determined according to the availability of observed data.

(3) Loss ratio calculation

In this step, given the above conditional probability, the joint probability (Eq. 12) is used to estimate the posteriori probability of the target loss ratio:

$$\begin{aligned} p(l_i, v_1, v_2, \ldots v_k \ldots v_s) &= p(l_i) p(v_1, v_2, \ldots v_k \ldots v_s | l_i) \\ &= p(l_i) \prod_{k=1}^{s} p(v_k | l_i) \end{aligned} \tag{12}$$

where l_i represents the i state of loss ratio l, and v_k is the vulnerability-related indicator or hazard-related indicator k, $k = 1,2,\ldots s$.

When the states of all vulnerability-related indicator and hazard-related indicators are given as j, the probability of loss ratio l_i occurring can be calculated based on the posteriori probability of the target loss ratio:

$$P(l_i) = \frac{p(l_i, v_{1j}, v_{2j}, \ldots v_{kj} \ldots v_{sj})}{\sum_{i=1}^{m} p(l_i, v_{1j}, v_{2j}, \ldots v_{kj} \ldots v_{sj})} \tag{13}$$

where l_i represent the i state of loss ratio l, $i = 1,2,\ldots m$, and v_{kj} represents the j state of vulnerability-related indicator or hazard-related indicator k, $k = 1,2, \ldots s$, $j = 1, 2,\ldots n$.

The final loss ratio L, with given all vulnerability-related indicator and hazard-related indicator states j, can then be calculated as:

$$L = \sum_{i=1}^{m} l_i \times P(l_i) \tag{14}$$

where l_i is the i state loss ratio with given all vulnerability-related indicator and hazard-related indicator states j, $i = 1,2,\ldots m$, and $P(l_i)$ is the corresponding probability of the target loss ratio l_i occurred.

The loss ratio with other states of vulnerability-related and hazard-related indicator can be calculated in the same way. This module can thus calculate the loss ratio induced by multi-hazards of different degree (different states in hazard-related indicators), whilst also addressing vulnerability using vulnerability-related indicators from physical, social, economic and environmental domains.

3.4 Multi-hazard risk assessment

At this point, and based on the hazard identification and hazard interaction analysis modules, the exceedance probability of multiple hazards can be determined. For any given exceedance probability of multiple hazards, the corresponding loss can be calculated by the value of exposure (identified in the exposure module), and the corresponding loss ratio induced by these hazards (measured in the loss ratio assessment module). With the help of ArcGIS software, the possible loss caused by multi-hazard with different exceedance probabilities in each spatial assessment unit can then be mapped. These maps can be used to identify high risk (large potential loss) areas. Furthermore, this model can help to identify what underlies these large losses. This is significant, as such information supports, guides and targets the development of appropriate loss prevention and risk mitigation measures.

4. MODEL APPLICATION: A CASE STUDY IN NORTHEAST ZHEJIANG

4.1 Study area and data

Northeast Zhejiang (Fig. 5), as part of the Yangtze River Delta, is one of China's main economic regions. This region comprises seven prefecture-level cities in Zhejiang provinces (hereafter "city"), and comprises 55 county level cities and counties (hereafter "county"). With both population density and economic activity growing, this already vulnerable region is increasingly at risk from natural disasters (Liu et al. 2013). Northeast Zhejiang, facing the Pacific to the east, is a typical floodplain with numerous rivers, lakes and canals, and is highly prone to natural hazards. This region is downstream of the Yangtze and Qiantang Rivers and their many tributaries, with channel density >0.5 km of river per km^2, factors that make it liable to frequent flooding (Li et al. 2013). The region is coastal and an oceanic landform between Eurasia and the Pacific, so is susceptible to typhoons and storm surges. In addition, some hilly areas are likely to be influenced by landslides. Hence, due to its geographical location and topography, multiple hazards, particularly typhoons, floods, landslides and storm surges are evident in Northeast Zhejiang.

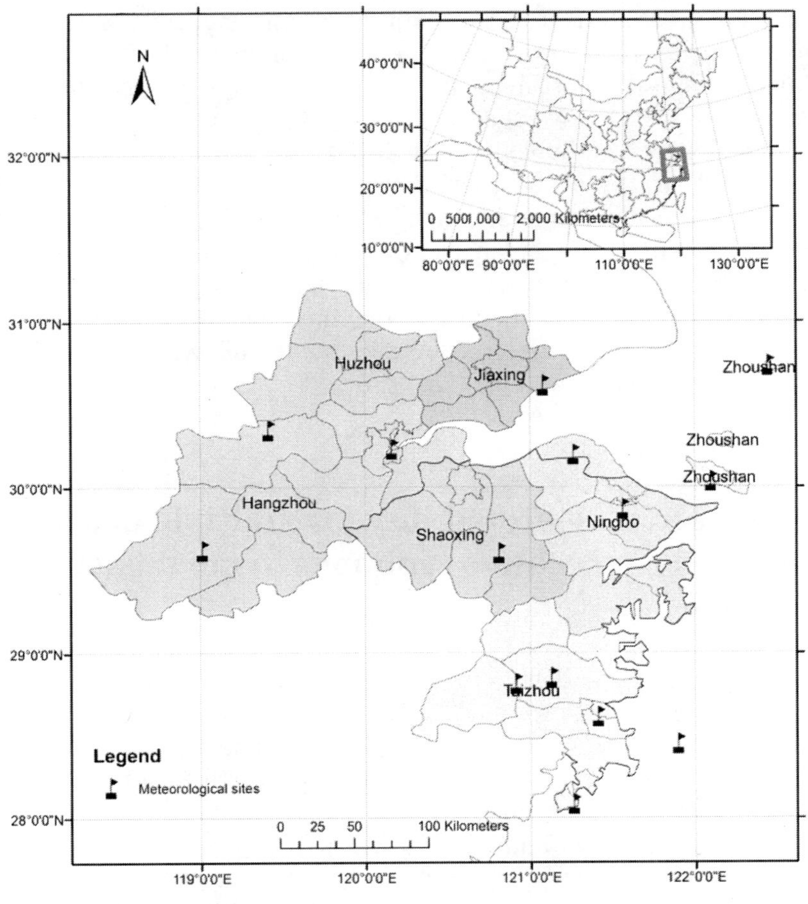

Figure 5. Northeast Zhejiang, China

In addition, according to historical observations, being struck by two consecutive typhoons (within 60 days) is the most common multi-hazard scenario in Northeast Zhejiang. The first and second typhoons have an independent relationship, and these two typhoons could induce floods, landslides and storm surges respectively. Typhoons, floods, landslides and storm surges have a strong relationship with each other in this multi-hazard scenario. Hence, Northeast Zhejiang being struck by typhoons twice consecutively (within 60 days) is a suitable case to show the application of the proposed model.

Three types of data are needed to implement the proposed model: environmental data, disaster data and socioeconomic data. Environmental data mainly comprises meteorological data, downloaded for 15 recording stations (Fig. 5) in the region, with daily data from 1980 to 2013, a suitably lengthy period for hazard-forming environment analysis. Disaster data for this period, collected from the Meteorological Department and the Civil Administration Department in China, includes the disaster type, time, place, and direct economic loss for each disaster in Northeast Zhejiang. Socioeconomic data, derived from statistics yearbooks (1980–2013) in each city includes GDP, income (of rural and urban residents), population density, gender ratio, age structure, telecommunication infrastructure (number of mobile phone users, fixed line phone users, and internet users), transport route (road length), medical service provision (number of medical institutions, beds and medical personnel), and social dependency (number of residents covered by subsistence allowances, number of employed) in each county.

4.2 Exceedance probability for multiple hazards from two consecutive typhoons in northeast Zhejiang

Typhoons, floods, landslides and storm surges are the main hazards in the region. In contrast to other hazards, typhoons can move thousands of kilometres accompanied by strong winds and heavy rain, and a series of hazards (strong winds, floods, landslides and storm surges) induced by changes in wind and rainfall are the reasons for losses in the typhoon track (Lee et al. 2012; Smith 2013). Thus typhoon is viewed as changes of wind speed and rainfall, with these changes used as trigger factors to measure the magnitude of the series of hazards in the track. Hence, the relationships among multiple hazards of two consecutive typhoons in northeast Zhejiang is shown by Fig. 6.

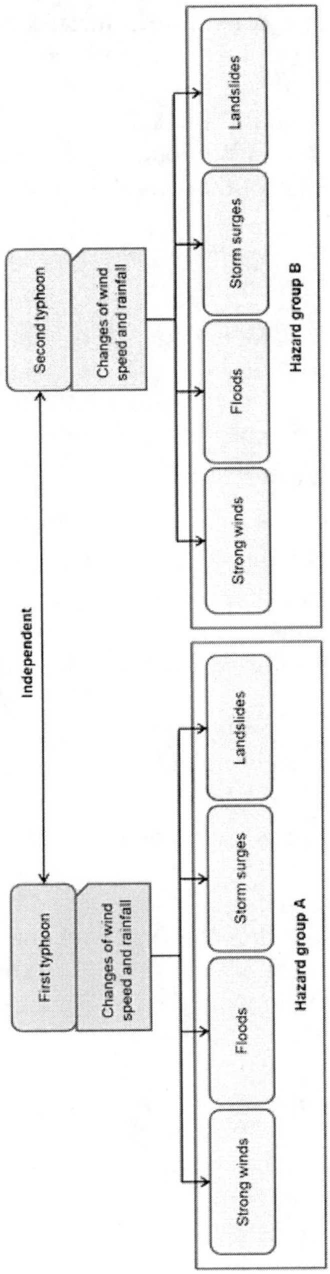

Figure 6. Relationships among multiple hazards of two consecutive typhoons in northeast Zhejiang

Both typhoons are viewed as the trigger factors, with changes of wind speed and rainfall, which induce strong winds, floods, landslides and storm surges. During each typhoon, these four hazard types are in a parallel relationship and constitute a hazard group with each hazard induced by common trigger factors (wind speed and rainfall). Hence, the frequency and magnitude of this hazard group are determined by changes in wind speed and rainfall. The exceedance probability of this hazard group (strong winds, floods, landslides and storm surges) occurring with different magnitudes can be expressed as:

$$EP(H_w \cap H_f \cap H_l \cap H_s) = EP(wind\ speed,\ rainfall) \tag{15}$$

where H_w is strong wind, H_f is flood, H_l is landslide, H_s is storm surge, and *EP(wind speed, rainfall)* is the exceedance probability of the corresponding maximum daily rainfall and maximum daily wind speed sets during the typhoon, calculated using the mathematical statistics approach with maximum daily rainfall and maximum wind speed during each historical typhoon.

According to the trigger factors for hazard occurrence, two typhoons are in an independent relationship, so the hazard groups A and B are also in an independent relationship. Hence, the exceedance probability for multiple hazards of two consecutive typhoons can be calculated as:

$$\begin{aligned} EP(MH) = {} & EP_f(wind\ speed,\ rainfall) \\ & \times EP_s(wind\ speed,\ rainfall) \end{aligned} \tag{16}$$

where *EP(MH)* is the exceedance probability for multiple hazards of two consecutive typhoons, EP_f *(wind speed, rainfall)* is the exceedance probability of the corresponding maximum daily rainfall and maximum daily wind speed sets during the first typhoon, and EP_s *(wind speed, rainfall)* is the exceedance probability of the corresponding maximum daily rainfall and maximum daily wind speed sets during the second typhoon.

In our analysis, we adopted the two dimension information diffusion method (Huang 1997) to calculate the exceedance probability of the corresponding maximum daily rainfall and maximum daily wind speed sets. Within this method, maximum daily wind speed and maximum daily rainfall are viewed as two associated factors in one set, and the coefficient of correlation between them is considered during the diffusion. The results in 3 meteorological sites are used as cases to be shown in Fig. 7. Then, a spatial interpolation technique is used to estimate the rainfall and wind distribution in the whole region. The results, as distribution of maximum daily rainfall and maximum wind speed with different exceedance probabilities, are shown in Fig. 8.

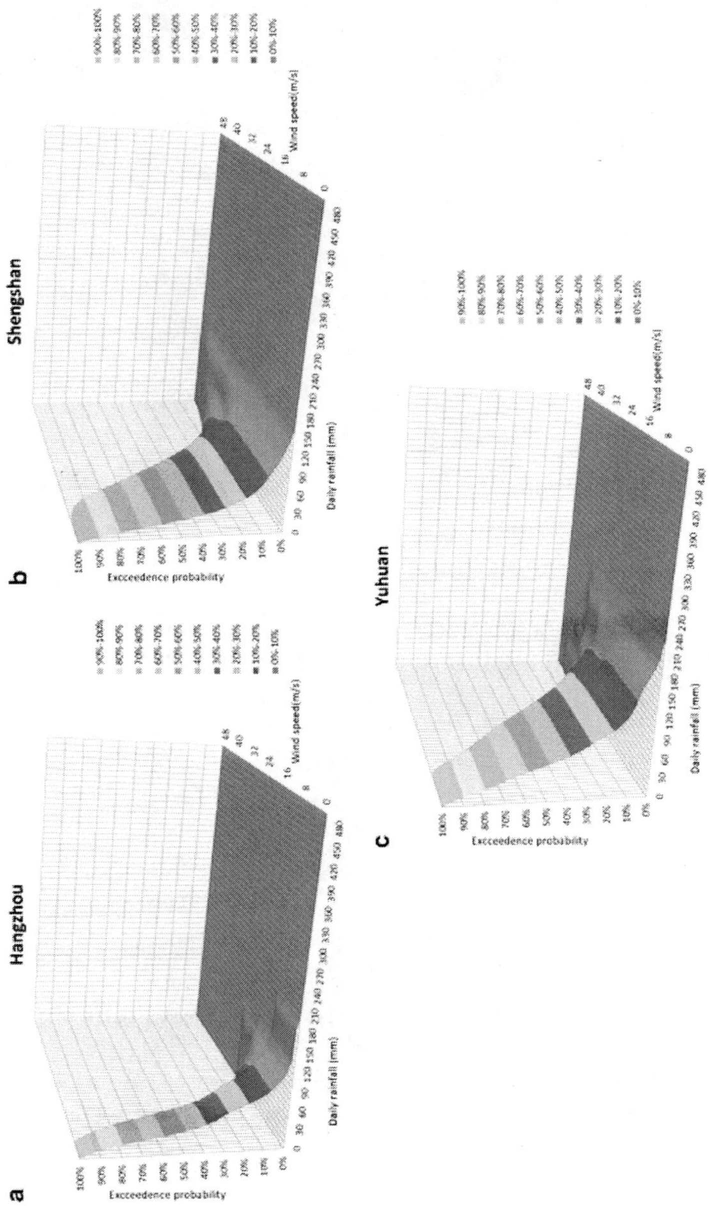

Figure 7. Exceedance probability distribution of rainfall and wind speed sets **a** Hangzhou, **b** Shengshan, **c** Yuhuan

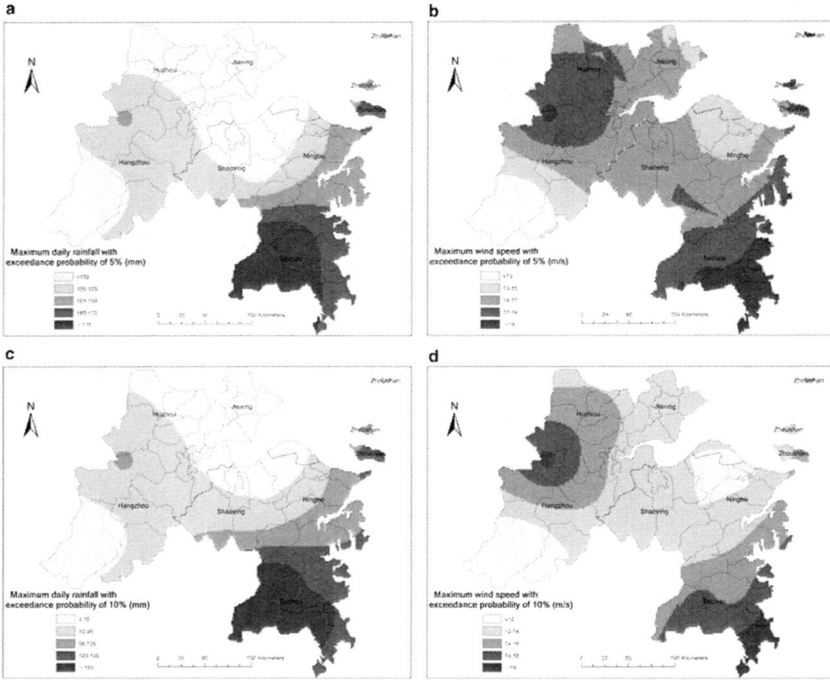

Figure 8. Distribution of maximum daily rainfall and maximum wind speed with different exceedance probabilities **a**Maximum daily rainfall distribution with exceedance probability of 5 %, **b** Maximum wind speed distribution with exceedance probability of 5 %, **c** Maximum daily rainfall distribution with exceedance probability of 10 %, **d** Maximum wind speed distribution with exceedance probability of 10 %

4.3 Possible loss caused by multiple hazards from two consecutive typhoons in northeast Zhejiang

4.3.1 Exposure distribution in northeast Zhejiang

With respect to losses, our case study takes economic loss as an example, with GDP selected as the exposure indicator. The assessment unit in northeast Zhejiang is the county level (government administrative division), so the official statistics data analysis method is used. From these statistics, GDP in each county was obtained, and mapped using ArcGIS. Figure 9 shows that countries with higher GDP in 2013 are mainly located in the north eastern part of the region.

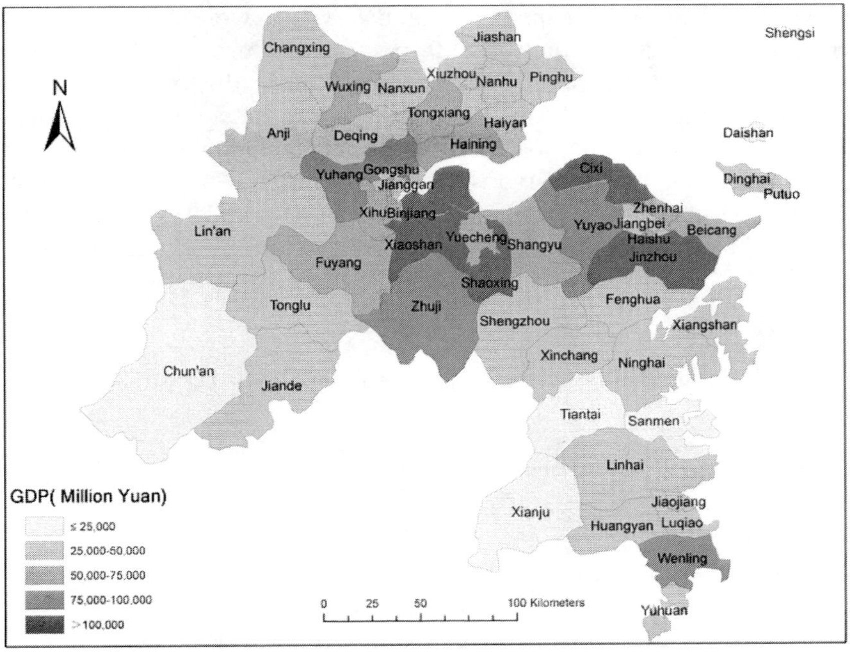

Figure 9. GDP distribution in northeast Zhejiang in 2013

4.3.2 Loss ratio assessment in northeast Zhejiang

For northeast Zhejiang, the relevant vulnerability-related indicators are selected as shown in Table 2. Among these indicators, GDP per km^2, population density and percentage of residents covered by subsistence allowances show the same trend direction with vulnerability, that is, as the value of these indicators increases, the value of vulnerability increases. The other indicators show the opposite directional trend with vulnerability. In order to unify the directional trend of these indicators with vulnerability, the reciprocal of the GDP per km^2, population density, and 1-percentage of residents covered by subsistence allowances are used in Factor analysis. PCA is adopted to make distinct the principal component. Table 3 shows eight principal components selected based on the cumulative variance, then a varimax rotation strategy is used to calculate the factor loading in each principal component. Variables with the highest loading in each principal component (bold figures in Table 3), are selected as the vulnerability-related indicators to construct the BN. These were: the number of mobile phone users per 10,000 persons (a proxy for income of residents, and telecommunication condition), doctors per 10,000 persons (a proxy for hospital beds and access to medical services), reciprocal of the population density (a proxy for population density and road length per 10,000 persons), reciprocal of the GDP per km^2, number of medical institutions per km^2, percentage of population age >15 and <65, and percentage of male residents and percentage of employed.

Table 2. Vulnerability-related indicators in northeast Zhejiang

Domain	Indicator	Indicator in northeast Zhejiang
Economic	GDP/capita	GDP per km^2
	Income of residents	Income of urban residents
Social		Income of rural residents
	Population density	Population density
	Gender ratio	Percentage of male residents
	Age structure	Percentage of population with age above 15 and under 65
	Telecommunication	Number of mobile phone users per 10,000 persons
		Number of fixed line phone users per 10,000 persons
		Number of internet users per 10,000 persons
	Transport route	Road length (km) per km^2
		Road length (km) per 10,000 persons
	Medical condition	Number of medical institutions per km^2
		Number of hospital beds per 10,000 persons
		Number of doctors per 10,000 persons
	Social dependency	Percentage of employed
		Percentage of residents covered by subsistence allowances
	Risk perception	–
	Warning systems	–
	Institutional preparedness	–
	Educational achievement	–
Physical	Technical infrastructure	–
Environmental	Significant natural areas	–
	Fragmented natural areas	–

– Represents the data is not available in these indicators. These indicators should be considered if such data are available

Table 3. Factor loadings in each principal component in northeast Zhejiang

Vulnerability-related indicators	Component							
	1	2	3	4	5	6	7	8
Reciprocal of the GDP per km^2	−0.294	−0.151	0.137	**−0.865**	−0.072	0.121	0.158	−0.068
Income of urban residents	0.849	0.125	0.128	0.325	0.035	−0.243	−0.070	0.166
Income of rural residents	0.829	0.190	0.105	0.328	0.079	−0.213	−0.174	0.161
Reciprocal of the population density	−0.146	−0.021	**0.884**	−0.303	−0.209	−0.036	0.110	−0.011
Percentage of male residents	−0.270	−0.237	0.226	−0.212	−0.220	−0.018	**0.815**	−0.083
Percentage of population with age above 15 and under 65	−0.226	0.046	−0.002	−0.103	0.063	**0.954**	−0.001	−0.061
Number of mobile phone users per 10,000 persons	**0.916**	0.254	0.005	0.066	0.058	−0.081	−0.069	0.190
Number of fixed line phone users per 10,000 persons	0.851	0.263	0.099	0.177	0.056	−0.162	−0.191	0.182
Number of internet users per 10,000 persons	0.850	0.355	−0.086	−0.059	0.042	0.054	−0.113	0.150
Road length (km) per km^2	0.793	−0.001	0.094	0.274	0.384	−0.002	0.023	0.079
Road length (km) per 10,000 persons	0.428	−0.010	0.859	0.164	0.036	0.029	0.079	0.011
Number of medical institutions per km^2	0.128	0.211	−0.147	0.056	**0.916**	0.066	−0.155	0.005
Number of hospital beds per 10,000 persons	0.347	0.807	−0.066	0.057	0.220	−0.023	−0.189	0.015
Number of doctors per 10,000 persons	0.248	**0.889**	0.020	0.133	0.073	0.068	−0.066	0.082
Percentage of employed	0.485	0.093	−0.006	0.085	0.011	−0.085	−0.075	**0.853**
1-Percentage of residents not covered by subsistence allowances	−0.692	−0.412	−0.051	0.048	0.018	0.145	0.391	−0.005
Cumulative variance (%)	49.4	63.1	72.0	77.8	82.8	86.6	89.7	92.8

Bold values represent the highest loading in each principal component

Bold values represent the highest loading in each principal component

Maximum daily rainfall and maximum daily wind speed in each typhoon are selected as trigger factors to construct the set of hazard-related indicators which represent the magnitudes of multiple hazards. The first and second typhoons have an independent relationship, hence based on the hazard interaction analysis (Sect. 4.2), the BN framework in northeast Zhejiang can be constructed as shown in Fig. 10.

In our case study, the loss ratio is divided into six states, the eight vulnerability-related indicators are each divided into five states, and the hazard-related indicators (maximum daily rainfall and maximum daily wind speed sets) are divided into eight states (Table 4). From the 1980–2012 historic disaster records, aggregate loss data from two consecutive typhoons were collected, along with corresponding data for vulnerability-related indicators (from the relevant statistics yearbook) to construct a complete observed data set. Then maximum-likelihood estimation (MLE) was then used to provide estimates of the conditional probabilities (Grossman and Domingos 2004). With these conditional probabilities, the loss ratio with different states of vulnerability-related and hazard-related indicator was calculated based on Eqs. 12–14. Taking the maximum daily rainfall distribution and maximum wind speed distribution of the first typhoon with exceedance probability of 10 %, and the second with exceedance probability of 5 % as example, the results are shown in Fig. 11.

4.4 Multi-hazard risk map in northeast Zhejiang

Taking northeast Zhejiang in 2013, influenced by consecutive typhoons (maximum daily rainfall distribution and maximum wind speed distribution of the first typhoon with exceedance probability of 10 %, and the second with exceedance probability of 5 %) as an example, and according to the hazard identification and interaction analysis described above, the magnitudes of multiple hazards can be expressed by the maximum daily rainfall distribution and maximum wind speed distribution (Fig. 8). With these rainfall and wind speed distributions, the corresponding loss can be calculated by the exposure distribution (Fig. 9) and corresponding loss ratio distribution (Fig. 11). The final risk map is shown in Fig. 12.

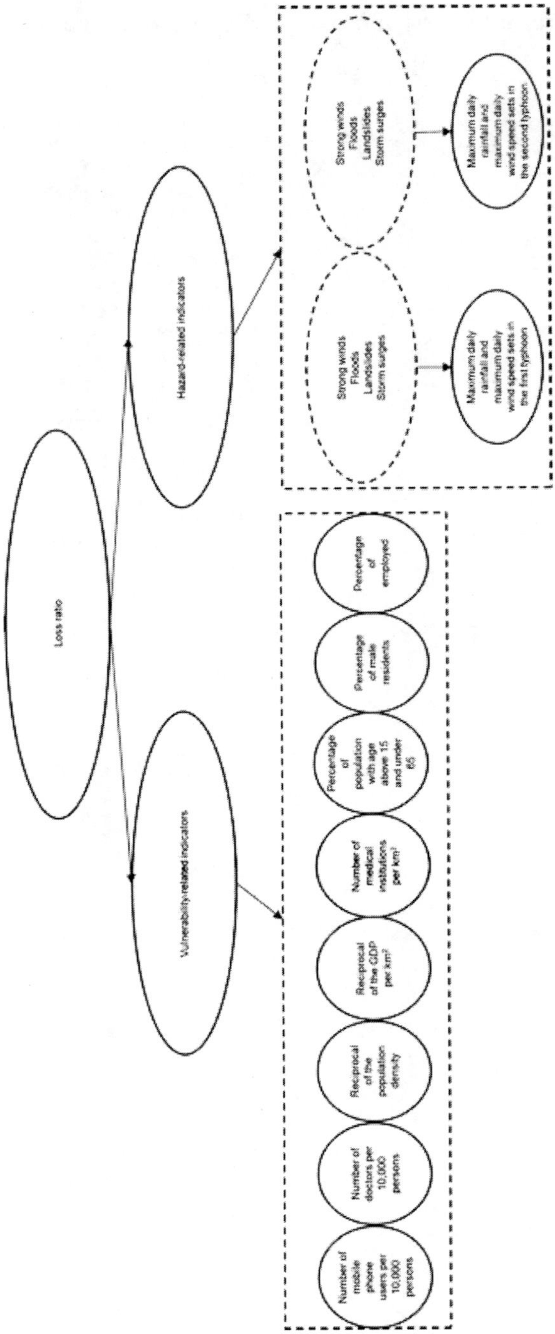

Figure 10. The basic structure of Bayesian Networks for loss ratio assessment in northeast Zhejiang

Table 4. Different states of factors in Bayesian network

Factor	State
Number of mobile phone users per 10,000 persons (M)	1. $M_1 < 2500$ phone users/10,000 persons
	2. 2500 phone users/10,000 persons $\leq M_2 <$ 5000 phone users/10,000 persons
	3. 5000 phone users/10,000 persons $\leq M_3 <$ 7500 phone users/10,000 persons
	4. 7500 phone users/10,000 persons $\leq M_4 <$ 10,000 phone users/10,000 persons
	5. $M_5 \geq$ 10,000 phone users/10,000 persons
Number of doctors per 10,000 persons (D)	1. $D_1 < 10$ doctors/10,000 persons
	2. 10 doctors/10,000 persons $\leq D_2 <$ 15 doctors/10,000 persons
	3. 15 doctors/10,000 persons $\leq D_3 <$ 20 doctors/10,000 persons
	4. 20 doctors/10,000 persons $\leq D_4 <$ 25 doctors/10,000 persons
	5. $D_i \geq 25$ doctors/10,000 persons
Reciprocal of the population density (Pd)	1. $Pd_1 <$ (1/1000) km^2/persons
	2. (1/1000) km^2/persons $\leq Pd_2 <$ (1/750) km^2/persons
	3. (1/750) km^2/persons $\leq Pd_3 <$ (1/500) km^2/persons
	4. (1/500) km^2/persons $\leq Pd_4 <$ (1/250) km^2/persons
	5. $Pd_5 \geq$ (1/250) km^2/persons
Reciprocal of the GDP per km^2 (G)	1. $G_1 <$ (1/30) km^2/million yuan
	2. (1/30) km^2/million yuan $\leq G_2 <$ (1/20) km^2/million yuan
	3. (1/20) km^2/million yuan $\leq G_3 <$ (1/10) km^2/million yuan
	4. (1/10) km^2/million yuan $\leq G_4 <$ (1/5) km^2/million yuan
	5. $G_5 \geq$ (1/5) km^2/million yuan
Number of medical institutions per km^2 (Mi)	1. $Mi_1 < 0.02$ medical institutions/km^2
	2. 0.02 medical institutions/km^2 $\leq Mi_2 <$ 0.03 medical institutions/km^2
	3. 0.03 medical institutions/km^2 $\leq Mi_3 <$ 0.04 medical institutions/km^2
	4. 0.04 medical institutions/km^2 $\leq Mi_4 <$ 0.05 medical institutions/km^2
	5. $Mi_5 \geq$ 0.05 medical institutions/km^2
Percentage of population with age above 15 and under 65 (Pa)	1. $Pa_1 < 72$ %
	2. 72 % $\leq Pa_2 < 73.5$ %
	3. 73.5 % $\leq Pa_3 < 75$ %
	4. 75 % $\leq Pa_4 < 76.5$ %
	5. $Pa_5 \geq 76.5$ %
Percentage of male residents (Ma)	1. $Ma_1 < 50$ %
	2. 50 % $\leq Ma_2 < 50.5$ %
	3. 50.5 % $\leq Ma_3 < 51$ %
	4. 51 % $\leq Ma_4 < 51.5$ %
	5. $Ma_5 \geq 51.5$ %
Percentage of employed (E)	1. $E_1 < 50$ %
	2. 50 % $\leq E_2 < 60$ %
	3. 60 % $\leq E_3 < 70$ %
	4. 70 % $\leq E_4 < 80$ %
	5. $E_5 \geq 80$ %
Maximum daily rainfall and maximum daily wind speed sets in the first typhoon (WRf)	1. WRf_1($W <$ 10 m/s, $R <$ 50 mm)
	2. WRf_2($W <$ 10 m/s, 50 mm $\leq R$)
	3. WRf_3(10 m/s $\leq W <$ 20 m/s, $R <$ 50 mm)
	4. WRf_4(10 m/s $\leq W <$ 20 m/s, 50 mm $\leq R <$ 150 mm)
	5. WRf_5(10 m/s $\leq W <$ 20 m/s, $R \geq$ 150 mm)
	6. WRf_6($W \geq$ 20 m/s, $R <$ 50 mm)
	7. WRf_7($W \geq$ 20 m/s, 50 mm $\leq R <$ 150 mm)
	8. WRf_8($W \geq$ 20 m/s, $R \geq$ 150 mm)

Maximum daily rainfall and maximum daily wind speed sets in the second typhoon (WRs)	1. WRs_1 ($W < 10$ m/s, $R < 50$ mm)
	2. WRs_2 ($W < 10$ m/s, 50 mm $\leq R$)
	3. WRs_3 (10 m/s $\leq W < 20$ m/s, $R < 50$ mm)
	4. WRs_4 (10 m/s $\leq W < 20$ m/s, 50 mm $\leq R < 150$ mm)
	5. WRs_5 (10 m/s $\leq W < 20$ m/s, $R \geq 150$ mm)
	6. WRs_6 ($W \geq 20$ m/s, $R < 50$ mm)
	7. WRs_7 ($W \geq 20$ m/s, 50 mm $\leq R < 150$ mm)
	8. WRs_8 ($W \geq 20$ m/s, $R \geq 150$ mm)
Loss ratio (l)	1. $l_1 = 0$ %
	2. $0\ \% < l_2 < 0.5$ %
	3. $0.5\ \% \leq l_3 < 1$ %
	4. $1\ \% \leq l_4 < 5$ %
	5. $5\ \% \leq l_5 < 10$ %
	6. $l_6 \geq 10$ %

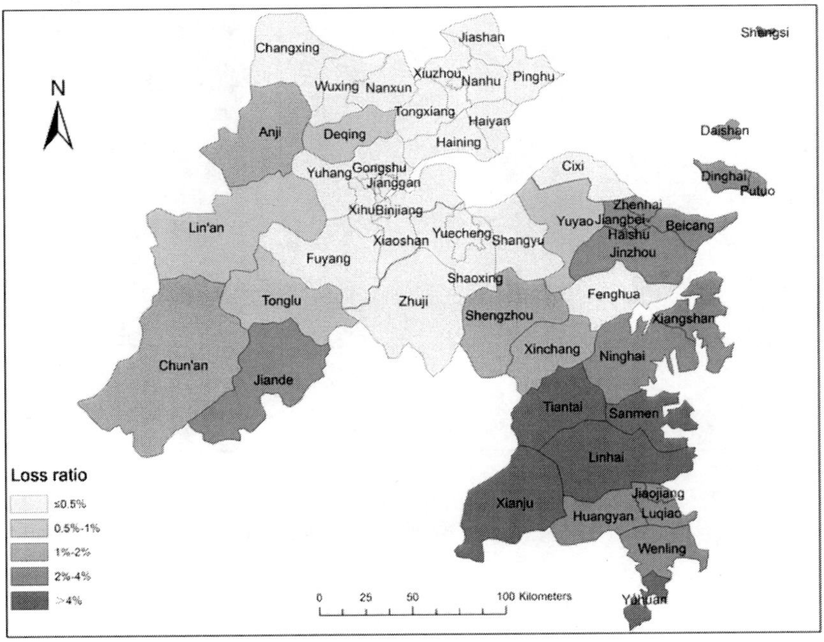

Figure 11. Loss ratio distribution influenced by two typhoons with exceedance probability of 10 % and exceedance probability of 5 %

This analysis shows that higher loss counties are mainly in the south eastern part of the region, with Linhai, Tiantai, Xianju, Sanmen and Jinzhou counties in the highest risk area. Risk is determined by the magnitudes of multiple hazards, vulnerability and value of exposure. Jinzhou is in the highest risk area due to its highest exposure value. The high risk at Linai, Tiantai, Xianju and Sanmen is due to the interaction of the highest magnitudes of multiple hazards and the highest vulnerability. Thus the relative importance of factors that underlies observed high risk varies geographically. The model can be used to estimate the loss distribution influenced by typhoon with other exceedance probabilities, and also through other hazard combinations.

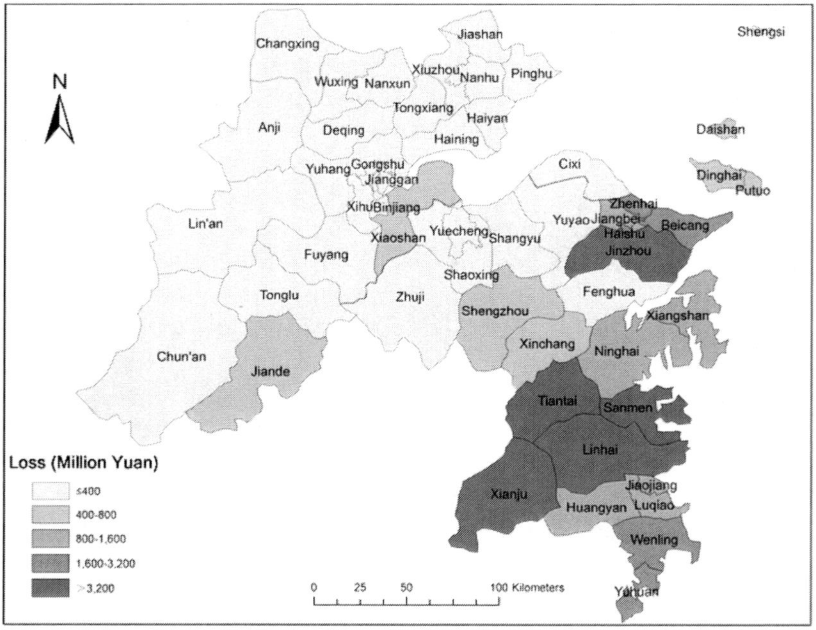

Figure 12. Multi-hazard risk map for two consecutive typhoons with exceedance probability of 10 % and exceedance probability of 5 %

5. MODEL VALIDATION

Model validation is used to check how well the model represents reality. In our example application, MmhRisk-HI is applied to estimate potential loss caused by multiple hazards in northeast Zhejiang. To test the effectiveness of the model, the hazards that occurred in 2013 were simulated, and calculated losses compared to observed losses. In 2013, northeast Zhejiang was influenced by typhoon Trami (21st August) then typhoon Fitow (7th October). According to the hazard identification and hazard interaction analysis, the magnitudes of multiple hazards induced by typhoons in northeast Zhejiang can be expressed by the maximum daily rainfall and maximum wind speed. Data for these variables was collected from 15 meteorological stations in northeast Zhejiang. With these hazard-related indicators (maximum daily rainfall and maximum wind speed) and vulnerability-related indicators described, using data for 2013, the loss ratio assessment module (based on historical data from 1980 to 2012) was used to estimate the probability of loss ratio in each county induced by these typhoons Trami and Fitow. The modelled and observed loss ratio in 55 counties in northeast Zhejiang are shown in Table 5.

Bold values represent that the real loss ratio falls into the loss ratio state which has the highest estimated probability

As shown in Table 5, among these 55 counties, the real loss ratio in 42 counties (76.36 %) falls into the loss ratio state (l_i) which has the highest estimated probability (bold figures in Table 5). Taking Shangcheng as an example, the real loss ratio in this county is 0.01 %, which falls into the loss ratio state 2 (0 % < l_2 < 0.5 %). In the corresponding estimated results, the estimated probability of l_2 occurring in Shangcheng is 98 %, which is the highest among all six loss ratio states. In addition, the total estimated loss in these 55 counties is 51,893.39 million yuan compared to the actual loss of 50,485.43 million yuan; a deviation of estimated from actual value of less than 2.79 %. Hence the MmhRisk-HI model is shown to effectively represent the real system, with estimated results reflecting the real loss situation.

6. CONCLUSION AND DISCUSSION

MmhRisk-HI fills a key research gap in existing MHRA methods as it goes beyond the simple consideration of multi-hazard interaction as domino effect, to calculate possible loss with an explicit consideration of all possible hazard relationships. In our case study example, whilst the typhoons are independent, the short time period between them means the area is more vulnerable as it has not recovered immediately from the first typhoon. Previous MHRA methods assume there is no change in vulnerability, and so would calculate the loss in each typhoon individually assuming the same vulnerability, then the aggregates the losses. Thus such results cannot reflect the real loss situation. In MmhRisk-HI, the loss ratio assessment module addresses this issue by considering the magnitudes of these two typhoons together in hazard-related indicators. These two typhoons are treated as a multiple hazards group, and the relevant vulnerability-related indicators correspond to this group rather than a single typhoon. Hence, the results obtained in the model are more reliable.

Table 5. The estimated results and observed real loss ratio

	Estimated probability of loss ratio l_i occurring						Real loss ratio
	$l_1 = 0\ \%$	$0\ \% < l_2 < 0.5\ \%$	$0.5\ \% \leq l_3 < 1\ \%$	$1\ \% \leq l_4 < 5\ \%$	$5\ \% \leq l_5 < 10\ \%$	$l_6 \geq 10\ \%$	
Shangcheng	1.62	**98.00**	0.31	0.07	0.00	0.00	0.01
Xiacheng	1.62	98.00	0.31	0.07	0.00	0.00	0.00
Jianggan	0.79	**99.08**	0.11	0.02	0.00	0.00	0.02
Gongshu	1.62	**98.00**	0.31	0.07	0.00	0.00	0.03
Xihu	1.62	**98.00**	0.31	0.07	0.00	0.00	0.28
Binjiang	1.46	97.80	0.53	0.21	0.00	0.00	0.00
Xiaoshan	0.61	**99.27**	0.10	0.02	0.00	0.00	0.49
Yuhang	1.03	**98.61**	0.28	0.08	0.00	0.00	0.49
Tonglu	10.07	**86.07**	1.56	2.29	0.00	0.00	0.16
Chun'an	**93.64**	4.78	0.74	0.41	0.02	0.41	0.00
Jiande	**55.55**	38.22	1.16	4.57	0.13	0.36	0.00
Fuyang	3.76	**95.13**	0.51	0.59	0.00	0.00	0.27
Lin'an	33.57	**64.42**	1.27	0.73	0.00	0.00	0.41
Haishu	0.75	98.82	0.36	0.06	0.00	0.00	1.06
Jiangdong	0.75	**98.82**	0.36	0.06	0.00	0.00	0.44
Jiangbei	0.00	0.00	0.00	**99.57**	0.43	0.00	4.04
Beicang	0.50	**97.90**	0.88	0.72	0.00	0.00	0.10
Zhenhai	1.10	**98.49**	0.29	0.11	0.00	0.00	0.33

	Estimated probability of loss ratio l_i occurring						Real loss ratio
	$l_1 = 0\ \%$	$0\ \% < l_2 < 0.5\ \%$	$0.5\ \% \leq l_3 < 1\ \%$	$1\ \% \leq l_4 < 5\ \%$	$5\ \% \leq l_5 < 10\ \%$	$l_6 \geq 10\ \%$	
Jinzhou	0.00	0.00	0.00	**98.01**	1.99	0.00	4.91
Yuyao	0.00	0.00	0.00	97.63	2.37	0.00	26.62
Cixi	0.00	0.00	0.00	**98.56**	1.44	0.00	1.96
Fenghua	0.00	0.00	0.00	99.77	0.23	0.00	6.76
Xiangshan	0.00	0.00	0.00	**97.99**	2.01	0.00	1.58
Ninghai	0.00	0.00	0.00	**98.87**	1.13	0.00	1.08
Nanhu	0.00	0.00	0.00	89.84	10.16	0.00	0.83
Xiuzhou	0.00	0.00	0.00	**89.87**	10.13	0.00	4.73
Pinghu	0.00	0.00	0.00	**91.64**	8.36	0.00	1.16
Haining	0.00	0.00	0.00	**73.04**	26.96	0.00	1.92
Tongxiang	0.00	0.00	0.00	**90.58**	9.42	0.00	1.37
Jiashan	0.00	0.00	0.00	**90.58**	9.42	0.00	2.28
Haiyan	0.00	0.00	0.00	97.56	2.44	0.00	5.63
Wuxing	4.66	94.92	0.37	0.04	0.00	0.00	0.66
Nanxun	0.00	0.00	0.00	**98.75**	1.25	0.00	1.50
Deqing	0.00	0.00	0.00	**98.28**	1.72	0.00	1.42
Changxing	0.00	0.00	0.00	**99.07**	0.93	0.00	1.07

Anji	0.00	0.00	0.00	93.37	6.63	0.00	7.54
Yuecheng	1.45	**98.18**	0.33	0.05	0.00	0.00	0.08
Shaoxing	0.00	0.00	0.00	96.16	3.84	0.00	0.55
Shangyu	0.00	0.00	0.00	**92.00**	8.00	0.00	2.75
Zhuji	7.03	**89.86**	1.44	1.67	0.00	0.00	0.02
Shengzhou	8.77	**63.45**	6.65	20.78	0.35	0.00	0.09
Xinchang	5.95	**52.34**	3.17	37.97	0.57	0.00	0.27
Dinghai	3.01	**93.09**	2.07	1.79	0.03	0.00	0.10
Putuo	4.08	**94.02**	1.43	0.46	0.01	0.00	0.18
Daishan	7.29	**77.66**	5.81	8.96	0.29	0.00	0.12
Shengsi	6.72	**71.91**	4.05	16.44	0.88	0.00	0.02
Jiaojiang	3.58	**92.16**	3.41	0.83	0.01	0.00	0.19
Huangyan	5.20	4.99	7.74	**58.89**	5.30	17.89	1.19
Luqiao	3.58	**92.16**	3.41	0.83	0.01	0.00	0.17
Wenling	8.42	**60.32**	5.23	25.13	0.90	0.00	0.21
Linhai	9.61	31.25	16.01	35.75	5.17	2.21	0.66
Yuhuan	3.74	69.94	1.98	13.96	10.38	0.00	0.95
Sanmen	0.00	0.00	0.00	**54.90**	4.99	40.11	1.89
Tiantai	3.81	3.97	13.91	65.10	3.52	9.70	0.90
Xianju	3.16	2.53	6.62	**67.25**	5.92	14.51	1.16

Bold values represent that the real loss ratio falls into the loss ratio state which has the highest estimated probability

Model validation is a highly desirable step in model development, but a process that has previously proved intractable in MHRA due to the nature of the existing models. MmhRisk-HI can be validated through comparison of modelled and observed data, with the model used to simulate different multiple natural hazards scenarios and estimate the corresponding loss. In the case study, MmhRisk-HI was used to simulate consecutive events, typhoons Trami and Fitow which struck northeast Zhejiang within a few months in 2013. The simulated results were then compared to the observed data, with good agreement. The validation, although based on a necessarily limited set of actual multi-hazard interaction, does indicate that MmhRisk-HI can represent the real world risk situation with greater confidence.

In conclusion, MmhRisk-HI provides more reliable results (possible loss caused by multiple hazards) with an explicit consideration of interaction between different hazards, and can also be used to explore and better explain what underpins large potential losses (high risk). Hence, the model is a useful tool which can provide better information in risk mitigation planning.

Further improvements to MmhRisk-HI are possible. First, a change in one (or several) trigger factors may induce more than one hazard at the same time. In MmhRisk-HI, these hazards are treated as a multiple hazards group, with all hazards in the group induced by the same trigger factor(s). These trigger factors can be used as hazard-related indicators to represent the intensity of this hazard group in loss ratio assessment. In this way, the results obtained in this model are more reliable. However, these results cannot show how much loss is induced by each single hazard in the hazard group. In reality, it is also hard to distinguish how much loss is induced by each single hazard. For example, during a typhoon, it is hard to distinguish how much loss can be attributed to strong winds and how much to floods. Indeed, in the historical disaster record, only records of loss induced by the whole typhoon are made, rather than for the constituent hazards. Understanding the loss induced by each single hazard could help decision-makers take more targeted mitigation measures. Addressing this issue, without historical loss data, will be challenging. Equally challenging will be the inclusion of uncertainty quantifiers in the MmhRisk-HI model. Uncertainty is inherent in natural disaster risk, and needs to be better addressed for effective risk management, yet uncertainty analysis in MHRA remains rare.

REFERENCES

1. Alexander D (1993) Natural disasters. UCL Press, London

2. Alexander D (2000) Confronting catastrophe, new perspectives on natural disaster. Terra, Harpenden

3. Armas I (2006) Earthquake risk perception in Bucharest Romania. Risk Anal 26(5):1223–1234

4. Armonia (Applied Multi-Risk Mapping of Natural Hazards for Impact Assessment) (2006) Applied multi-risk mapping of natural hazards for impact assessment. Report on new methodology for multi-risk assessment and the harmonisation of different natural risk maps. Armonia, European Community, Genova

5. Bayraktarli YY et al (2006) Capabilities of the Bayesian probabilistic networks approach for earthquake risk management. First European Conference on Earthquake Engineering and Seismology. Geneva

6. Blaikie P, Cannon T, Davis I, Wisner B (1994) At risk: natural hazards, people's vulnerability, and disasters. Routledge, New York

7. Blanchard W (2005) Select emergency management-related terms and definitions, vulnerability assessment techniques and applications (VATA). NOAA Coastal Services Center. http://www.csc.noaa.gov/vata/glossary.html. Accessed 12 May 2015

8. Brooks N (2003) Vulnerability, risk and adaptation: a conceptual framework. Tyndall Centre for Climate Change Research Working Paper 38:1–16

9. Burton I, Kates RW, White GF (1993) The environment as hazard, 2nd edn. The Guildford Press, New York

10. Carpignano A, Golia E, Di Mauro C, Bouchon S, Nordvik JP (2009) A methodological approach for the definition of multi-risk maps at regional level: first application. J Risk Res 12(3–4):513–534

11. Cutter SL, Mitchell JT, Scott MS (2000) Revealling vulnerability of people and place: a case study of Geogretown county, South Carolina. Annal Assoc Am Geogr 90(4):713–737

12. Cutter SL, Boruff BJ, Shirley WL (2003) Social vulnerability to environmental hazards. Soc Sci Quart 84(2):242–261

13. Di Mauro C, Bouchon S, Carpignana A, Golia E, Peressin S (2006) Definition of multi-risk maps at regional level as management tool: experience gained by civil protection authorities of Piemonte region. 5th conference on risk assessment and management in the civil and industrial settlements. University of Pisa

14. Dilley M et al (2005) Natural disaster hotspots, a global risk analysis. World Bank, Washington, DC

15. Eshrati L, Mahmoudzadeh A, Taghvaei M (2015) Multi hazards risk assessment, a new methodology. Int J Health Syst Disaster Manag 3(2):79–88

16. FEMA (2004) Using HAZUS-MH for risk assessment. http://www.fema.gov. Accessed 11 March 2012

17. Frigerio S, Kappes MS, Glade T, Malet J-P (2012) MultiRISK: a platform for Multi-Hazard Risk Modelling and Visualisation. http://www.mae-srl.it/allegati/3_22_692.pdf Accessed 18 March 2014

18. Ge Y, Dou W, Gu Z, Qian X, Wang J, Xu W, Shi P, Ming X, Zhou X, Chen Y (2013) Assessment of social vulnerability to natural hazards in the Yangtze River Delta, China. Stoch Environ Res Risk Assess 27(8):1899–1908

19. Gill JC, Malamud BD (2014) Reviewing and visualizing the interactions of natural hazards. Rev Geophys 52(4):680–722

20. Granger K, Trevor J (2000) A multi-hazard risk assessment. In: Middelmann M, Granger K (eds) Community risk in MacKay, a multi-hazard risk assessment. Australian geological survey organization, Canberra

21. Grossman D, Domingos P (2004) Learning Bayesian network classifiers by maximizing conditional likelihood. Proceedings of the twenty-first international conference on Machine learning. ACM, New York

22. Grünthal G et al (2006) Comparative risk assessment for the city of Cologne-storms, floods, earthquake. Nat Hazards 38:21–44

23. Gumbel EJ (1958) Statistics of extremes. Columbia University Press, New York

24. Haque CE (2000) Risk assessment, emergency preparedness and response to hazards: the case of the 1997 Red River Valley flood, Canada. In: Natural Hazards. Springer, Dordrech

25. Heckerman D, Geiger D, Chickering DM (1995) Learning Bayesian networks: the combination of knowledge and statistical data. Mach Learn 20(3):197–243

26. Hewitt K (1997) Regions of risk: a geographical introduction to disasters. Longman, Essex

27. Huang CF (1997) Principle of information diffusion. Fuzzy Sets Syst 91(1):69–90

28. ISDR (Intentional Strategy for Disaster Reduction) (2004) Living with risk. A global review of disaster reduction initiatives. United Nations publication, Geneva

29. Jensen FV, Nielsen TD (2007) Bayesian networks and decision graphs, 2nd edn. Springer, New York

30. Jolliffe I (2002) Principal component analysis. John Wiley & Sons Ltd, New York

31. Kappes MS, Keiler M, von Elverfeldt K, Glade T (2012) Challenges of analyzing multi-hazard risk: a review. Nat Hazards 64(2):1925–1958

32. Komendantova N et al (2014) Multi-hazard and multi-risk decision-support tools as a part of participatory risk governance: feedback from civil protection stakeholders. Int J Disaster Risk Reduct 8:50–67

33. Lauritzen SL (1995) The EM algorithm for graphical association models with missing data. Comput Stat Data Anal 19(2):191–201

34. Lee YH, Yang HD, Chen CS (2012) Spatial risk assessment of typhoon cumulated rainfall: a case study in Taipei area. Stoch Environ Res Risk Assess 26(4):509–517

35. Li LF et al (2010) Assessment of catastrophic risk using Bayesian network constructed from domain knowledge and spatial data. Risk Anal 30(7):1157–1175

36. Li GF, Xiang XY, Tong YY, Wang HM (2013) Impact assessment of urbanization on flood risk in the Yangtze River Delta. Stoch Environ Res Risk Assess 27(7):1683–1693

37. Liao W, Ji Q (2009) Learning Bayesian network parameters under incomplete data with domain knowledge. Pattern Recogn 42(11):3046–3056

38. Linares-Rivas A (2012) CAPRA initiative: integrating disaster risk into development policies in Latin America and the Caribbean. http://www.ecapra.org/capra-initiative-integrating-disaster-risk-development-policies-latam. Accessed Oct 2013

39. Liu B, Siu YL, Mitchell G, Xu W (2013) Exceedance probability of multiple natural hazards: risk assessment in China's Yangtze River Delta. Nat Hazards 69(3):2039–2055

40. Liu R, Chen Y, Wu J, Gao L, Barrett D, Xu T, Li L, Huang C, Yu J (2015) Assessing spatial likelihood of flooding hazard using naïve Bayes and GIS: a case study in Bowen Basin, Australia. Stoch Environ Res Risk Assess: 1–16.

41. Liu B, Siu YL, Mitchell G (2016) Hazard interaction analysis for multi-hazard risk assessment: a systematic classification based on hazard-forming environment. Nat Hazards Earth Syst Sci 16(2):629–642

42. Maldonado AD, Aguilera PA, Salmerón A (2015) Continuous Bayesian networks for probabilistic environmental risk mapping. Stoch Environ Res Risk Assess: 1–15.

43. Marulanda MC et al (2013) Probabilistic earthquake risk assessment using CAPRA: application to the city of Barcelona, Spain. Nat Hazards 69(1):59–84

44. Marzocchi W et al (2009) Principles of multi-risk assessment. Interaction amongst natural and man-induced risks. European Communities, Brussels

45. Marzocchi W et al (2012) Basic principles of multi-risk assessment: a case study in Italy. Nat Hazards 62(2):551–573

46. McGuire B, Mason I, Kilburn C (2002) Natural hazards and environmental change. Arnold, London

47. Ming X, Xu W, Li Y, Du J, Liu B, Shi P (2015) Quantitative multi-hazard risk assessment with vulnerability surface and hazard joint return period. Stoch Environ Res Risk Assess 29(1):35–44

48. Morrow BH (1999) Identifying and mapping community vulnerability. Disasters 23(1):11–18

49. Munich Re (Munich Reinsurance Company) (2003) Topics—annual review: natural catastrophes 2002. Munich Re Group, Munich

50. Ngo EB (2001) When disasters and age collide: reviewing vulnerability of the elderly. Nat Hazards Rev 2(2):80–89

51. Osborne JW (2008) Best practices in quantitative methods. Sage, London

52. Park C (1994) Environment issues. Prog Phys Geogr 18(3):411–424

53. Pelling M (2003) The vulnerability of cities. Natural disasters and social resilience. Earthscan Publications, London

54. Platt RH (1991) Lifelines: an emergency management priority for the United States in the 1990s. Disasters 15(2):172–176

55. Puente S (1999) Social vulnerability to disaster in Mexico City. In: Mitchell JK (ed) Crucibles of Hazard: Mega-cities and disasters in transition. United Nations University Press, Tokyo, pp 295–334

56. Redner RA, Walker HF (1984) Mixture densities, maximum likelihood and the EM algorithm. SIAM Rev 26(2):195–239

57. Russell DW (2002) In search of underlying dimensions: the use (and abuse) of factor analysis in Personality and Social Psychology Bulletin. Pers Soc Psychol Bull 28(12):1629–1646

58. SCEMDOAG (South Carolina Emergency Management Division Office of the Adjutant General) (2009) State of South Carolina hazards assessment 2008. University of South Carolina, South Carolina

Emergency Management Division Office of the Adjutant General, Hazards Research Lab, Department of Geography, South Carolina

59. Schmidt J et al (2011) Quantitative multi-risk analysis for natural hazards: a framework for multi-risk modelling. Nat Hazards 58(3):1169–1192

60. Schmidt-Thomé P (2006) The spatial effects and management of natural and technological hazards in Europe. European Spatial Planning and Observation Network (ESPON) project 1.3.1. Geological Survey of Finland, Luxembourg

61. Shi PJ (1996) Theory and practice of disaster study. J Nat Disaster 5(4):6–17

62. Slovic PE (2000) The perception of risk. Earthscan Publications, London

63. Smith K (2000) Environmental hazards: assessing risk and reducing disaster, 3rd edn. Routledge, New York

64. Smith K (2013) Environmental hazards: assessing risk and reducing disaster, 6th edn. Routledge, New York

65. Straub D (2005) Natural hazards risk assessment using bayesian networks. The 9th international conference on structural safety and reliability. Rome

66. Tong ZJ, Zhang JQ, Liu XP (2009) GIS-based risk assessment of grassland fire disaster in western Jilin province, China. Stoch Environ Res Risk Assess 23(4):463–471

67. Villagrán de León JC (2006) Vulnerability: a Conceptual and methodological review. UNU-EHS (The United Nations University, Institute for Environment and Human Security), Bonn

68. Wisner B, Blaikie P, Cannon T, Davis I (2004) At risk: natural hazards, people's vulnerability and disasters, 2nd edn. Routledge, London

69. Zschau J, Küppers AN (eds) (2003) Early warning systems for natural disaster reduction. Springer Science & Business Media, New York

2

A NEURAL NETWORK BASED GENERAL RESERVOIR OPERATION SCHEME

Nima Ehsani[1,2] • Balazs M. Fekete[1,2] • Charles J. Vo¨ro¨smarty[1,2] • Zachary D. Tessler[2]

[1]Civil Engineering Department, City College of New York (CUNY), 160 Convent Ave, Marshak Rm. 925, New York, NY 10031, USA

[2] CUNY Environmental CrossRoads Initiative, CUNY Advanced Science Research Center, New York, NY, USA

ABSTRACT

Construction of dams and the resulting water impoundments are one of the most common engineering procedures implemented on river systems globally; yet simulating reservoir operation at the regional and global scales remains a challenge in human–earth system interactions studies. Developing a general reservoir operating scheme suitable for use in large-scale hydrological models can improve our understanding of the broad impacts of dams operation. Here we present a novel use of artificial neural networks to map the general input/output relationships in actual operating rules of real world dams. We developed a new general reservoir operation scheme (GROS) which may be added to daily hydrologic routing models for simulating the releases from dams, in regional and global-scale studies. We show the advantage of our model in distinguishing between dams with various storage capacities by demonstrating how it modifies the reservoir operation in respond to changes in capacity of dams. Embedding GROS in a water balance model, we analyze the hydrological impact of dam size as well as their distribution pattern within a drainage basin and conclude that for large-scale studies it is generally acceptable to aggregate the capacity of smaller dams and instead model a hypothetical larger dam with the same total storage capacity; however we suggest limiting the aggregation area to HUC 8 sub-basins (approximately equal to the area of a 60 km or a 30 arc minute grid cell) to avoid exaggerated results.

KEYWORDS

Dams, Reservoir operation, Neural network, Hydrological alteration, Hydrological models

1 INTRODUCTION

Construction of dams and the resulting water impoundments are one of the most common engineering procedures implemented on river systems. Half of the major global river systems are affected by dams (Dynesius and Nilsson 1994). There are over 45,000 operational large dams globally with an estimated aggregate storage capacity of over 6000 km^3 (Vörösmarty et al. 1997; WCD 2000; ICOLD 2011; Lehner et al.2011) trapping over 17 % of global annual runoff (Nilsson et al. 2005; Piao et al. 2007). Dams increase the storage of water in river systems by 700 % and triple the mean residence time of water in the rivers (Vörösmarty et al. 1997). Dams impact ecosystems by flow regulation, upstream flooding, change in sedimentation patterns, draining floodplain wetlands and altering water temperature patterns (Vörösmarty and Sahagian 2000; Kingsford 2000; Syvitski et al. 2005). Overlooking those impacts may significantly affect modelling results and influence decisions addressing water management issues.

In the context of this paper, reservoir or dam operation refers to alteration of the outgoing flow regime via accumulation of the incoming flow and delayed release of water over time. One major problem in dams' impact studies is the lack of reliable methods for simulating reservoir operation. In reality, dams are regulated in different ways and virtually each dam has a unique operating rule (Simonovic 1992; Wurbs1993). Actual reservoir operating rules are not available to the public, thus their direct use in models is infeasible, especially for macro-scale applications where hundreds or thousands of dams exist in the study domain.

To address this problem, metrics have been proposed to assess the aggregate hydrologic behavior of dam regulated rivers. For example, Graf (1999) used the total reservoir storage in a watershed as a measure of changes in flow regimes and associated downstream effects. Nilsson et al. (2005) used the percentage of the annual discharge of a river system that can be contained by the reservoirs within that system as a measure to quantify flow regulation by dams (Dynesius and Nilsson 1994). Vörösmarty et al. (1997) used the ratio of aggregated reservoir storage along river networks and the mean annual river discharge to calculate the mean local aging of water and showed how large dams might change the residency time in rivers. They used a similar approach to study the impact of reservoir construction on sediment transport to the ocean (Vörösmarty et al. 2003).

Conceptual or empirical relationships have also been used to model reservoir operation. Meigh et al. (1999) and Döll et al. (2003) used an empirical relationship to simulate the monthly release from dams based on reservoir water storage (Qout~SbQout~Sb). Coe (2000) modelled dam operation by assuming that a reservoir is full in the month of average maximum inflow and is at its minimum storage for the month of average minimum inflow. He calculated the storage for other months assuming a linear relation between those upper and lower limits. Wisser et al. (2010a) used a relationship between daily inflow and average long-term inflow to calculate the daily release from a reservoir. Haddeland et al. (2006) calculated reservoir storage and release to meet monthly irrigation and hydropower demand. Hanasaki et al. (2006) predicted monthly release based on reservoir characteristics, river discharge and water use information. Their model tries to meet the industrial, domestic and irrigation water demand (Yoshikawa et al. 2013). Recent studies have applied remote sensing applications to model dam characteristics like storage, surface area and water level (Coe and Birkett 2004; Peng et al. 2006; Gao et al. 2012).

1.1 Artificial neural networks in hydrology

Artificial neural networks (ANN) have been successfully used in various water resources studies (Hsu et al.1995; Maier and Dandy 2000; Govindaraju 2000a, b) notably in river flow forecasting (Karunanithi et al. 1994; Zealand et al. 1999; Coulibaly et al. 2000; Coulibaly and Baldwin 2008) and also to solve reservoir operation optimization problems through dynamic programming (Raman and Chandramouli 1996; Fontane et al. 1997; Jain et al. 1999; Rani and Moreira 2010). It has been demonstrated that performances of ANN-based models are similar or superior over conventional statistical and stochastic models for prediction of river flows (Abrahart and See 2000; Cigizoglu 2003; Rani and Moreira 2010).

Here for the first time we use ANN to map the general input/output relationships in actual operating rules of real world reservoirs. Our goal is to parameterize actual dam operation data by using ANN and develop a general reservoir operation scheme (GROS) that is suitable for use in large-scale hydrological models and is sufficiently accurate in simulating the operation of existing reservoirs. GROS enables us to collectively investigate the impact of dams and reservoirs operation on hydrological systems. We seek to limit the input requirements to essential and conveniently calculable data. For regional and global studies, accurate water demand data with applicable resolution is rarely available; therefore, we do not use water demand as an input to GROS. We compare the performance of GROS with a monthly reservoir model developed by Hanasaki et al. (2006) and a daily reservoir model developed by Wisser et al. (2010a).

1.2 Assessment of hydrologic alteration caused by dams

Dams alter the frequency, duration and timing of annual flooding and drought events. While this has beneficiary effects on human water security, aquatic biota is distressed by these changes as they rely on the natural hydrologic cycle for food and reproduction (Pringle et al. 2000; Kingsford 2000). Earlier works have analyzed the impact of dams on natural hydrology by comparing pre and post-dam flow regime from gage station data (Thoms and Sheldon 2000; Magilligan and Nislow 2005). This approach is only valid if no substantial change exists in other factors affecting the hydrologic regime between two periods. Impacts of climate variability and other anthropogenic disturbances such as water use, land cover change, and water transfer projects between the two periods should be accounted for (Yang et al. 2008; Chen et al. 2010). Moreover when comparing data from two periods, it is impossible to see how a different damming strategy could have affected the hydrology of that region.

2 DEVELOPMENT OF GROS

We are using ANN to develop our GROS for use in large-scale hydrological models to simulate the operation of existing reservoirs. The architecture of the ANN used in this study (Fig. 1) is determined by trial and error (Basheer and Hajmeer 2000; Rafiq et al. 2001). Each input set consists of three vectors; if t represents the daily time step, the three input vectors are: Inflow = $[I_t, I_{t-1}, I_{t-2}]$, Release = $[R_{t-1}, R_{t-2}]$ and Storage = $[S_{t-1}]$, which are used to calculate the daily release (output) from reservoir $[R_t]$. Based on our analysis of significance of inputs (discussed in Sect. 2.1), using inflow and release data from previous time steps (i.e. $t-3$, $t-4$, etc.) does not have a significant impact on performance of the ANN and since storage levels do not fluctuate substantially every day, we limited the inputs to daily reservoir storage volume, three consecutive days of inflow, and release in past 2 days. Redundant inputs increase the likelihood of overfitting and decrease ANN prediction ability (Tetko et al. 1996; Maier et al. 2010).

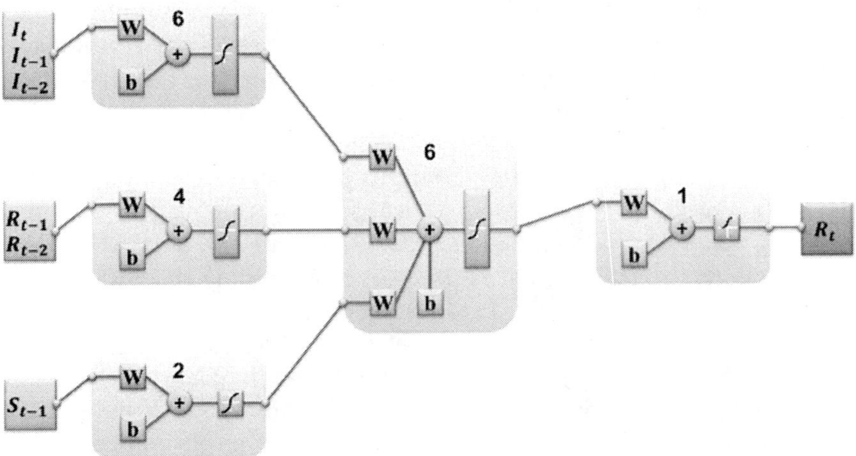

Figure 1. Architecture of the artificial neural network used in developing GROS. t stands for daily time step, I for inflow, R for release and S for Storage. W and b represent weights and biases. Rectangles in *green* are input layers, rectangles in *gray* are hidden layers, and rectangle in *orange* is the output layer. Numbers shown for each hidden layer show the number of nodes in that layer

Inflow, release and storage data are each handled by a separate input layer. The three elements of the inflow vector $[I_t, I_{t-1}, I_{t-2}]$ are connected to a hidden layer with six nodes. The two elements of the release vector $[R_{t-1}, R_{t-2}]$ are connected to a hidden layer with four nodes, and the storage $[S_{t-1}]$ is connected to a hidden layer with two nodes. Outputs from these layers are connected to the fourth layer with six nodes which are connected to the fifth layer with one node. The sigmoid hyperbolic tangent was selected as the transfer function for all the hidden layers since it provides better accuracy and faster learning speed compared to other sigmoid transfer functions (Adeloye and De Munari 2006; Taormina et al. 2012). The Log-sigmoid transfer function was used in the output layer to ensure that the output is always between 0 and 1. Testing with multiple learning algorithms, the Levenberg–Marquardt algorithm had the best performance (Maier and Dandy 2000; Kişi 2007).

Daily inflow, release and storage data of 12 dams (Table 1) are used in calibration and validation of the ANN. On average each dam has 23 years of daily data. These dams represent a wide range of storage sizes (1.5–32.3 km^3), inflows (38.5–2993.5 m^3/s), residence time (45–1089 days) and purpose. Two of the 12 dams (Sirikit and Bhumibol) are in Thailand where the climate is different compared to the other sites, which are all in the USA. Data from these two dams were excluded from the ANN training process to be used exclusively as independent validation datasets to ensure that the final ANN has not overfitted to the specific training sites. The input/output pairs from the other 10 dams were randomly divided into three subsets of training (60 %), cross training (20 %) and validation (20 %) (Table 1). In ANN specific terminology, the training set is used for determining the ANN weights and biases to minimize the error function and maximize accuracy in each iteration. The cross training set is used to oversee the training process and improve ANN generalization by minimizing overfitting. An overfitted ANN yields high accuracy on training data but fails to generalize from the training data, thus has poor performance on new, independent input data. The validation dataset provides an unbiased estimate of the generalization error (Govindaraju 2000a).

Table 1. List of the dams used in this study

Name of dam	Location	Capacity (km^3)	Mean flow (m^3/s)	Residence time (day)	Purpose	Data period	Source	Training (%)	Cross training (%)	Validation (%)
Palisades	Idaho	1.5	193.3	90	F–I–H	1970–2000	a	60	20	20
American Falls	Idaho	2.1	207.1	117	I–F–H–R	1978–1995	a	60	20	20
Navajo	New Mexico	2.1	38.5	631	I–H–F–R	1962–2002	b	60	20	20
Trinity	California	3.0	55.3	628	I–H–F	1970–2000	c	60	20	20
Falcon	Texas-Mexico	3.3	110.8	345	I–F–H–R	1958–1995	b	60	20	20
Hungry Horse	Montana	4.3	100.5	495	H–I–F	1970–2000	d	60	20	20
Oroville	California	4.4	190	268	H–F–I–S	1995–2004	c	60	20	20
Flaming Gorge	Utah	4.7	64.4	845	I–F–H–S	1962–2002	a	60	20	20
Sirikit	Thailand	9.5	166.4	661	I–F–H	1980–1996	e	0	0	100
Grand Coulee	Washington	11.6	2993.5	45	F–I–H	1978–1990	d	60	20	20
Bhumibol	Thailand	13.5	143.5	1089	I–F–H	1980–1996	e	0	0	100
Glen Canyon	Arizona	32.3	453.1	825	I–H–R	1970–2002	a	60	20	20

I, F, H, R and S stand for irrigation, flood control, hydropower, recreation and water supply

a US Bureau of Reclamation, *b* International Border and Water Commission, *c* California Department of Water Resources, *d* US Army Corp of Engineers, *e* Energy Generation Agency of Thailand

Data used in this study comes from 12 dams with different operating rules, spanning several orders of magnitude. If the actual data is used directly, the training process will fail as no significant pattern will be detectable between inputs and outputs. To avoid this problem, the data were rescaled to be dimensionless, ranging from 0 to 1. For each dam, flow (m^3/s) was converted to daily volume (m^3) to be comparable to storage (m^3); then all the data for each dam were divided by the maximum capacity of that dam. To use outputs of ANN in a hydrological model, the scaling procedure should be reversed (Appendix 1) by multiplying it by the maximum capacity of the dam to get the daily volume of released water (m^3) and then converting it to flow rate (m^3/s).

This trained ANN is the core of our GROS. Appendix 1 contains a simplified algorithm that explains how GROS simulates the daily reservoir release. Using GROS in a water balance model (WBM_{plus}) (Wisser et al.2010b), we isolated reservoir effects from other disturbances (i.e. water withdrawal, flow diversions and climate variability) and studied the impact of dams with various storage sizes and distribution patterns on river system dynamics. We calculated Colwell's parameters (Colwell 1974) and used the Indicators of hydrological alteration (Richter 1996) and Range of Variability Approach (Richter et al. 1997) to show how dams impact rivers hydrology under different scenarios (Mathews and Richter 2007).

2.1 Significance of ANN inputs

We used an Improved Stepwise method (Gevrey et al. 2003) to identify the importance of inputs used in developing GROS. The change in the sum of square errors (SSE) is calculated when that input and its corresponding weights are removed from the Neural Network. The most important variables are those which cause the largest change in SSE.

Table 2 lists the SSE and coefficient of determination (R^2) for 10 experiments that show how eliminating inputs from ANN training affects its performance. From the six single inputs used in this study (I_t, I_{t-1}, I_{t-2}, R_{t-1}, R_{t-2}, S_{t-1}), daily storage (S_{t-1}) is the most significant input. S_{t-1} represents the fraction of dam filled with water in each day. Inputs from older times have a smaller impact on ANN performance. Among the inputs, inflow at time $t-2$ (I_{t-2}) had the smallest impact on SSE.

Table 2. Significance of ANN inputs using the improved stepwise method

Experiment	Input variables	SSE	R^2
Control	ANN (I_t, I_{t-1}, I_{t-2}, R_{t-1}, R_{t-2}, S_{t-1})	0.102	0.976
Exclude inflow at $t-2$	ANN (I_t, I_{t-1}, R_{t-1}, R_{t-2}, S_{t-1})	0.106	0.975
Exclude inflow at $t-1$	ANN (I_t, I_{t-2}, R_{t-1}, R_{t-2}, S_{t-1})	0.114	0.973
Exclude inflow at t	ANN (I_{t-1}, I_{t-2}, R_{t-1}, R_{t-2}, S_{t-1})	0.119	0.972
Exclude inflow at $t-2$ and $t-1$	ANN (I_t, R_{t-1}, R_{t-2}, S_{t-1})	0.126	0.971
Exclude release at $t-2$	ANN (I_t, I_{t-1}, I_{t-2}, R_{t-1}, S_{t-1})	0.173	0.960
Exclude release at $t-1$	ANN (I_t, I_{t-1}, I_{t-2}, R_{t-2}, S_{t-1})	0.181	0.959
Exclude storage	ANN (I_t, I_{t-1}, I_{t-2}, R_{t-1}, R_{t-2})	0.188	0.958
Exclude release	ANN (I_t, I_{t-1}, I_{t-2}, S_{t-1})	0.818	0.808
Exclude inflow	ANN (R_{t-1}, R_{t-2}, S_{t-1})	2.844	0.351

2.2 Model performance

We evaluated the performance of GROS against observed time series and compared the outcomes against existing daily and monthly reservoir models (Hanasaki et al. 2006; Wisser et al. 2010a). Wisser et al. (2010a) use a simple relationship linking daily reservoir inflow I_t (m^3/s) and long-term mean inflow I_m (m^3/s) to calculate daily reservoir release R_t (m^3/s) as:

$$R_t = \begin{cases} kI_t & I_t \geq I_m \\ \lambda I_t + (I_m - I_t)I_t & I_t < I_m \end{cases} \tag{1}$$

where k and λ are empirical constants set to 0.16 and 0.6, respectively.

Hanasaki et al. (2006) categorized reservoirs as irrigation and non-irrigation and developed a set of rules for monthly release from reservoirs based on their intended purpose and water demand. For a non-irrigation reservoir, monthly release was parameterized as:

$$r'_{m,y} = i_{mean} \tag{2}$$

where r′m,yrm,y′ is the provisional monthly release (m^3/s), and i_{mean} is the mean annual inflow (m^3/s).

For an irrigation reservoir, monthly release was parameterized as:

$$r'_{m,y} = \begin{cases} \dfrac{i_{mean}}{2} \times \left(1 + \dfrac{\sum_{area}\{k_{alc} \times (d_{irg,m,y} + d_{ind} + d_{dom}\}}{d_{mean}}\right), & (d_{mean} \geq 0.5 \times i_{mean}) \\ i_{mean} + \sum\limits_{area}\{k_{alc} \times (d_{irg,m,y} + d_{ind} + d_{dom}\} - d_{mean}, & (d_{mean} \geq 0.5 \times i_{mean}) \end{cases}$$

$$d_{mean} = \sum_{area}\{k_{alc} \times (d_{irg,m,y} + d_{ind} + d_{dom}\} \tag{3}$$

where k_{alc} is an allocation coefficient for grids with more than one reservoir upstream. $d_{irg,m,y}$ is the monthly irrigation water withdrawal (m^3/s); d_{dom} is the domestic water withdrawal (m^3/s); d_{ind} is the industrial water withdrawal (m^3/s); and d_{mean} is the mean annual total water demand of the reservoir (m^3/s). The subscripts*m*, *y* and *mean*, indicate month, year and annual mean, respectively. The term $\sum_{area}$ indicates integration over the basin downstream reservoir.

The monthly release $r_{m,y}$ (m^3/s) was calculated as

$$r_{m,y} = \begin{cases} k_{rls,y} \times r'_{m,y}, & (c \geq 0.5) \\ \left(\frac{c}{0.5}\right)^2 k_{rls,y} \times r'_{m,y} + \left\{1 - \left(\frac{c}{0.5}\right)^2\right\} i_{m,y}, & (0 \leq c \leq 0.5) \end{cases} \quad (4)$$

where c is the storage capacity to mean total annual inflow ratio ($c = C/I_{mean}$); and $k_{rls,y}$ is the release coefficient, which reflects water storage at the beginning of the operational year; r′m,yrm,y′ is the provisional monthly release.

We used the observed inflow time series as inputs to GROS and calculated the daily release from the reservoirs (Appendix 1). Based on analysis of observed data, the minimum storage level for reservoirs is set to 10 % of capacity. Nash–Sutcliffe efficiency coefficient (E) (B1), coefficient of determination (R^2) (B2) and normalized root mean square error (NRMSE) (B3) are calculated for simulated dam releases (Fig. 2).

Simulation results from GROS are significantly more accurate than the outputs from the other two models. On average, GROS reduces the NRMSE (and RMSE) for simulated daily release by 72 % compared to Wisser et al. (2010a) method. For monthly simulated release, the average NRMSE (and RMSE) for Wisser et al. (2010a) and Hanasaki et al. (2006) models are comparable to each other at around 0.65 but the average NRMSE (and RMSE) for GROS is 0.11, nearly 80 % smaller. The average daily Nash efficiency coefficient is 0.86, and R-squared is 0.85 using GROS.

Accuracy of GROS outputs for simulated release from Bhumibol and Sirikit dams which were excluded from the development of ANN to be used as independent validation datasets, confirms that the ANN used in GROS is not overfitted to its training datasets.

3 METHODS FOR ASSESSING ALTERATION OF FLOW REGIMES BY DAMS

We used Colwell's parameters (Colwell 1974), a suite of indicators of hydrological alteration (Richter 1996) and range of variability approach (Richter et al. 1997) to demonstrate how changes in storage size and distribution pattern of dams in a drainage basin alter their hydrological impacts.

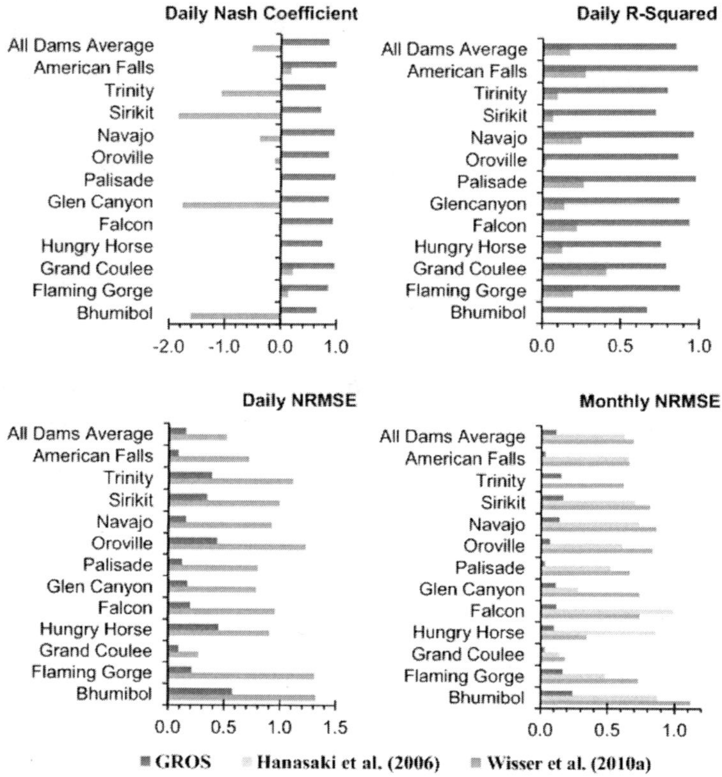

Figure 2. RMSE, E and R^2 of Daily simulated release using GROS and Wisser et al. (2010a). NRMSE of monthly release simulation using GROS, Hanasaki et al. (2006) and Wisser et al. (2010a). All dams average shows the arithmetic mean for all the 12 dams

Colwell (1974) proposed three parameters of predictability, constancy and contingency to describe patterns of temporal fluctuation in physical and biological phenomena. Predictability is a measure of temporal uncertainty across successive time domains spanning a periodic phenomenon. Maximum predictability occurs when the state of a phenomenon is known with absolute certainty in time. Maximum constancy happens when the state of a phenomenon is always constant. Contingency shows to what degree state of the phenomena depends on time. Values of these parameters range from 0 to 1 (Colwell 1974; Poff and Ward 1989).

Richter (1996) proposed the indicators of hydrologic alteration (IHA) for assessing the degree of hydrologic alteration attributable to human influence within an ecosystem. IHA consists of 32 indices (Table 3) selected to quantitatively describe hydrological alterations caused by anthropogenic disturbances between pre-impact and post-impact time periods (Richter 1996; TNC 2009). The range of variability approach (RVA) defines a range of variation (i.e. mean ± 1SD) for IHA parameters from pre-impact period data (i.e. before dam construction). The degree to which the RVA target range is not attained at the post-impact period (i.e. after dam construction) is a measure of Hydrologic Alteration (Richter et al. 1997).

We implemented GROS in WBM_{plus} and used various scenarios to understand how variation in size and location of dams in a drainage basin changes their hydrological impact. Size and location of dams in each scenario is allocated based on analysis of national inventory of dams (NID) database (USACE 2013) to be representative of real-world conditions (Appendix 4).

Discharges were simulated for a length of 25 years (1986–2010) for each scenario using MERRA precipitation and temperature data (Rienecker et al. 2011) and a gridded 3-min (longitude × latitude) simulated topological network (Vörösmarty et al. 2000) of a large arbitrary drainage basin with 36,450 km^2area (Fig. 3). The average discharge at the mouth of this drainage basin is 690 m^3/s and low pulse and high pulse thresholds (mean ± 1SD for IHA and RVA analysis) are at 86.3 and 1293 m^3/s.

3.1 The significance of water storage capacity of a single dam

Analysis of the NID database (USACE 2013) reveals that real world dams, with different purposes and significantly different storage capacities, may be built on rivers with statistically similar flow conditions (Appendix 4). Therefore, storage capacity is a critical input to reservoir models and models that do not use storage values for simulation of release cannot reliably capture the impact of dam operation. For example Wisser et al. (2010a) reservoir operation model (Eq. 1) does not use water storage in its calculations thus simulated release will be identical for dams with significantly different storage capacities. Hanasaki et al. (2006) use ratio of storage capacity to mean total annual inflow (Eq. 4); but the way it is formulated does not enable the model to respond to change in storage capacity properly.

Table 3. Summery of IHA hydrological parameters

IHA statistics group	Parameter number	Hydrological parameters
Group 1: magnitude of monthly water conditions	1–12	Mean value for each calendar month
Group 2: magnitude and duration of annual extreme water conditions	13–23	Annual min 1-day means
		Annual min 3-day means
		Annual min 7-day means
		Annual min 30-day means
		Annual min 90-day means
		Annual max 1-day means
		Annual max 3-day means
		Annual max 7-day means
		Annual max 30-day means
		Annual max 90-day means
		Base flow index: 7-day min flow/mean flow for year
Group 3: timing of annual extreme water conditions	24 & 25	Julian date of each annual 1-day max
		Julian date of each annual 1-day min
Group 4: frequency and duration of high/low pulses	26–29	No. of high pulses each year
		No. of low pulses each year
		Mean duration of high pulses in each year (days)
		Mean duration of low pulses in each year (days)
Group 5: rate/frequency of water condition changes	30–32	Rise rates: means of all positive differences between consecutive daily values
		Fall rates: means of all negative differences between consecutive daily values
		Number of hydrologic reversals

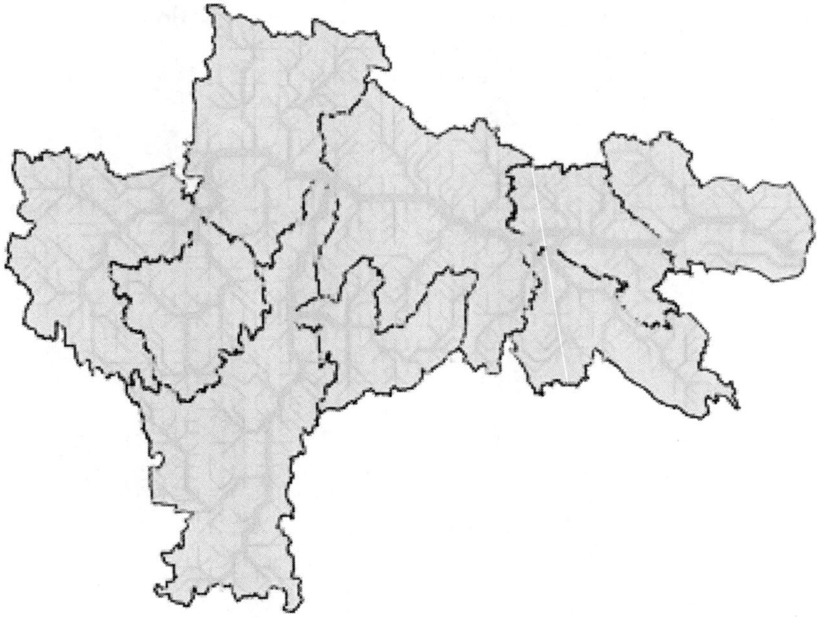

Figure 3. HUC 8 sub-basins contained in the study domain

We assumed five scenarios to show the advantage of GROS compared to other models and to quantify how variation in the size of a dam changes its hydrological impact. In the first scenario, flow was simulated without any dams. For the other scenarios, operation of a dam with 5, 10, 15 and 20 km^3 storage volume was simulated at the mouth of the drainage basin (Fig. 6, S4).

Figure 4 shows how variation in storage capacity of a dam affects its impact on extreme flow conditions. The IHA and RVA analysis results are presented in Table 7 and Fig. 5. As expected, the impact of dams on natural hydrology increases as storage capacity increases.

Dams decrease the range of fluctuations in the magnitude of monthly flows (IHA parameter 1–12). Magnitude, frequency and duration of extreme water conditions were considerably affected by dams operation (IHA parameter 13–23). Dams shifted the date of the occurrence of the minimum and maximum flows by up to 2 months (IHA parameter 24–25). There were no low pulses and the number of high pulses was reduced (IHA parameter 26–29). Because of the flow regulation by dams, the frequency and rate of daily flow change were also significantly reduced (IHA parameter 30–32).

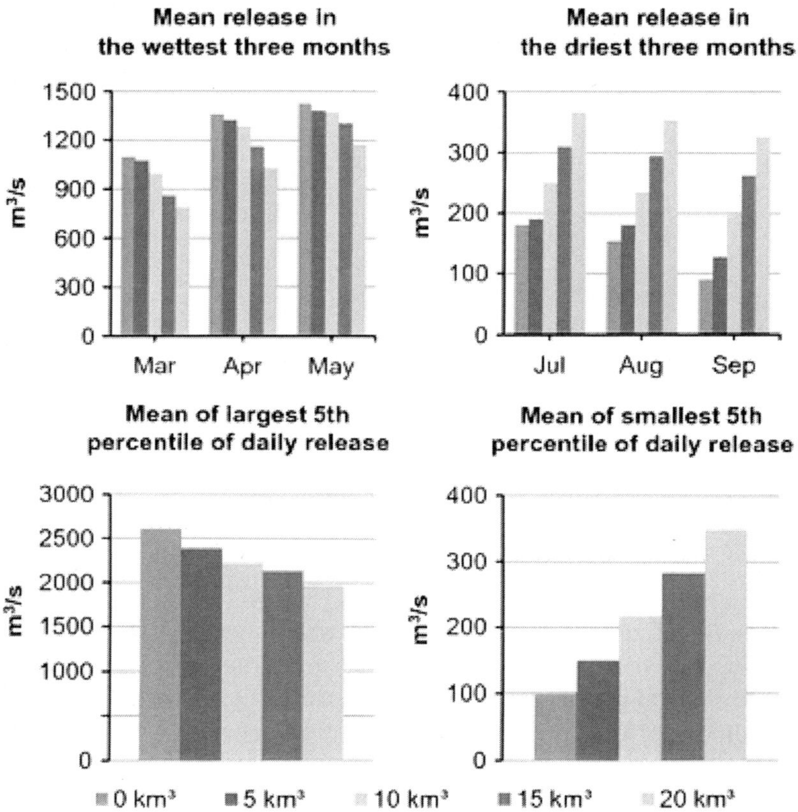

Figure 4. Impact of change in storage size of a dam on natural flow

The Colwell's parameters also consistently change with the increase in dam capacity (Table 4). Dams reduce the range of flow compared to natural conditions by reducing the magnitude of high flows and increasing the magnitude of low flows (Table 7; Fig. 5). As the storage capacity of dams increases, the standard deviation and coefficient of variation ($CV = \sigma/\mu$) of flows decrease. Limiting the range of flow variation also results in an increase in the value of Constancy, which respectively increases Predictability (Table 4).

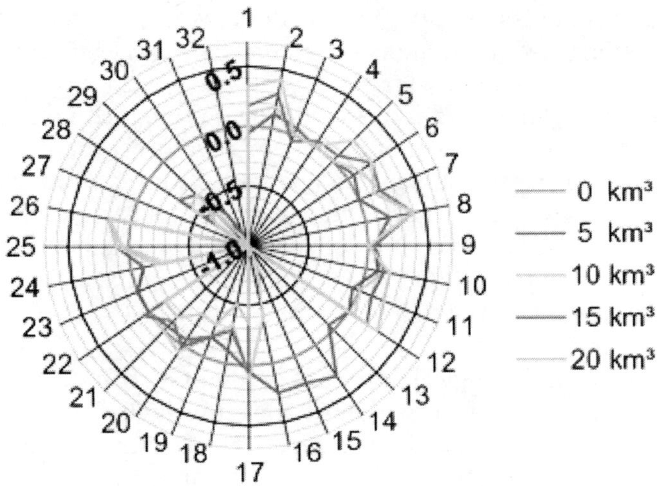

Figure 5. RVA analysis for the 32 IHA parameters using GROS for dams with different storage capacities

Table 4. Colwell's parameters for dams with different storage sizes

Parameters	0 km^3 No dam	5 km^3 GROS	10 km^3	15 km^3	20 km^3
Annual C. V.	0.87	0.82	0.77	0.71	0.63
Predictability	0.5	0.54	0.58	0.62	0.72
Constancy	0.36	0.40	0.46	0.51	0.64
Contingency	0.14	0.13	0.12	0.11	0.08

3.2 The significance of distribution patterns of water storage capacity of multiple dams in a basin

Working with incomplete reservoir databases that do not list relatively small reservoirs is a concern in reservoirs impacts studies (ICOLD 2011; Lehner et al. 2011). This is especially important as most of the world's dams are relatively small structures (Rosenberg et al. 2000).

Hydrological impacts of individual small dams may be relatively small, but the aggregate effects of numerous small dams may be substantial.

In regional and global studies it is challenging to match reservoirs with the correct rivers in the model due to inaccurate or missing geo-referencing information in many databases. To address these issues, it is customary to aggregate the storage capacity of multiple reservoirs in a watershed (Graf 1999; Nilsson et al.2005) or a grid cell (Vörösmarty et al. 1997; Hanasaki et al. 2010) and assume only one larger reservoir exists on the main river of that watershed or grid cell.

By studying the hydrological impact of different distribution patterns of dams in a drainage basin, we show how the hydrological impact of numerous, dispersed, small dams compares to the impact of a few larger ones. We also investigate if this is a valid approach to aggregate the capacity of smaller dams and instead model a hypothetical larger dam with the same total storage capacity.

We used GROS in WBM_{plus} to model river flow and operation of reservoirs under four scenarios (Table 5; Fig. 6). Based on our analysis (Appendix 4) the total storage volume of dams in the basin was set to 20 km^3. In scenario S1, 475 relatively small dams are on lower order streams of tributary branches. For scenario S2, there is a single dam in each HUC 12 sub-watersheds with the storage volume equal to aggregated storage volume of smaller dams from scenario S1 in that HUC 12 sub-watershed. Scenario S3 is similar to scenario S2 except that storage volumes are aggregated to HUC 8 sub-basins. In scenario S4 storage volume of all the dams in the basin is aggregated into one 20 km^3 dam.

Figure 7 illustrates parts of the drainage basin disturbed by reservoirs operation and the accumulated upstream storage volume at each grid cell. Figure 8 shows how variation in the distribution pattern of water storage capacity in a basin impacts some of the extreme flow conditions. Analysis results for IHA and RVA methods are summarized in Table 8 and Fig. 9, and results for the Colwell's parameters are presented in Table 6. In general, the hydrological impact on the water flow leaving the basin increases from scenario S1 to scenario S4; larger dams have a larger hydrological impact and they are generally located on larger streams, and thus regulate larger amounts of water.

Table 5. Five scenarios to study the impact of dams' distribution patterns

Scenario	Description
S0	The base scenario to model flow conditions without dams
S1	475 Dams with average capacity of 0.04 km^3 and total capacity of 20 km^3
S2	Storage volume of S1 dams aggregated to HUC 12 sub-watersheds. 217 dams with average capacity of 0.09 km^3 and total capacity of 20 km^3
S3	Storage volume of S1 (or S2) dams aggregated to HUC 8 sub-basins. 9 dams with average capacity of 2.2 km^3 and total capacity of 20 km^3
S4	One large dam with 20 km^3 capacity located at the mouth of the drainage basin

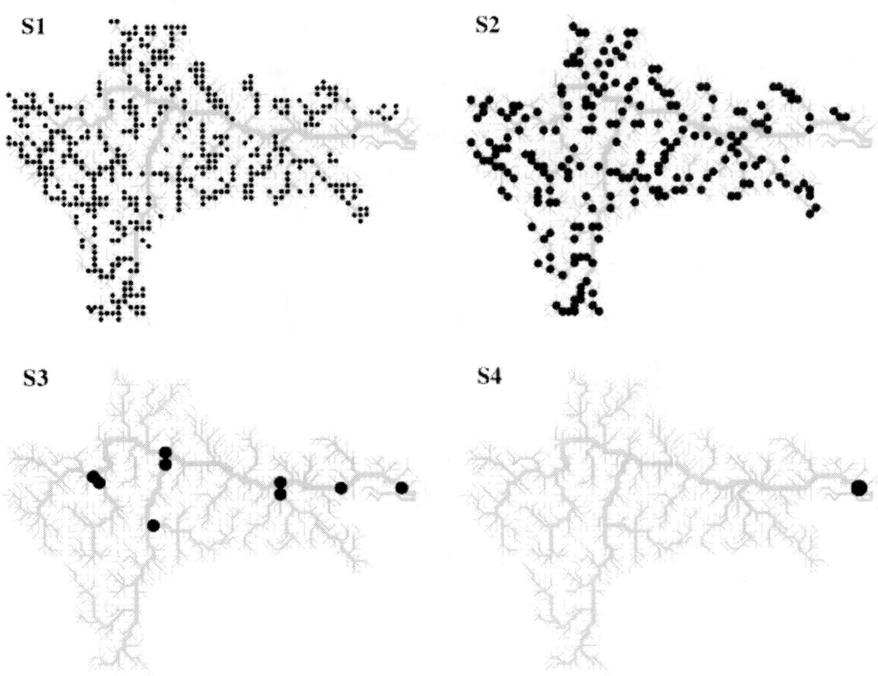

Figure 6. Location of dams in each scenario

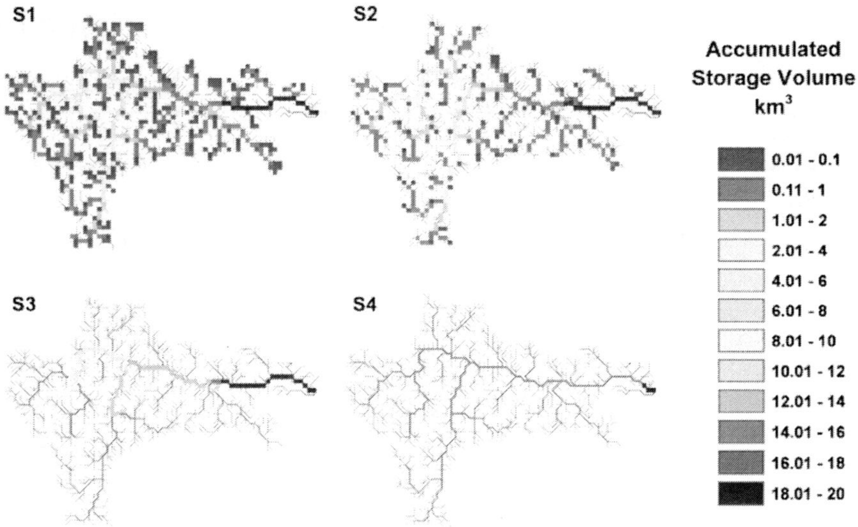

Figure 7. Parts of the basin disturbed by the dams and the accumulated upstream storage volume (km^3)

Magnitudes of monthly flows (IHA parameter 1–12) are very similar for scenarios S1–S3. From scenario S1–S4, as the number of dams decreases and their capacity increases, they have a larger impact on magnitude, frequency and duration of extreme water conditions (IHA parameter 13–23). Date of the minimum and maximum flows (IHA parameter 24–25) is approximately the same for scenarios S1–S4 and is not sensitive to the distribution pattern of dams. The same is true for low and high pulses (IHA parameter 26–29). The frequency and rate of flow changes (IHA parameter 30–32) do not show a significant difference between different scenarios.

Colwell's parameters for scenarios S1 and S2 are similar. Generally as the number of dams decreases and their size increases and they move from small streams to larger ones, their impact on Colwell's parameters increases.

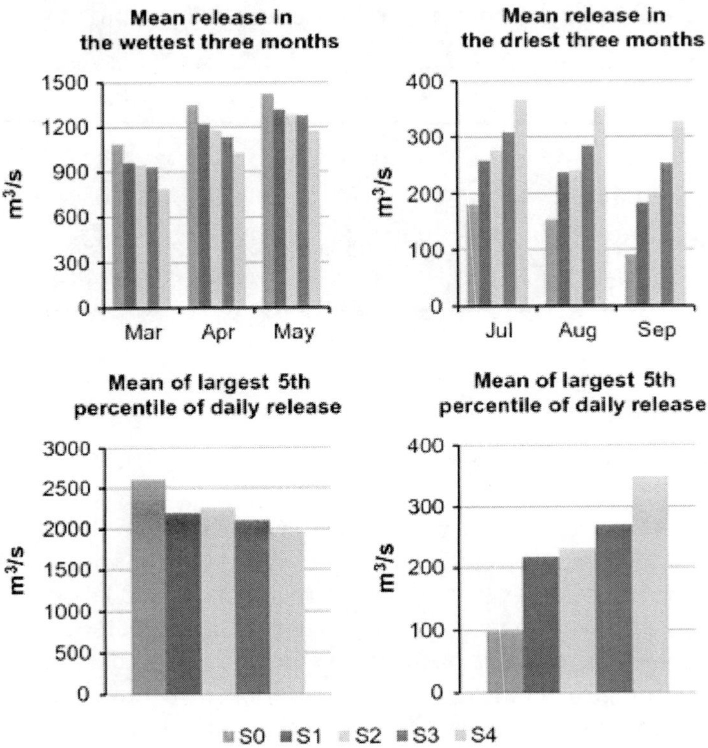

Figure 8. Impact of change in storage distribution pattern of dams on natural flow

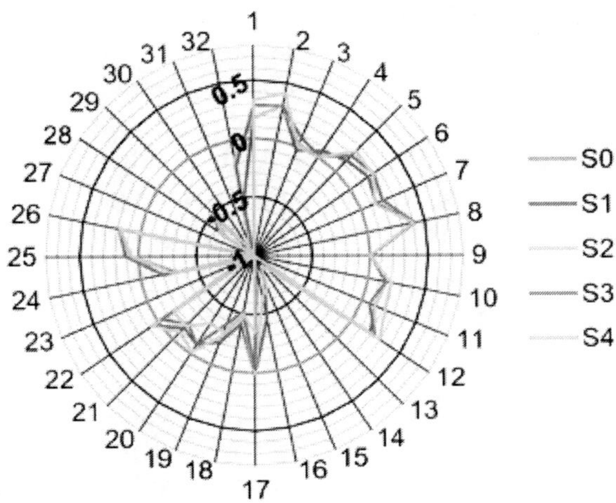

Figure 9. RVA analysis for the 32 IHA parameters using GROS for different water storage distribution patterns

Table 6. Colwell's parameters for different dam distribution patterns

Parameters	S0	S1	S2	S3	S4
Annual C. V.	0.87	0.70	0.70	0.65	0.63
Predictability	0.50	0.60	0.59	0.63	0.72
Constancy	0.36	0.49	0.48	0.52	0.64
Contingency	0.14	0.11	0.11	0.11	0.08

4 SUMMARY OF RESULTS AND DISCUSSION

By using ANN, we developed a new general reservoir operation scheme (GROS) which may be added to daily hydrologic routing models for simulating the releases from dams, in regional and global-scale studies. GROS is specifically designed to provide a broad perspective of the general behavior of dams and improve our understanding of the large-scale hydrological impact of dams operation in a relatively easy and efficient way. Comparisons with two other models using a variety of performance metrics verify that using GROS to model the operation of reservoirs can significantly improve the accuracy of the simulation of daily and monthly reservoir release time series.

Analysis of the NID database shows that dams with significantly different storage capacities may be located on rivers with similar flow characteristics. General reservoir models should be tested for their response to changes in storage capacity of dams before being implemented into hydrological models. One advantage of GROS over other models is that it properly responds to changes in storage capacity of dams and therefore can be reasonably used for simulating reservoir releases in regional and global domains where hundreds and thousands of dams with various storage capacities exist.

Using GROS in WBM_{plus} we investigated the practice of aggregating the storage capacity of multiple reservoirs in a watershed and simulating the operation of a hypothetical larger reservoir with the same total storage capacity. For this purpose we aggregated the storage capacity of dams in three scales, HUC 12 sub-watersheds (scenario S2), HUC 8 sub-basins (scenario S3) and basin level (scenario S4) and compared their hydrological impact on the water flow that leaves the basin. Based on our analysis results, hydrological impact of the original condition (scenario S1) is almost identical to scenario S2 (HUC 12 level aggregation) and very similar to scenario S3 (HUC 8 level aggregation). We conclude that for large-scale studies it is generally acceptable to aggregate the capacity of smaller dams and model a hypothetical larger dam with the same total storage capacity; however we suggest limiting the aggregation area to HUC 8 sub-basins (average of 3861 km^2 in this experiment) or a grid cell of approximately 60 km or 30 arc minute resolution to avoid exaggerated results.

Based on analysis of significance of capacity and distribution pattern of dams in the way they alter water flow out of a basin, hydrological parameters are mostly affected by the total storage capacity in the basin rather than the pattern in which storage is distributed in the basin. However, it should be noted that a few large dams have a greater impact on hydrological parameters compared to numerous smaller dams with the same total storage capacity. In our experiment, hydrologic alteration of the flow leaving the basin caused by a single large dam was greater than the combined impact of 475 relatively smaller dams with the same cumulative storage. This means that a single large reservoir is a more effective structure to regulate water compared to numerous smaller reservoirs with the same cumulative water storage capacity. Having only one large dam on the main stream of a basin decreases the level of river fragmentation as tributary branches are not affected by the dam operation. These points should be considered for both cases of building new dams and restoration of rivers by dam removal.

ACKNOWLEDGMENTS

We are grateful to Naota Hanasaki for sharing datasets and giving us access to results from his previously published work. The authors wish to express their gratitude to Reza Khanbilvardi, Eugene Z. Stakhiv, Michael Piasecki and Matthew C. Larsen for providing valuable insights. We extend our gratitude to Katherine M. Jensen and the anonymous reviewers for their valuable comments that shaped the final form of this article. This work was supported by the National Science Foundation, Award No. 1049181.

REFERENCES

1. Abrahart RJ, See L (2000) Comparing neural network and autoregressive moving average techniques for the provision of continuous river flow forecasts in two contrasting catchments. Hydrol Process 14:2157–2172.

2. Adeloye AJ, De Munari A (2006) Artificial neural network based generalized storage-yield-reliability models using the Levenberg–Marquardt algorithm. J Hydrol 326:215–230.

3. Basheer IA, Hajmeer M (2000) Artificial neural networks: fundamentals, computing, design, and application. J Microbiol Methods 43:3–31.

4. Chen YD, Yang T, Xu CY et al (2010) Hydrologic alteration along the middle and upper East River (Dongjiang) basin, South China: a visually enhanced mining on the results of RVA method. Stoch Environ Res Risk Assess 24:9–18.

5. Cigizoglu HK (2003) Estimation, forecasting and extrapolation of river flows by artificial neural networks. Hydrol Sci J 48:349–361

6. Coe MT (2000) Modeling terrestrial hydrological systems at the continental scale: testing the accuracy of an atmospheric GCM. J Clim 13:686–704.

7. Coe MT, Birkett CM (2004) Calculation of river discharge and prediction of lake height from satellite radar altimetry: example for the Lake Chad basin. Water Resour Res 40:1–12

8. Colwell RK (1974) Predictability, constancy, and contingency of periodic phenomena. Ecology 55:1148–1153

9. Coulibaly P, Baldwin CK (2008) Dynamic neural networks for nonstationary hydrological time series modeling. Pract Hydroinf.

10. Coulibaly P, Anctil F, Bobée B (2000) Daily reservoir inflow forecasting using artificial neural networks with stopped training approach. J Hydrol 230:244–257.

11. Döll P, Kaspar F, Lehner B (2003) A global hydrological model for deriving water availability indicators: model tuning and validation. J Hydrol 270:105–134.

12. Dynesius M, Nilsson C (1994) Fragmentation and flow regulation of river systems in the northern third of the world. Science 266:753–762.

13. Fontane DG, Gates TK, Moncada E (1997) Planning reservoir operations with imprecise objectives. J Water Resour Plan Manag 123:154–162

14. Gao H, Birkett C, Lettenmaier DP (2012) Global monitoring of large reservoir storage from satellite remote sensing. Water Resour Res. doi:10.

15. Gevrey M, Dimopoulos I, Lek S (2003) Review and comparison of methods to study the contribution of variables in artificial neural network models. Ecol Model 160:249–264.

16. Govindaraju RS (2000a) Artificial neural networks in hydrology. I: preliminary concepts. J Hydrol Eng 5:115–123.

17. Govindaraju RS (2000b) Artificial neural networks in hydrology. II: hydrologic applications. J Hydrol Eng 5:124–137.

18. Graf WL (1999) Dam nation: a geographic census of American dams and their large-scale hydrologic impacts. Water Resour Res 35:1305–1311.

19. Haddeland I, Lettenmaier DP, Skaugen T (2006) Effects of irrigation on the water and energy balances of the Colorado and Mekong river basins. J Hydrol 324:210–223.

20. Hanasaki N, Kanae S, Oki T (2006) A reservoir operation scheme for global river routing models. J Hydrol 327:22–41.

21. Hanasaki N, Inuzuka T, Kanae S, Oki T (2010) An estimation of global virtual water flow and sources of water withdrawal for major crops and livestock products using a global hydrological model. J Hydrol 384:232–244.

22. Hsu K, Gupta HV, Sorooshian S (1995) Artificial neural network modeling of the rainfall-runoff process. Water Resour Res 31:2517.

23. ICOLD (2011) The world registers of dams. International Commission on Large Dams, Paris

24. Jain SK, Das A, Srivastava DK (1999) Application of ANN for reservoir inflow prediction and operation. J Water Resour Plan Manag 125:263–271.

25. Karunanithi N, Grenney WJ, Whitley D, Bovee K (1994) Neural networks for river flow

26. Kingsford RTT (2000) Ecological impacts of dams, water diversions and river management on floodplain wetlands in Australia. Austral Ecol 25:109–127.

27. Kişi Ö (2007) Streamflow forecasting using different artificial neural network algorithms. J Hydrol Eng 12:532–539.

28. Lehner B, Liermann CR, Revenga C et al (2011) High-resolution mapping of the world's reservoirs and dams for sustainable river-flow management. Front Ecol Environ 9:494–502.

29. Magilligan FJ, Nislow KH (2005) Changes in hydrologic regime by dams. Geomorphology 71:61–78.

30. Maier HR, Dandy GC (2000) Neural networks for the prediction and forecasting of water resources variables: a review of modelling issues and applications. Environ Model Softw 15:101–124.

31. Maier HR, Jain A, Dandy GC, Sudheer KP (2010) Methods used for the development of neural networks for the prediction of water resource variables in river systems: current status and future directions. Environ Model Softw 25:891–909.

32. Mathews R, Richter BD (2007) Application of the indicators of hydrologic alteration software in environmental flow setting. J Am Water Resour Assoc 43:1400–1413.

33. Meigh JR, McKenzie AA, Sene KJ (1999) A grid-based approach to water scarcity estimates for eastern and southern Africa. Water Resour Manag 13:85–115

34. Nilsson C, Reidy CA, Dynesius M, Revenga C (2005) Fragmentation and flow regulation of the world's large river systems. Science 308:405–408.

35. Peng D, Guo S, Liu P, Liu T (2006) Reservoir storage curve estimation based on remote sensing data. J Hydrol Eng 11:165–172.

36. Piao S, Friedlingstein P, Ciais P et al (2007) Changes in climate and land use have a larger direct impact than rising CO_2 on global river runoff trends. Proc Natl Acad Sci U S A 104:15242–15247.

37. Poff NL, Ward JV (1989) Implications of streamflow variability and predictability for lotic community structure: a regional analysis of streamflow patterns. Can J Fish Aquat Sci 46:1805–1818.

38. Pringle CM, Freeman MC, Freeman BJ (2000) Regional effects of hydrologic alterations on riverine macrobiota in the new world: tropical-temperate comparisons. Bioscience 50:807.

39. Rafiq MY, Bugmann G, Easterbrook DJ (2001) Neural network design for engineering applications. Comput Struct 79:1541–1552.

40. Raman H, Chandramouli V (1996) Deriving a general operating policy for reservoirs using neural network. J Water Resour Plan Manag 122:342–347.

41. Rani D, Moreira MM (2010) Simulation-optimization modeling: a survey and potential application in reservoir systems operation. Water Resour Manag 24:1107–1138.

42. Richter BD (1996) A method for assessing hydrologic alteration within ecosystems. Conserv Biol 10:1163–1174.

43. Richter BD, Baumgartner JV, Wigington R, Braun DP (1997) How much water does a river need? Freshw Biol 37:231–249.

44. Rienecker MM, Suarez MJ, Gelaro R et al (2011) MERRA: NASA's modern-era retrospective analysis for research and applications. J Clim 24:3624–3648.

45. Rosenberg DM, McCully P, Pringle CM (2000) Global-scale environmental effects of hydrological alterations: introduction. Bioscience 50:746

46. Simonovic SP (1992) Reservoir systems analysis: closing gap between theory and practice. J Water Resour Plan Manag 118:262–280.

47. Syvitski JPM, Vörösmarty CJ, Kettner AJ, Green P (2005) Impact of humans on the flux of terrestrial sediment to the global coastal ocean. Science 308:376–380.

48. Taormina R, Chau K, Sethi R (2012) Artificial neural network simulation of hourly groundwater levels in a coastal aquifer system of the Venice lagoon. Eng Appl Artif Intell 25:1670–1676.

49. Tetko IV, Villa AEP, Livingstone DJ (1996) Neural network studies. 2. variable selection. J Chem Inf Model 36:794–803.

50. Thoms MC, Sheldon F (2000) Water resource development and hydrological change in a large dryland river: the Barwon–Darling River, Australia. J Hydrol 228:10–21.

51. TNC (2009) Indicators of hydrologic alteration version 7.1 user's manual. The Nature Conservancy, Charlottesville

52. USACE (2013) National inventory of dams (NID). In: U.S. Army Corps Eng. http://nid.usace.army.mil

53. Vörösmarty CJ, Sahagian D (2000) Anthropogenic disturbance of the terrestrial water cycle. Bioscience 50:753.

54. Vörösmarty CJ, Sharma KP, Fekete BM et al (1997) The storage and aging of continental runoff in large reservoir systems of the world. Ambio 26:210–219

55. Vörösmarty CJ, Fekete BM, Meybeck M, Lammers RB (2000) Geomorphometric attributes of the global system of rivers at 30-minute spatial resolution. J Hydrol 237:17–39.

56. Vörösmarty CJ, Meybeck M, Fekete B et al (2003) Anthropogenic sediment retention: major global impact from registered river impoundments. Glob Planet Chang 39:169–190.

57. WCD (2000) Dams and development: a new framework for decision-making: the report of the world commission on dams. Earthscan Publications Ltd, London and Sterling

58. Wisser D, Fekete BM, Vörösmarty CJ, Schumann AH (2010a) Reconstructing 20th century global hydrography: a contribution to the global terrestrial network-hydrology (GTN-H). Hydrol Earth Syst Sci 14:1–24.

59. Wisser D, Frolking S, Douglas EM et al (2010b) The significance of local water resources captured in small reservoirs for crop production—a global-scale analysis. J Hydrol 384:264–275.

60. Wurbs RA (1993) Reservoir-system simulation and optimization models. J Water Resour Plan Manag 119:455–472.

61. Yang T, Zhang Q, Chen YD et al (2008) A spatial assessment of hydrologic alteration caused by dam construction in the middle and lower Yellow River, China. Hydrol Process 22:3829–3843.

62. Yoshikawa S, Cho J, Yamada HG et al (2013) An assessment of global net irrigation water requirements from various water supply sources to sustain irrigation: rivers and reservoirs (1960–2000 and 2050). Hydrol Earth Syst Sci Discuss 10:1251–1288.

63. Zealand CM, Burn DH, Simonovic SP (1999) Short term streamflow forecasting using artificial neural networks. J Hydrol 214:32–48.

3

DYNAMICS OF CARBON DIOXIDE CONCENTRATION IN INDOOR AIR

Andrzej Szczurek 1• Monika Maciejewska1 • Rafał Połoczan´ski2 • Marek Teuerle 2• Agnieszka Wyłoman´ska2

[1] Institute of Air Conditioning and District Heating, Wroclaw University of Technology, Wybrzeze Wyspian _ ´skiego 27, 53-007 Wrocław, Poland

[2] Hugo Steinhaus Center, Institute of Mathematics and Computer Science, Wroclaw University of Technology, Wybrzeze, Wyspian´skiego 27, 53-007 Wrocław, Poland

ABSTRACT

Carbon dioxide is an indicator of indoor air quality. A number of factors influence its concentration. Due to the fact that they all present time variability, CO_2 concentration indoors considerably varies over time. In this work we focus on the dynamics of indoor CO_2 concentration changes. We examine the dynamics of CO_2variation and use it as a source of information on the character of collective impact of factors on indoor air. The proposed method is based on mean square displacement analysis (MSD) applied to the segments of the time series of CO_2 monitoring data. The segments are determined based on the introduced criterion for optimal sample size selection. The method was validated by showing that it reproduces the known stochastic dynamics of the simulated data set properly. From the real data analysis, we found that indoors the stochastic dynamics of CO_2 concentration in time was mainly nonlinear. Moreover, it exhibited a cycle of change which could be associated with the daily variation of the collective influence of factors on indoor air. We intend to apply the method to other parameters of indoor air, aiming at developing a capability of describing the dynamics of indoor air as a complex system.

KEYWORDS

Indoor air quality, Mean square, displacement, Time series, Variability

1. INTRODUCTION

It is widely recognized that the concentration of CO_2 is an indicator of indoor air quality. Although this gas is not considered to pose serious health risks to occupants, the experience indicates that elevated CO_2 cause drowsiness, lethargy and a general sense that the air is stale. Researchers demonstrate correlations between concentrations of this gas and reduced productivity as well as the overall performance decrease (Seppanen et al. 1999; Seppanen and Fisk 2004; Apte et al. 2000a, b; Daisey et al. 2003; Mendell and Heath2005; Haverinen-Shaughnessy et al. 2011). For these reasons, people are frequently interested in checking whether the actually observed CO_2 concentrations exceed limit or guideline values. However, this type of interest is quite limited. It completely overlooks the fact that the data on CO_2 concentration indoors are much more informative. In particular, they may be employed as a source of information on the driving forces of indoor microclimate.

The residential CO_2 is affected by many factors. Major ones are associated with wide range of indoor sources of this compound, their type e.g. metabolic or combustion and their intensity e.g. human activity. Highly meaningful are the operation of heating, ventilation and air conditioning system (HVAC) as well as the overall building maintenance. Other contributions come from ambient conditions and the interaction between building and its surroundings. Essentially, all these factors present time variability and in various manner they induce perturbations of CO_2 concentration. Indeed, continuous measurements demonstrate there exists a considerable temporal variation of CO_2 in indoor air.

Yet, the focus of this work is not the temporal variation of CO_2. We propose to make one step forward and to focus on the dynamics of indoor CO_2 concentration changes. While the variability of CO_2 concentration in time is able to reflect the influence of factors on indoor air, the dynamics of changes may provide the information on the character of their impact. This statement is fundamental for our work. The proposal to study the dynamics of CO_2 variation, and use it as a source of information on the type of collective impact of factors on indoor air, is a major novelty in this work. We focused on the collective impact, because in practice it is difficult to separate the influences originating from individual factors.

The traditional temporal analysis of time series refers to the variation in the mean and other standard sample statistics such as e.g. variance and standard deviation, range. A major shortcoming of this approach lies in the fact that the methods do not allow to study the dynamics of change. In this work, we propose a method which allows to recognize the character of dynamics of temporal variation of CO_2 concentration. It is our original approach.

The method is based on three major assumptions. (1) The dynamics of indoor CO_2 concentration is determined on the basis of univariate time series of measurement data. They reflect the stochastic dynamics of the values observed over time. The term univariate time series means the set of well-defined data of a single quantity (CO_2 concentration) which have been measured at regular time intervals. The data shall be collected for a relatively long time. (2) The dynamics maintains common character over some period of time, i.e. in time intervals. The size of intervals as well as their position in time domain is unknown prior to the analysis. For this reason, the method shall allow to identify segments of the time series reflecting one kind of dynamics of CO_2 concentration. (3) The objective is to distinguish between three types of dynamics of indoor CO_2 concentration, namely: linear, sub-linear and super-linear. For this reason it is proposed to apply for characterization the sample mean square displacement (MSD) analysis. More specifically, the individual categories of dynamics are indicated using β parameter.

In this paper we demonstrate the feasibility of the approach for studying the temporal dynamics of CO_2concentration. In our future work the method will be applied to investigate other parameters of indoor air. Ultimately, we would like to achieve the capability of characterizing the dynamics of an entire complex system which is indoor air. It may become very useful in diagnosing the microclimate as well as designing a proper one.

2. METHODOLOGY

In this section we present major elements involved when analyzing the stochastic dynamics of the time series of indoor CO_2 concentration. The key element of our proposal is to characterize the dynamics in time intervals, which delimit adjacent, non overlapping fragments of the time series. For those data segments, the dynamics may be described in a consistent manner, with a single value of an indicative parameter.

The main tool which we apply to describe the stochastic dynamics for a fragment of the time series is the time-averaged mean squared displacement (MSD). The choice is based on the fact that for samples being realizations of processes with stationary increments, the MSD behaves like power function τβτβ, for some parameter β>0β>0. The value of this parameter characterizes the stochastic dynamics of the process.

In this paper we are primarily interested in the problem of choosing a proper size of data segment for which the MSD behaves like the mentioned power function.

In order to solve this problem we introduce an information criterion. It takes under consideration the quality of MSD fitting with the power function τβτβ, but it also penalizes for the excessive number of observations involved in calculations. On the basis of this criterion we can conclude how many observations should be taken in order to achieve a reliable estimate of β. In other words, we obtain the size of time window which may be applied to time series when calculating β. Hence, there is obtained the time resolution adequate for analyzing the stochastic dynamics of the time series of indoor air parameter.

2.1 Mean square displacement

The MSD for the sample {Xi,i=1,2,...,n} of length n with stationary increments is defined as follows (Burnecki and Weron 2010):

$$M_n(\tau) = \frac{1}{n-\tau}\sum_{k=1}^{n-\tau}(X_{k+\tau} - X_k)^2, \tag{1}$$

and it is a function of ττ which indicates the difference between observations. Let us mention, the MSD defined as in Eq. (1) is a random variable.

In general, the MSD can be obtained either by performing an average over an ensemble of particles, or by taking the temporal average over a single trajectory. The second approach is used in this paper. Some examples of using the MSD measure one can find in Caspi et al. (2000), Golding and Cox (2006), He et al. (2008), Lubelski et al. (2008).

One of important properties of stochastic process {Xt} which can be recognized by using MSD is its dynamics. For sample being a realization of process {Xt} with stationary increments, the MSD defined in (1) behaves like:

$$M_n(\tau) \sim \tau^{\beta}. \tag{2}$$

We classify the process with β=1 as exhibiting linear dynamics. If β<1, the process may be described as featured by sub-linear dynamics, and consequently β>1β>1 points at the super-linear dynamics of the stochastic process. Some examples of processes that exhibit non-linear dynamics one can find in Barkai et al. (2012), Burov et al. (2001), Hoefling and Franosch (2013), Sokolov (2012).

The origin of nonlinear dynamics in a given system is often unknown (Guigas et al. 2007; Szymanski and Weiss 2009). However, (1) sub-linear (β<1) dynamics is usually observed in disordered systems, which are in stable conditions (Cushman et al. 2011). They possess memory. In such systems, long-range correlations are valid. System variables are typically described by Lévy-type statistics and heavy tailed distributions. Upon sub-linear dynamics, gross system changes are retarded by various factors and their interactions. More specifically, shifts of the system in one direction are quickly compensated by changes in the opposite one. (2) If β=1 sample MSD is proportional to time. It is characteristic for systems with stationary conditions and random fluctuations. The processes which have this type of dynamics are local in space and do not have memory. The random variations can be described as a series of transitions characterized by finite length scales for each step and finite time scales between transitions. Mathematically, this represents a sequence of steps done one after another, where each step follows a random direction which does not depend on one of the previous step (Markov property). One may attribute linear dynamics to neutral conditions in the system, i.e. the situation when various influences are well balanced. (3) The super-linear growth of MSD with time (β>1) is observed when the system is accelerated. In other words, when it is unstable or in transient state. Such conditions may be caused by imbalanced influences, e.g. resulting from active transport processes. From statistical point of view, the super-linear dynamics is typical for standard random walks with heavy tailed jump length distributions. There are many examples of processes with non-linear dynamics. We should mention here Lévy walks (Froemberg and Barkai 2013; Godec and Metzler 2013; Magdziarz and Teuerle 2015), fractional Brownian motion (Deng and Barkai 2009; Jeon and Metzler 2012), continuous time random walk (Barkai et al. 2000), heterogenous diffusion process (Cherstvy et al. 2013; Fulinski 2011), scaled Brownian motion (Lim and Muniandy 2002). For additional examples we refer the reader to (Harder et al. 1987; Iomin 2011; Jeon et al. 2013; Meroz et al.2010).

2.2 A method to select optimal sample size for MSD analysis

There is given a time series with removed deterministic trend and seasonal components.

First, for the window length *m* we calculate the sample MSD according to formula (1). More precisely, we divide the sample {Xi,i=1,2,...,n} of length *n* to sub-samples of lengths *m* (non-overlapping sub-samples) and for each subsample we calculate MSD. As a result we obtain ⌊n/m⌋ MSD functions, one for each sub-sample. We call them Mi,m(τ)Mi,m(τ) for i=1,2.,,,⌊n/m⌋i=1,2.,,,⌊n/m⌋. Next, to each Mi,m(τ) we fit the power function τβmiτβim by using least squares method. As a result we obtain the parameter βmiβim corresponding to each sub-sample. The fitting procedure is described in details in Maciejewska et al. (2012).

In order to check the quality of fit, we introduce the measure of fitting accuracy. This measure is defined for i=1,2,3,...,⌊n/m⌋i=1,2,3,...,⌊n/m⌋ as a mean square error between the log(Mi,m(τ))log and function log($(\tau^{\beta_i^m})$

$$e(i,m) = \frac{1}{k}\sum_{j=1}^{k}\left(\log(M_{i,m}(\tau_j)) - \log(\tau_j^{\beta_i^m})\right)^2, \qquad (3)$$

where τ1,...,τkτ1,...,τk are arguments for which the functions are calculated.

In order to find the optimal window length for MSD analysis, we introduce the information criterion. For a particular window length *m* we define IC_m as follows:

$$IC_m = \frac{\frac{1}{\lfloor n/m \rfloor}\sum_{i=1}^{\lfloor n/m \rfloor} e(i,m)}{\max_i\{e(i,m)\}} + \frac{m}{n}. \qquad (4)$$

The criterion is a sum of two components. The first refers to the fitting error when using sub-samples of length *m* and the second accounts for the size of the time window itself. As we observe, the value of the criterion is small when the fitting error is small and when the time window is short. Hence, the minimum value of the statistics IC_m is considered as indicative for the optimal window length in MSD calculations. We denote the optimal window length as m_{opt}. The βmoptβmopt calculated for the time windows of that size may be taken into consideration when analyzing the stochastic dynamics of the time series of indoor air parameter.

3. EXPERIMENTAL

Measurements of carbon dioxide concentration were performed in a lecture room with an amphitheatric layout. Room dimensions are 19 × 8 m × (4–2.9 m). It has only one external wall, which is fitted with huge, openable windows. Despite availability of mechanical ventilation, air exchange is realized predominantly via the natural ventilation. Teaching hours extend from 7:30 to 21:00. Classes are held during all working days and on majority of weekends (part time studies). Teaching blocks are typically 1.5 h long with the brakes of 15 min in-between. Although designed for 90 students, the lecture room is hardly ever occupied to that extent. In the examined period, the number of listeners changed considerably within a single day as well as from day to day.

CO_2 measurements were performed the central part of the room at the height of about 1 m. The measuring device was separated from the direct influence of the emission sources (students and the teacher).

The monitoring was realized with the instrument dedicated to continuous measurements and data logging. It is based on the NDIR sensor and it offers the following measuring characteristics: measuring range 0–5000 ppm; accuracy 50 ppm +3 % of measured value and the measurement data resolution 1 ppm. This level of performance may be currently considered as a standard in indoor air quality studies.

The measurement results were recorded with time resolution of 15 s. In this work we analyzed the data collected during eight consecutive days, between 6th and 14th June 2012.

4. ANALYSIS OF SIMULATED DATA

In this section we test whether the proposed methodology recovers the true dynamics of an exemplary time series. We present the results of applying the methods presented in Sect. 2 to simulated data with known characteristics. The simulated time series was characterized by the time varying β. More specific, we assumed that simulated data consisted of subsamples of equal length with the parameter β constant within every subsample and changing on consecutive subsamples. Due to the periodicity observed in real-world data from Sect. 5 we included in our simulated data a periodic parameter β on subsequent subsamples. Nevertheless, our method is capable to cope with a nonperiodic case as well.

In order to obtain the requested data we use the fractional Brownian motion (fBM), since it is a process with the MSD being asymptotically like a power law function, see Eq. (2), with the power law exponent controlled by the model's parameter. Namely, the fBM is a mean zero Gaussian process {BH(t)} defined as follows (Mandelbrot and Van Ness 1968):

$$B_H(t) = \int_{-\infty}^{\infty} \left((t-u)_+^{H-1/2} - (-u)_+^{H-1/2} \right) dB(u), \quad t \geq 0,$$

where {B(t)} is the classical Brownian motion, (x)+=max(x,0) and parameter HH (H∈(0;1)) is called Hurst exponent. What is important, the second moment of the process {BH(t)} is expressed as $\left\langle B_H(t)^2 \right\rangle = \sigma^2 t^{2H}$ (for some parameter σ>0σ>0) and it coincides with its sample MSD, therefore we have that β=2Hβ=2H.

The simulation was performed according to algorithm presented in Stoev and Taqqu (2004). We simulated 12,000 observations. This set consisted of 2424 subsets of length 500 (k=500) and was characterized by varying parameter *H*. Namely, the entire data set was

$$\left(\left(B_{H(1)}(i)\right)_{i=1}^{k}, \left(B_{H(2)}(i)\right)_{i=k+1}^{2*k}, \ldots, \left(B_{H(24)}(i)\right)_{i=23*k+1}^{24*k} \right),$$

with the parameter H(·) being periodic:

$$H(j) = \begin{cases} 0.2 & \text{for } mod(j,4) = 1, \\ 0.8 & \text{for } mod(j,4) = 2, \\ 0.6 & \text{for } mod(j,4) = 3, \\ 0.3 & \text{for } mod(j,4) = 0. \end{cases} \quad (5)$$

It is worth noticing that the increments of our data are stationary only within the subsamples of length 500 on which the parameter *H* is constant. Therefore our simulated data are of different type than so-called multifractal fBM considered in Ralchenko and Shevchenko (2010).

The trajectory of simulated data is presented in Fig. 1. One can notice that on subsequent intervals of length 500, we observe trajectories of fBM with parameter *H* given by formula (5). The symptoms of time-varying parameter *H* are visible in Fig. 2, where the amplitude of the increments for the presented trajectories varies over time.

The simulated data was analyzed using an approach described in Sect. 2. The goal was to select optimal time window size for MSD analysis. In Fig. 3 we present the information criterion plotted versus time window length. From the obtained results it appears that the optimal window length is mopt=502mopt=502 observations. This value almost replicates the true value (k = 500) for the simulated data.

As the final step we attempted to reconstruct the values of parameter β in consecutive time windows of length moptmopt. In Fig. 4 we plot the estimated values of the parameter β together with true values. There we find almost perfect agreement. This result confirms that the approach proposed in Sect. 2 can be used to such data. In next section we present our tools applied to real-world data.

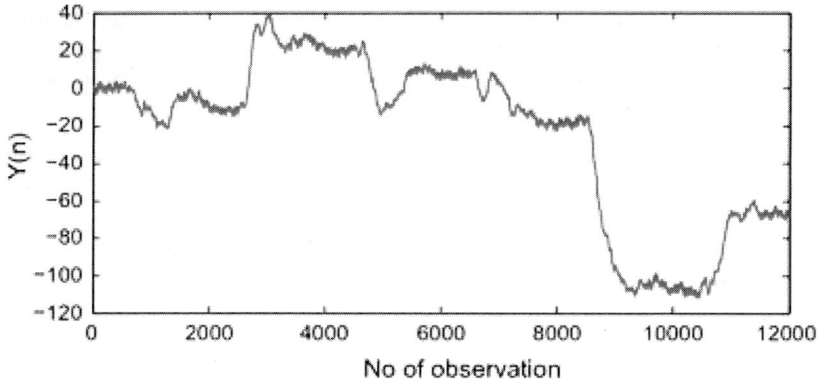

Figure. 1

A trajectory of simulated data considered in Sect. 4

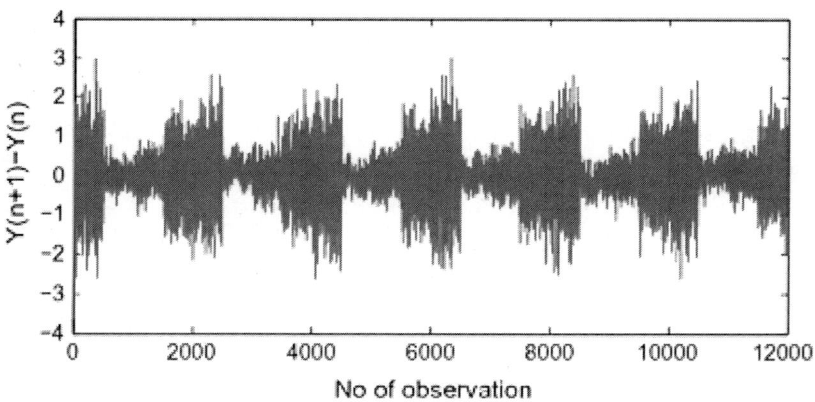

Figure 2. The increments of a trajectory presented in Fig. 1—one can observe typical periodic variance change between subsequent subsamples of length 500

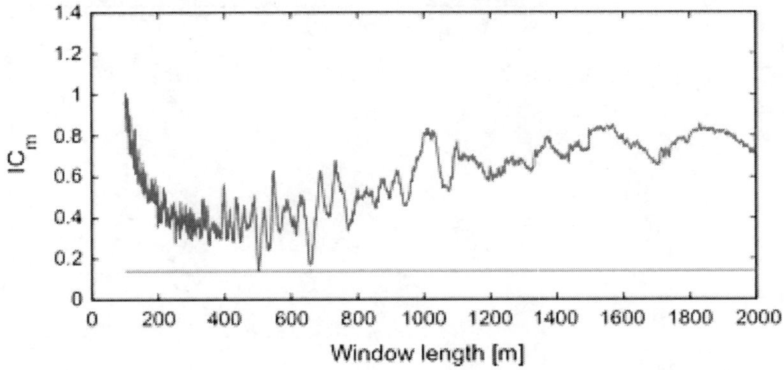

Figure 3. The values of information criterion for different window sizes. The minimum value of the criterion is obtained at m=502m=502, which indicates the optimal window size

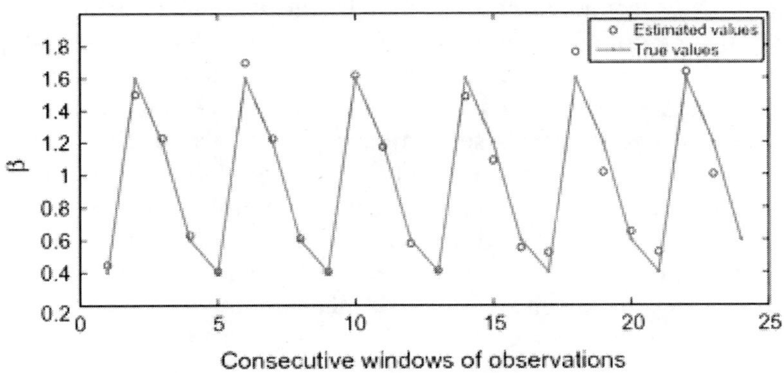

Figure 4. The comparison between true and estimated values of parameter β on consecutive windows of observations (of length mopt=502mopt=502) for simulated data

5. ANALYSIS OF CARBON DIOXIDE TIME SERIES

The recorded time series of CO_2 concentration is displayed in Fig. 5. We analyzed the stochastic dynamics of this data.

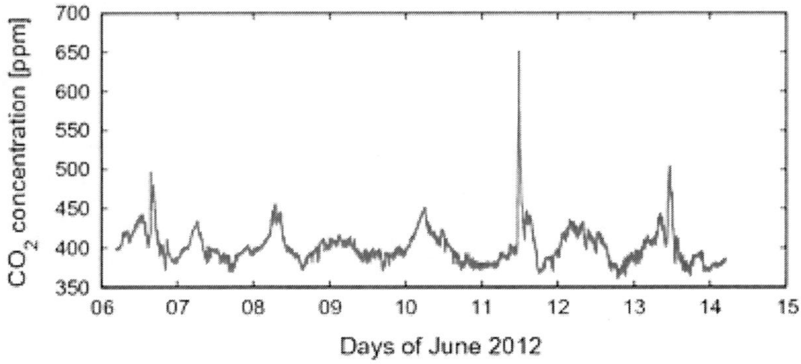

Figure 5. Time series of CO_2 concentration

In the initial step, we performed data transformations which allowed us to obtain the process with stationary increments. For that sake, all deterministic components observed in the data had to be removed. First, the polynomial trend was fitted to the data by using least squares method. The proper order of polynomial was calculated on the basis of mean square error measure. After removing the deterministic polynomial, one should also eliminate the deterministic seasonality. Using periodogram we have identified the dominant cycle which was equal to 5,760 points, i.e. number of data points in a single day. We proposed to fit the data with the function which is a sum of sine waves. Similar as previously, we could find the proper number of sinusoidal functions by using mean square error measure.

After the mentioned transformations the data exhibited behavior adequate to a process with stationary increments. The stationarity was checked using autocorrelation and partial autocorrelation function of the differentiated time series. Respective plots are displayed in Fig. 6. It is easy to observe that the values of ACF and PACF tend to 0 with larger lags, indicating that the analyzed time series is stationary.

Next, following the same steps as in Sect. 4, we determined the optimal window length. The results presented in Fig. 7 indicate that the optimal window length is 693 observations, which is equal to 3 h 5 min, approximately.

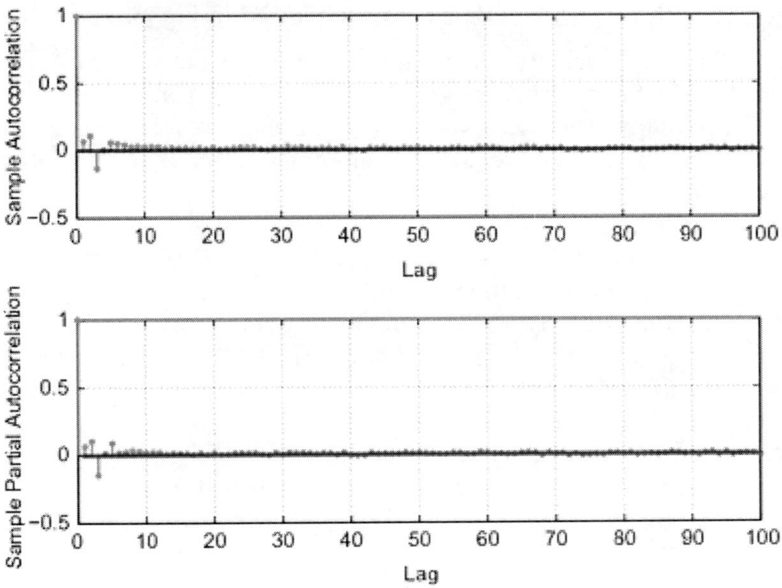

Figure 6. Stationarity check of differentiated CO_2 time series

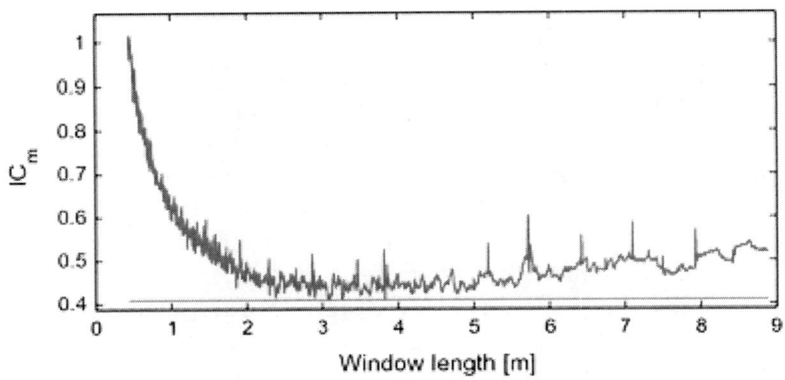

Figure 7. The values of information criterion for different window sizes. The minimum value of the criterion is obtained at m=693m=693 (approximately 3 h 5 min), which indicates the optimal window size

Having calculated the optimal window length, we estimated mean squared displacement in subsequent non-overlapping intervals, each consisting of 693 observations. The estimated values of β parameter together with respective error bars are plotted in Fig. 8. The *i*-th error bar in Fig. 8 is calculated using the root-mean-square error $\sqrt{e(i, 693)}$ related to the fitting accuracy given by Eq. (3). One can observe that the errors are in general of low values, therefore the resulting estimates of parameter β in consecutive non-overlapping intervals can be recognized as valid.

Based on the obtained results the dynamics of CO_2 concentration in indoor air was quite complex. First, in the analyzed time series there were recognized all types of dynamics, namely: sub-linear ($\beta<0.9$), linear ($0.9<\beta<1.1$) as well as super-linear ($\beta>1.1$) . Secondly, there were observed the periodic changes of the dynamics over time. Namely, the period of sub-linear dynamics was typically followed by the one of linear dynamics, next there appeared super-linear dynamics and via the subsequent time interval of linear dynamics the system returned back to the sub-linear one. In a more simple case sub-linear and linear dynamics were observed interchangeably. We have examined the periodicity of $\beta\beta$ time series by analyzing the autocorrelation function, as shown in Fig. 9. From it, the dominant cycle was identified, equal to 8.25. It is worth noticing that the calculated cycle is close to $243.08\approx7.8$. From this fact we might draw a conclusion that the estimated characteristic is periodic with cycle approximately equal to one day.

As displayed in Fig. 9 for the examined 1 week + 1 day long data, CO_2 concentration in indoor air most frequently exhibited nonlinear dynamics. Sub-linear one was typically associated with the nighttime. In these conditions we may view the air as a stable system. Positive changes of CO_2 concentration are quickly compensated by negative ones. Gas motion driving forces are largely inactive, similar as CO_2 sources. Super-linear dynamics was usually observed during midday. In this case, we may describe air as being unstable or in transient state. Once initiated, the changes of CO_2 concentration are further accelerated, irrespective of their direction. Gas motion driving forces are active and not balanced, similarly as CO_2delivery and removal processes. The linear dynamics was usually recognized in periods between the nighttime and midday, but sometimes, it extended over entire daytime. In circumstances of linear dynamics changes of CO_2 concentration are neither retarded not accelerated. Gas motion driving forces are active but balanced, similar to CO_2 delivery and removal processes. The air is in neutral state.

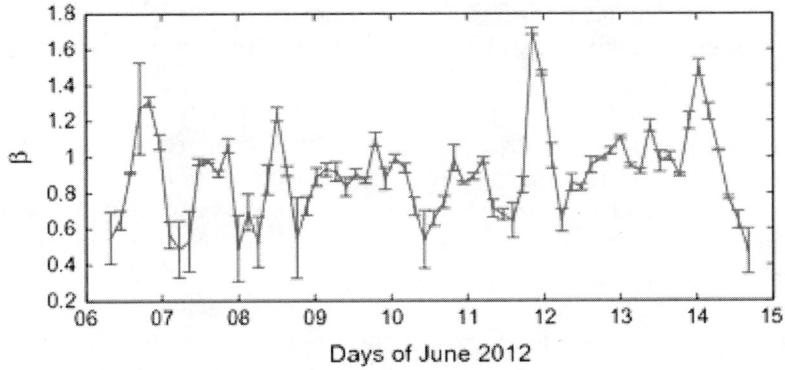

Figure 8. Estimated values of β parameter in consecutive windows of observations together with respective error bars (see Sect. 5 for details of *error bars* calculations)

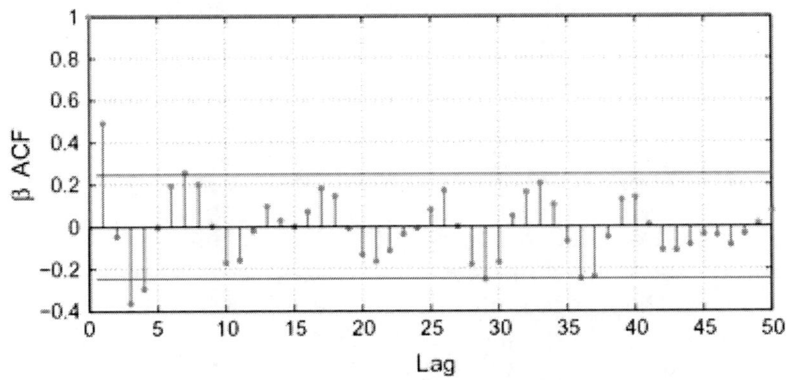

Figure 9. Autocorrelation function of β's estimates depicted in Fig. 8

6. CONCLUSIONS

In this work we proposed a method which allows to recognize the character of stochastic temporal dynamics of CO_2 concentration in indoor air.

The method focuses on the time series of measured CO_2 concentrations and it utilities MSD analysis as a tool to characterize the dynamics. More specifically, β parameter is applied as the dynamics descriptor.

The method was verified by demonstrating high accuracy of reproducing the known stochastic dynamics of the simulated data set.

We also showed that the approach was successful in studying the dynamics of CO_2 concentration based on real measurement data. In the examined time series there were identified segments featured by linear, sub-linear and super-linear dynamics. Their occurrence in time could be associated with the daily cycle of change of the collective influence of factors on indoor air.

Following, we intend to apply the method for investigating the occurrence of various types of CO_2concentration dynamics in long term perspective (a year or longer) in various enclosed spaces. Further, the study, utilizing this approach, will be extended to other parameters of indoor air. Ultimately, we would like to achieve the capability of characterizing the dynamics of an entire complex system which is indoor air. We believe it will become useful for understanding and designing proper microclimate in buildings. Other important area of application is the diagnostics of indoor environment.

ACKNOWLEDGMENTS

This paper is a result of the project: 'The variability of physical and chemical parameters in time as the source of comprehensive information about indoor air quality'. The research of A. S., M. M., R. P., M. T. and A. W. is co-financed by the National Science Center, Poland, under the contract No. UMO-2012/07/B/ST8/03031.

REFERENCES

1. Apte MG, Fisk WJ, Daisey JM (2000a) Associations between indoor CO2 concentrations and sick building syndrome symptoms in U.S. office buildings: an analysis of the 1994–1996 BASE study data. Indoor Air 10(4):246–257

2. Apte MG, Fisk WJ, Daisey JM (2000b) Indoor carbon dioxide concentrations and SBS in office workers. Proc Healthy Build 2000(1):133–138

3. Barkai E, Metzler R, Klafter J (2000) From continuous time random walks to the fractional Fokker–Planck equation. Phys Rev E 61:132

4. Barkai E, Garini Y, Metzler R (2012) Strange kinetics of single molecules in living cells. Phys Today 65(8):29–35

5. Burnecki K, Weron A (2010) Fractional Levy stable motion can model subdiffusive dynamics. Phys Rev E 82:021130

6. Burov S, Jeon J-H, Metzler R, Barkai E (2001) Single particle tracking in systems showing anomalous diffusion: the role of weak ergodicity breaking. Phys Chem Chem Phys 13:1800–1812

7. Caspi A, Granek R, Elbaum M (2000) Enhanced diffusion in active intracellular transport. Phys Rev Lett 85:5655–5658

8. Cherstvy AG, Chechkin AV, Metzler R (2013) Particle invasion, survival, and non-ergodicity in 2D diffusion processes with space-dependent diffusivity. New J Phys 15:083039

9. Cushman J, Park M, Moroni M, Kleinfelter-Domelle N, OMalley D (2011) A universal field equation for dispersive processes in heterogeneous media. Stoch Environ Res Risk Assess 25:1–10

10. Daisey JM, Angell WJ, Apte MG (2003) Indoor air quality, ventilation and health symptoms in schools: an analysis of existing information. Indoor Air 13:53–64

11. Deng W, Barkai E (2009) Ergodic properties of fractional Brownian–Langevin motion. Phys Rev E 79:011112

12. Froemberg D, Barkai E (2013) Time-averaged Einstein relation and fluctuating diffusivities for the Lévy walk. Phys Rev E 87:030104(R)

13. Fulinski A (2011) Anomalous diffusion and weak nonergodicity. Phys Rev= E 83:061140

14. Godec A, Metzler R (2013) Finite-time effects and ultraweak ergodicity breaking in superdiffusive dynamics. Phys Rev Lett 110:020603

15. Golding I, Cox EC (2006) Physical nature of bacterial cytoplasm. Phys Rev Lett 96:098102

16. Guigas G, Kalla C, Weiss M (2007) Probing the nanoscale viscoelasticity of intracellular fluids in living cells. Biophys J 93:316–323

17. Harder H, Havlin S, Bunde A (1987) Diffusion on fractals with singular waiting-time distribution. Phys Rev B 36:3874

18. Haverinen-Shaughnessy U, Moschandreas DJ, Shaughnessy RJ (2011) Association between substandard classroom ventilation rates and students academic achievement. Indoor Air 21:121–131

19. He Y, Burov S, Metzler R, Barkai E (2008) Random time-scale invariant diffusion and transport coefficients. Phys Rev Lett 101:058101

20. Hoefling F, Franosch T (2013) Anomalous transport in the crowded world of biological cells. Rep Prog Phys 76:046602

21. Iomin A (2011) Subdiffusion on a fractal comb. Phys Rev E 83:052106

22. Jeon J-H, Metzler R (2012) Inequivalence of time and ensemble averages in ergodic systems: exponential versus power-law relaxation in confinement. Phys Rev E 85:021147

23. Jeon J-H, Metzler R, Barkai E (2013) Noisy continuous time random walks. J Chem Phys 139:121916

24. Lim SC, Muniandy SV (2002) Self-similar Gaussian processes for modeling anomalous diffusion. Phys Rev E 66:021114

25. Lubelski A, Sokolov IM, Klafter J (2008) Nonergodicity mimics inhomogeneity in single particle tracking. Phys Rev Lett 100:250602

26. Maciejewska M, Szczurek A, Sikora G, Wyłomańska A (2012) Diffusive and subdiffusive dynamics of indoor microclimate: a time series modeling. Phys Rev E 86:031128

27. Magdziarz M, Teuerle M (2015) Asymptotic properties and numerical simulation of multidimensional Levy walks. Commun Nonlinear Sci Numer Simul 20:489–505

28. Mandelbrot BB, Van Ness JW (1968) Fractional Brownian motions, fractional noises and applications. SIAM Rev 10(4):422–437

29. Mendell MJ, Heath GA (2005) Do indoor pollutants and thermal conditions in schools influence student performance? A critical review of the literature. Indoor Air 15:27–52

30. Meroz Y, Sokolov IM, Klafter J (2010) Subdiffusion of mixed origins: when ergodicity and nonergodicity coexist. Phys Rev E 81:010101(R)

31. Ralchenko KV, Shevchenko GM (2010) Path properties of multifractal Brownian motion. Theory Probab Math Stat 80:119–130

32. Satish U, Mendell MJ, Shekhar K, Hotchi T, Sullivan D, Streufert S, Fisk WJ (2012) Impact of CO2 on human decision making and productivity. Environ Health Persp 120:1671–1677

33. Seppanen OA, Fisk WJ, Mendell MJ (1999) Association of ventilation rates and CO2 concentrations with health and other responses in commercial and institutional buildings. Indoor Air 9(4):226–252

34. Seppanen OA, Fisk WJ (2004) Effects of temperature and outdoor air supply rate on the performance of call center operators in the tropics. Indoor Air 14(Suppl 7):102–118

35. Sokolov IM (2012) Models of anomalous diffusion in crowded environments. Soft Matter 8:9043–9052

36. Stoev S, Taqqu M (2004) Simulation methods for linear fractional stable motion and FARIMA using the fast Fourier transform. Fractals 12:95–121

37. Szymanski J, Weiss M (2009) Elucidating the origin of anomalous diffusion in crowded fluids. Phys Rev Lett 103:038102

4

REGIONAL AND TEMPORAL CHARACTERISTICS OF *BOVINE TUBERCULOSIS* OF CATTLE IN GREAT BRITAIN

Aristides Moustakas · Matthew R. Evans

School of Biological and Chemical Sciences, Queen Mary University of London, Mile End Road, London E1 4NS, UK

ABSTRACT

Bovine tuberculosis (TB) is a chronic disease in cattle that causes a serious food security challenge to the agricultural industry in terms of dairy and meat production. Spatio-temporal disease analysis in terms of time trends and geographic disparities of disease dynamics can provide useful insights into the overall efficiency of control efforts as well as the relative efficiency of different management measures towards eradication. In GB, Scotland has had a risk based surveillance testing policy under which high risk herds are tested frequently, and in September 2009 was officially declared as TB free. Wales have had an annual or more frequent testing policy for all cattle herds since January 2010, while in England several herds are still tested every 4 years except some high TB prevalence areas where annual testing is applied. Time series analysis using publicly available data for total tests on herds, total cattle slaughtered, new herd incidents, and herds not TB free, were analysed globally for GB and locally for the constituent regions of Wales, Scotland, West, North, and East England. After detecting trends over time, underlying regional differences were compared with the testing policies in the region. Total cattle slaughtered are decreasing in Wales, Scotland and West England, but increasing in the North and East English regions. New herd incidents, i.e., disease incidence, are decreasing in Wales, Scotland, West English region, but increasing in North and East English regions. Herds not TB free, i.e., disease prevalence, are increasing in West, North, and East English regions, while they are decreasing in Wales and Scotland. Total cattle slaughtered were positively correlated with total tests in the West, North, and East English regions, with high slopes of regression indicating that additional testing is likely to facilitate the eradication of the disease. There was no correlation between total cattle slaughtered and total tests on herds in Wales indicating that herds are tested frequent enough in order to detect all likely cases and so control TB. The main conclusion of the analysis conducted here is that more frequent testing is leading to lower TB infections in cattle both in terms of TB prevalence as well as TB incidence.

KEYWORDS

Public data analysis, Spatio-temporal epidemiology, Time series decomposition, Geographic analysis, Cattle infections, *Mycobacterium bovis*

1. INTRODUCTION

Disease spread is a complex process involving several factors such as social structure of the individuals forming the population, biological and epidemiological characteristics of both the population and the disease, any economic and geographic factors associated with the mobility of the population as well as the spatial scale of that movement (Christakos et al. 2006). Therefore, understanding the spread of any disease is a highly interdisciplinary exercise as biological, social, geographic, economic, and medical factors may shape the way a disease moves through a population and options for its eventual eradication (Christakos et al.2006, 2014; Hethcote 2000). Historically, *bovine tuberculosis* (TB, also abbreviated as bTB) was almost eradicated from the GB, with a minimum in the number of cattle herds containing an individual that reacted positively to a test for TB being reached in the late 1970s (Krebs et al. 1997). Since this time the disease has been steadily increasing in both its prevalence in the cattle herd and its spread around the country (Abernethy et al. 2013; Gilbert et al. 2005).

Bovine TB is a chronic disease in cattle (Thrusfield 2013). Although TB only rarely poses a threat to human health it causes a serious food security challenge to the agricultural industry in terms of dairy and meat production (Marsden and Morley 2014; Tomlinson 2013): Cattle which are detected or thought to have a high chance of being TB infected are slaughtered and thus the farmer loses the value of the animal and its output, although in GB the government pays farmers compensation for slaughtered animals based on their market value and so a large part of the economic cost is borne by the taxpayer (DEFRA 2014b; Krebs et al.1997). In addition there are costs associated with testing animals for TB. When an animal in a herd tests positive for the disease, the entire herd is put under movement restrictions until all the remaining animals have tested repeatedly negative (DEFRA 2014b; Krebs et al. 1997). Until that point cattle cannot be moved to market, nor can animals be brought into the herd to replace slaughtered animals (DEFRA 2014b; Krebs et al. 1997). Further, TB infected cattle may facilitate the spread of infection to other domestic and wild animals (Caron et al. 2013; McCallan et al. 2014). Within the past 10 years over £0.5 billion on testing, compensation and research with further costs being borne by the agricultural industry (Godfray et al. 2013).

In order to control and eradicate TB from cattle, all cattle are regularly tested via a skin test at a frequency that is determined by the prevalence of TB within the local cattle population—the primary ante-mortem test for TB in the GB. In addition, all cattle moved to another farm are also tested prior to movement (Krebs et al.1997). According to EU Directives 64/432/EEC and 97/12/EC the minimum testing frequency for cattle depends on the percentage of infected cattle herds: Annual testing is required unless the percentage of infected herds in a state or region of the state is 1 % or less. When the percentage of infected herds is 0.2 % or less than 0.1 % testing may be conducted every 3 or 4 years respectively. In practice, most places in the GB tested cattle every 4 years during the 1990s (DEFRA 2014b; Krebs et al. 1997). Increasing test frequency would increase the annual cost of testing (Krebs et al. 1997). All cattle moving from one farm to another are tested prior to movement and the ones detected as infected are removed and slaughtered and the herd of origin is placed under movement restrictions (AHVLA 2013). The skin test on cattle is known to be imperfect—the test will not always detect an infected individual—and the presence of a common parasite*Fasciola hepatica* is reported to under-ascertain the rate of the skin test by about one-third (Claridge et al.2012). Studies evaluating the sensitivity of the test suggest that its sensitivity lies between 52 and 100 % with median values of 80 and 93.5 % for standard and severe interpretation, respectively (DEFRA

2009). As a result many positive TB cases are first discovered in slaughterhouses.

Spatio-temporal disease analysis (Lin et al. 2015; Moustakas and Evans 2015; Ruiz-Medina et al. 2014) in terms of time trends and geographic disparities of disease dynamics may provide useful insights into the overall efficiency of control efforts and eradication of the disease (Booth et al. 2005) as well as the relative efficiency of different management measures towards eradication (Yu et al. 2013). In the case of TB monthly statistics are recorded as part of legal requirements of the EU Member States (DEFRA 2014b) and these data are publicly available. In October 2009 Scotland was officially declared as TB free while the rest of GB was not (Abernethy et al. 2013). Scotland has had a risk based surveillance testing policy (Dorjee et al.2013) under which high risk herds are tested frequently while lower risk herds are tested every 4 years and since September 2009 has been designated as TB free. Wales has had an annual or even more frequent testing policy of all herds since January 2010 while in England the number of herds tested annually (instead of every 4 years) has increased since 2010 and expanded further in the East and North regions in 2011; however several areas are still tested every 4 years (DEFRA 2014b). Thus, regional differences in TB detections may provide insights of different policies against eradicating the disease. In the current study we sought to quantify seasonal trends of TB in cattle population across different regions of GB and link potential differences to the control strategies operated in the three countries and regions of England.

2. METHODS

2.1 Data

Publicly available time series of monthly bovine TB statistics in cattle in GB published by DEFRA were used. The bovine TB statistics are presented for GB by different geographical areas with regional-level statistics provided for Scotland, Wales, and North, East and West English region (DEFRA 2014a). Statistics for GB (without distinguishing between regions) are available from January 1996 to June 2014, while the regional statistics are available from January 2008 to June 2014, all on a monthly time step (DEFRA 2014a). Statistics are affected by seasonal patterns and variations in the frequency of testing. TB testing is seasonal as more herds are tested in the winter when more cattle are housed inside. The animals tested are not a random sample of the whole GB herd: Herds are tested more frequently in areas of higher TB incidence than in those with a historically low incidence and thus herds experiencing several separate incidents in the same reporting period will appear more often in the dataset (see below '*Herds not TB free*'). In general, as more tests are carried out, more TB infected herds are likely to be found (DEFRA 2014a). From the available datasets the following monthly statistics were analysed:

(1)*Herds not TB free*: The number of herds that did not have TB free status (i.e., were under movement restrictions) at some time during the period shown due to a TB incident. This figure may include herds experiencing a TB incident that began before (but continued into) the report period. Likewise, any herds experiencing two separate incidents in the same reporting period could appear twice under this data column.

(2)*Total cattle tests*: The count of the number of cattle for which a tuberculin skin test has been carried out. An individual animal could be tested more than once in each time period. All test types are included except for slaughterhouse cases.

(3)*Total tests on herds*: Herds for which tuberculin skin testing is carried out on at least one animal in a herd during the period listed. The following test types are not included in this measure: Pre- and post-movement, gamma interferon blood tests, and private tests.

(4)*New herd incidents*. The disclosure of at least one test reactor or a confirmed slaughterhouse case in a herd that was previously TB free during the period shown.

(5)*Total cattle slaughtered*: This is the number of infected cattle slaughtered and it is the sum of reactors slaughtered + inconclusive reactors slaughtered + direct contacts slaughtered. *Reactors slaughtered*include cattle which were compulsorily slaughtered by AHVLA because it responded to a relevant test for TB in a way that was consistent with it being TB infected. In any given period, the majority of animals are slaughtered as reactors to the tuberculin skin test. A much smaller number of reactors are those giving a positive result to the ancillary interferon-gamma blood test. *Inconclusive reactors slaughtered* are cattle showing a positive reaction to TB that was not strong enough for it to be deemed a reactor. However, instead of being tested again after 60 days (the normal procedure in this case) the animal was voluntarily slaughtered by the owner. *Direct contacts slaughtered* are cattle from an officially TB free status withdrawn herd that, although not a test reactor, was considered to have been exposed to TB and compulsorily slaughtered.

Details of the methodologies and revisions of the methodologies through time can be found at (DEFRA2014a) and references therein. All data are freely available at: https://www.gov.uk/government/ statistics/incidence-of-tuberculosis-tb-in-cattle-in-great-britain.

2.2 Time series decomposition

We analysed the five monthly variables (*Total cattle tests, Total tests on herds, Total cattle slaughtered, Herds not TB free*, and *New herd incidents*) using decomposition time series analysis (Ye et al. 2015). In order to disentangle the actual variance of the signal due to stochastic fluctuations in the time series data (Koutsoyiannis 2011) across time scales (Markonis and Koutsoyiannis 2013) qualitative diagnostics were used to make a deterministic attribution of linear trends (Koutsoyiannis 2005). In order to detect linear trends in time series data we estimate the variance change points of the time series using the method of Lavielle (1999, 2005). In brief the method identifies breaking points in the data such that linear models can be fit within these points (Lavielle 1999, 2005). The method assumes that the data follow a normal distribution and detects simultaneous changes in the mean and variance of the distribution but requires no assumptions about the dependence structure of the time series (i.e., linear, or non-linear trends). After the detection of breaks, the time series is split into a seasonal and trend component (Lavielle 1999, 2005). The method of Lavielle (1999) finds the best segmentation of a time series, given that it is built by K segments. It searches the segmentation for which a contrast function (measuring the contrast between the actual series and the segmented series) is minimized. The contrast function used for measuring different aspects of the variation of the series from one segment to the next was calculated accounting for both the mean and the variance varying between segments. It is required to specify a value for the minimum number of observations L_{min} in a segment (here L_{min} = 12 months), as well as the maximum number of segments K_{max} in the series. In order to estimate the best number of segments K_{max} to partition the time series the decrease of the contrast function with the number of segments is examined. An automated way to estimate the optimal number of segments relying on the presence of a 'break' in the decrease of the contrast function has been suggested (Lavielle 2005). According to this method the last value of K for which the second derivative of a standardized contrast function is greater than a threshold S is chosen—see (Lavielle 2005) for details. Based on numerical experiments, it has been proposed to choose the value $S = 0.75$ (Lavielle 2005). As for some short time series it has been indicated that this value may not be optimal and may depend on the value of K_{max} (Lavielle 2005), we also performed graphical examination of the decrease of the contrast function with the number of segments. See e.g. Sur et al. (2014) for an application of the method in ecology, and Lijffijt et al. (2014) for segmentation of time series.

Having determined segments (time periods within the data) in which linear trends exist, seasonal filters were used to de-seasonalize a time series using a multiplicative decomposition (Brockwell and Davis 2002). Decomposition in time series analysis is used in order to separate the time series into linear trend and seasonal components, as well as error and stochasticity. As the seasonal pattern in the data depended on the level of the data—more cattle tests are carried out during winter months and in general as more tests are carried out, more TB infected herds are likely to be found (DEFRA 2014a)—a multiplicative model structure was employed accounting for this effect. A multiplicative de-trending model is used when the size of the seasonal pattern in the data depends on the level of the data (Carstensen 2005; Gocheva-Ilieva et al. 2014). The model assumes that as the data increase, so does the seasonal pattern. Several time series exhibit such a pattern (Barucci et al. 2010). The multiplicative model is expressed as: $Y_t = Trend * Seasonal * Error$, where Y_t is the observation at time t. An approximate confidence interval (95 %) of the prediction using a rule of thumb *prediction* $\pm 2 \times$ *Standard Error* were also plotted. All data were $\log_{10}(x + 1)$ transformed prior to the analysis and successfully tested for normality.

In order to assess model fit to the data the mean absolute percentage error (MAPE) index was used. MAPE measures the accuracy as a percentage of the error (Armstrong and Overton 1977) and is expressed as: $MAPE=\sum||yt-\hat{y}t||n\times 100,(yt\neq 0)MAPE=\sum|yt-\hat{y}t|n\times 100,(yt\neq 0)$, where y_t is the actual value and $\hat{y}t\hat{y}t$ the fitted value. Because this number is a percentage, it may be easier to understand than other performance measures. For example, if the MAPE is 5, on average the prediction is out by 5 %.

2.3 Analysis between total tests on herds and total cattle slaughtered

In addition we sought to investigate the relation between the number of total tests on herds or total cattle tests and total cattle slaughtered. We sought to quantify whether there are geographic regions in GB where a higher number of tests does not lead to the detection of more infected herds/cattle. While in general as more tests are carried out, more TB infected herds are likely to be found (DEFRA 2014a), an absence of correlation between number of tests and number of detections can provide statistical inference that herds are tested sufficiently often so that additional testing would not facilitate the detection of TB infected herds/cattle and a higher slope in the regression implies that more frequent testing would lead to higher detection of TB infected herds/cattle (García-Portugués et al. 2014; Guo et al. 2006; Weisberg 2014). We therefore examined the slopes of linear regressions between total cattle slaughtered versus total tests on herds and total cattle tests. In addition a local regression line was fitted between total cattle slaughtered and total tests on herds. A local regression helps explore the potential relationships between two variables without fitting a specific model, such as a regression line or a theoretical distribution and thus deviations from linearities could be explored, as well as the effect of extreme values or outliers in the regression analysis (Abaurrea et al. 2015; Cleveland 1979). Local regressions performed here are calculated using the lowess smoothing method. Lowess (locally-weighted scatterplot smoother) works by selecting a fraction (here f = 0.4 & 0.6, and number of steps = 1 & 3 respectively) of all points, using the data closest in x-value on either side of the (x,y) point. For each data point a weighted linear regression was performed, giving points closest to each x-value the most weight in the smoothing and limiting the effect of outliers.

RESULTS

3.1 Trends in the raw data between and within years

Mean values and 95 % confidence intervals of the mean of the raw data (i.e., not de-trended and log-transformed data) are shown in Fig. 1 (values between years) and Fig. 2 (values per month). In terms of between years values, total tests on herds are increasing on an annual basis in all regions (Wales, West, North, and East of England), except Scotland where the total tests on herds are decreasing (Fig. 1a). Total cattle slaughtered are increasing in the North and in the past 3 years in the East English region (Fig. 1b). There are no consistently clear patterns in total cattle slaughtered in Scotland, West England, and Wales (Fig. 1b). Herds not TB free are increasing in the North and East English regions and oscillate in Wales, Scotland and West English region (Fig. 1c). New herd incidents are increasing in North England while no clear trend in mean values is recorded in any other region (Fig. 1d). Seasonal (within the year) patterns of total tests on herds and total cattle tests have similar patterns in all GB regions: Testing is peaking during winter months when winter housing is applied, declining until August and gradually increasing thereafter (Fig. 2a). Total cattle slaughtered does not follow this pattern as in most cases mean values do not exhibit a clear trend prior to de-trending (Fig. 2b). Herds not TB free in terms of mean values are peaking in late winter months (March, April) and declining thereafter across all regions (Fig. 2c). New herd incidents are consistently higher during winter months across all regions (Fig. 2d).

3.2 Patterns following de-trending time series analysis

3.2.1 Number of cattle slaughtered

Once the raw data have been de-trended it can be seen that total cattle slaughtered in GB are increasing; the rate of increase was higher during 1996–2001 (variance break point September 2001, MAPE = 0.89) than in from Sept 2001 to June 2014 (MAPE = 2.42); (Fig. 3a). Within the year variance as indicated from seasonal components was also larger during 1996–2001 (Fig. 3a). No break points were detected for the regional trends (all could be approximated with a single linear trend). Total cattle slaughtered are decreasing in Wales (MAPE = 2.92), Scotland (MAPE = 0.36) and West England (MAPE = 3.17); (Fig. 3b–d), but increasing in the North (MAPE = 3.62) and East (MAPE = 3.82) English regions (Fig. 3e, f). There are some cases where model residuals are autocorrelated as for example in the decomposition of total cattle slaughtered Scotland. As for the decomposition of regional data $\log_{10}(x + 1)$ transformation produced good model fit for most regions we decided against applying more stringent data transformations for the cases where autocorrelated residuals exist and proceed with a consistent analysis for all regions and data. Residuals of the decomposition are provided in Supp. 1 'Residuals of Total cattle slaughtered'.

a

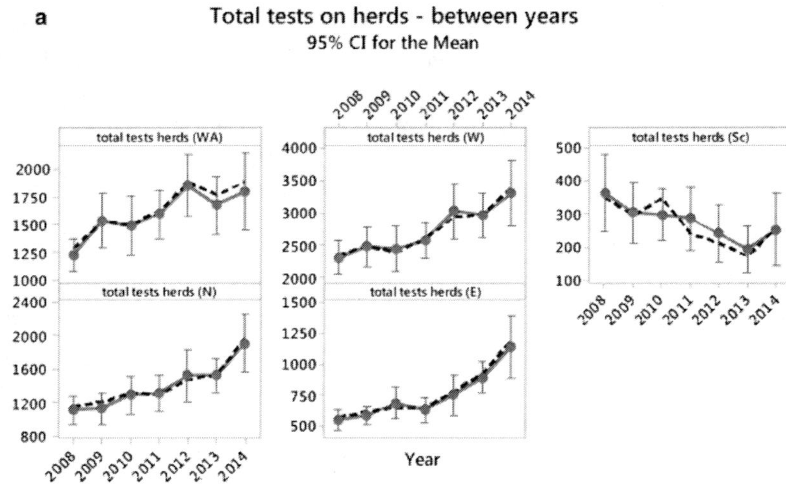
Total tests on herds - between years
95% CI for the Mean
total tests herds (WA)
total tests herds (W)
total tests herds (Sc)
total tests herds (N)
total tests herds (E)
2008 2009 2010 2011 2012 2013 2014
Year

b

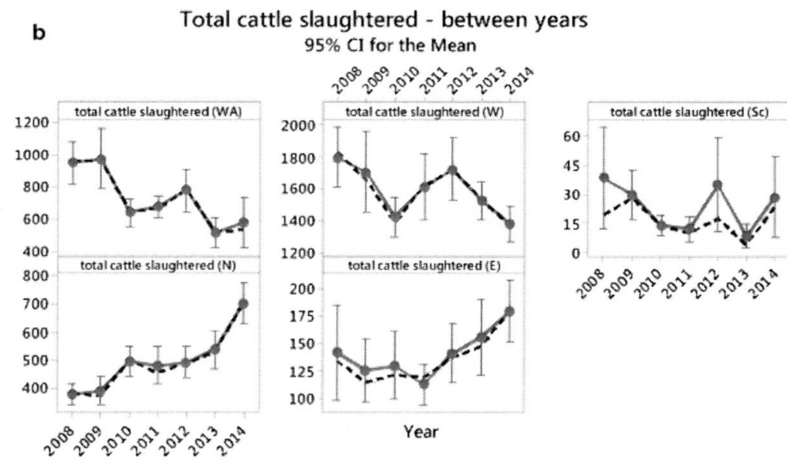
Total cattle slaughtered - between years
95% CI for the Mean
total cattle slaughtered (WA)
total cattle slaughtered (W)
total cattle slaughtered (Sc)
total cattle slaughtered (N)
total cattle slaughtered (E)
2008 2009 2010 2011 2012 2013 2014
Year

c

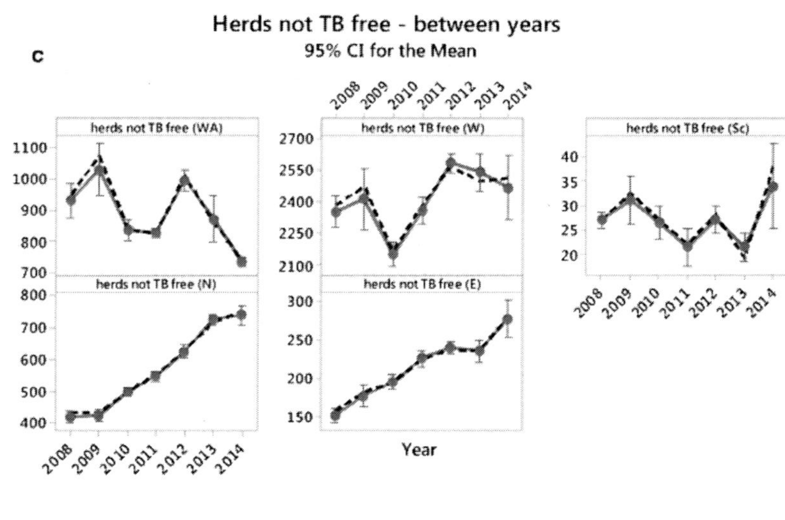
Herds not TB free - between years
95% CI for the Mean
herds not TB free (WA)
herds not TB free (W)
herds not TB free (Sc)
herds not TB free (N)
herds not TB free (E)
2008 2009 2010 2011 2012 2013 2014
Year

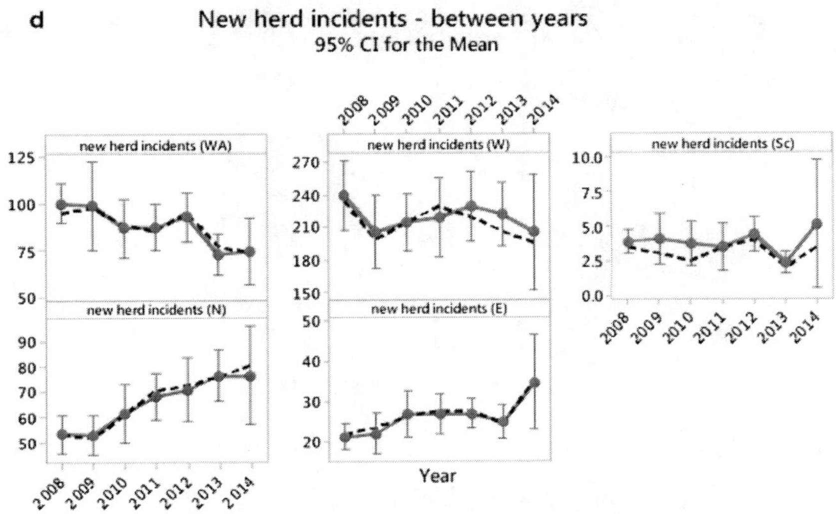

Figure 1. Confidence intervals of 95 % of the mean values of **a** total tests on hers, **b** total cattle slaughtered, **c** herds not TB free, and **d** new herd incidents per year. Data per region in the GB from Jan 2008 to June 2014 recorded on a monthly basis. *Wa* Wales, *W* West English Region, *Sc* Scotland, *N* North English Region, *E*East English Region. Intervals show the 95 % confidence envelope of the mean, *solid red lines* connect mean values, and *grey dotted lines* connect median values between years

3.2.2 New herd incidents: TB incidence

New Herd incidents are increasing in GB throughout from 1996 to 2014, and the within year variance has been lower during January 1996 to October 2001 (variance break point, MAPE = 0.79) and increasing from November 2001 to June 2014 (MAPE = 1.34); (Fig. 4a). No break points were detected for the regional trends (all could be approximated with a single linear trend). New herd incidents, and thus disease incidence, are decreasing in Wales (MAPE = 0.89), Scotland (MAPE = 0.91); (Fig. 4b, d), and marginally decreasing in West English region (MAPE = 1.86); (Fig. 4c), but increasing in North (MAPE = 1.42) and East (MAPE = 2.89) English regions (Fig. 4e, f). Residuals of the decomposition are provided in Supp. 1 'Residuals of New herd incidents'.

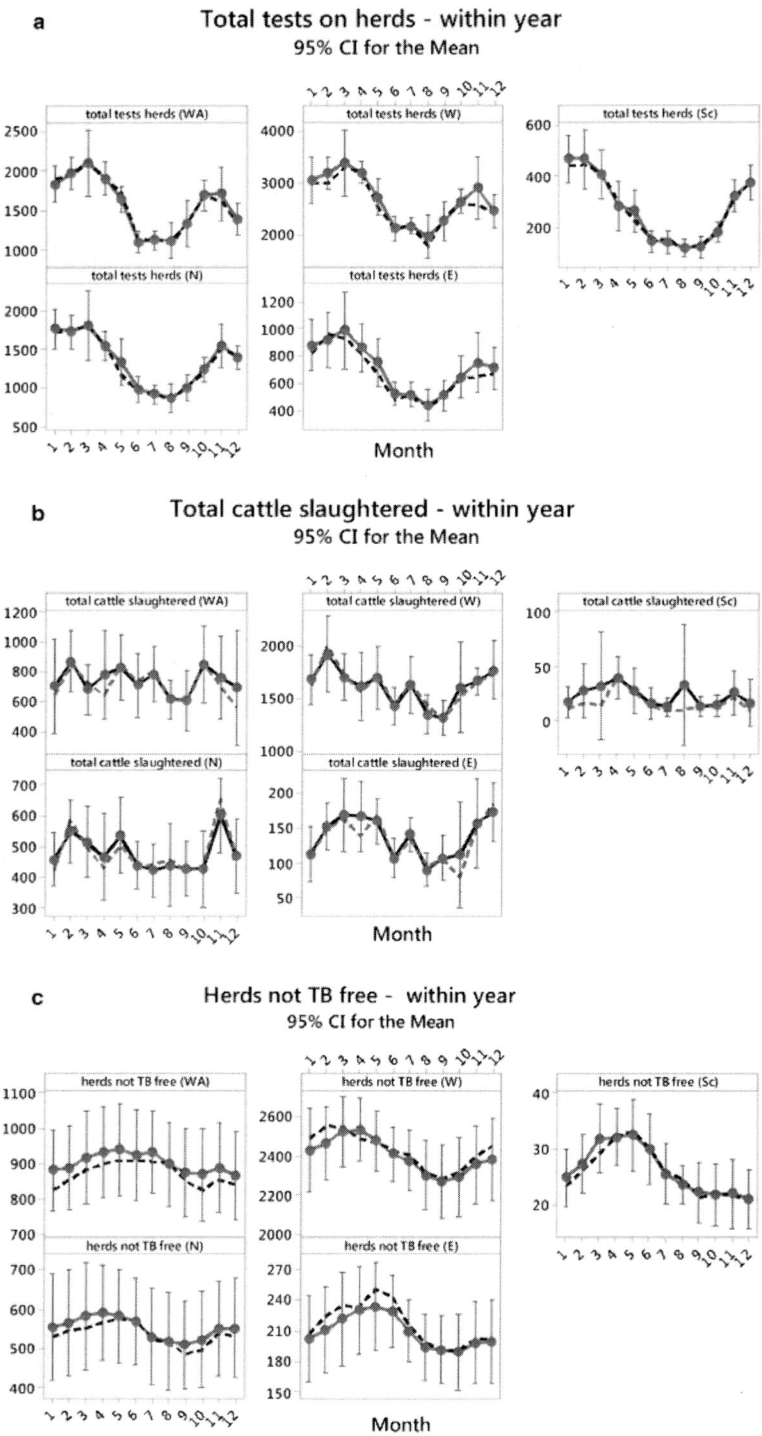
a
Total tests on herds - within year
95% CI for the Mean
total tests herds (WA)
total tests herds (W)
total tests herds (Sc)
total tests herds (N)
total tests herds (E)
Month
b
Total cattle slaughtered - within year
95% CI for the Mean
total cattle slaughtered (WA)
total cattle slaughtered (W)
total cattle slaughtered (Sc)
total cattle slaughtered (N)
total cattle slaughtered (E)
Month
c
Herds not TB free - within year
95% CI for the Mean
herds not TB free (WA)
herds not TB free (W)
herds not TB free (Sc)
herds not TB free (N)
herds not TB free (E)
Month

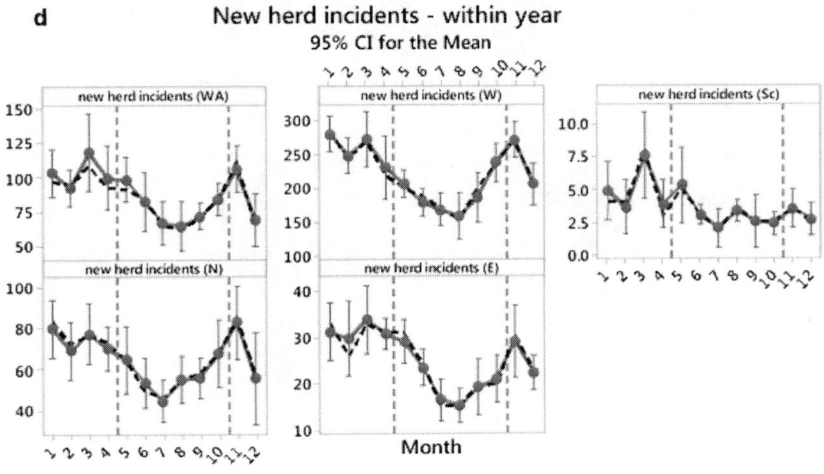

Figure 2. Confidence intervals of 95 % of the mean values of **a** total tests on hers, **b** total cattle slaughtered, **c** herds not TB free, and **d** new herd incidents per month. Data per region in the GB from Jan 2008 to June 2014 recorded on a monthly basis. *Wa* Wales, *W* West English Region, *Sc* Scotland, *N* North English Region, *E*East English Region. Intervals show the 95 % confidence envelope of the mean, *solid red lines* connect mean values, and *grey dotted lines* connect median values between months. *Dotted vertical lines* on graph d (new herd incidents per month) show winter months (January–April & November–December) and summer months (May–October)

3.2.3 Herds not TB free: TB prevalence

Herds not TB free were increasing in GB from January 1996 to January 2002 (break point, MAPE = 0.61), followed by a steeper increase during February 2002–December 2006 (break point, MAPE = 2.41), and from January 2007 to June 2014 (MAPE = 1.07) are also increasing but at a lower rate previously (Fig. 5a). During the last period of Jan 2007–June 2014 the within the year variance is also lower (Fig. 5a). No break points were detected for the regional trends (all could be approximated with a single linear trend). Herds not TB free, and thus disease prevalence, are increasing in West (MAPE = 1.98), North (MAPE = 0.99), and East (MAPE = 1.81) English regions (Fig. 5c, e, f), while they are decreasing in Wales (MAPE = 0.41) and Scotland (MAPE = 2.82); (Fig. 5b, d). In general there was autocorrelation in the residuals of the decomposition of herds not TB free in most regions. Residuals of the decomposition are provided in Supp. 1 'Residuals of Herds not TB free'.

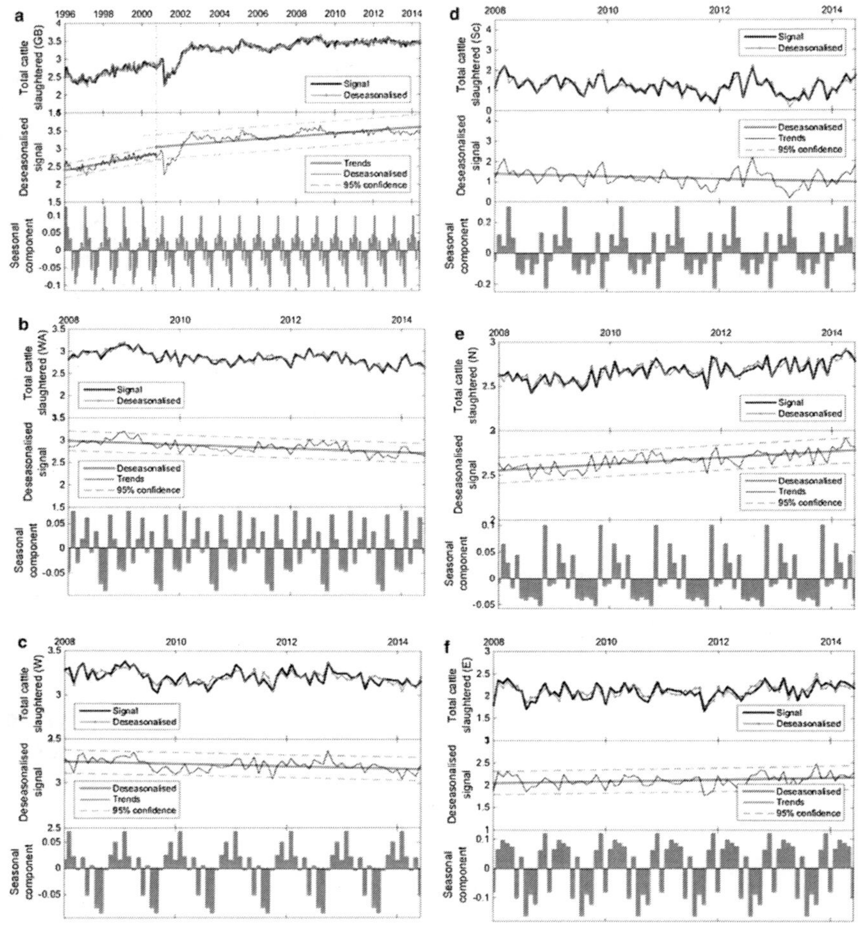

Figure 3. Time series decomposition of total cattle slaughtered (log-transformed data) in GB (data from Jan 1996 to June 2014) and in different regions in GB (data from Jan 2008 to June 2014) on a monthly time step. Upper panels of each graph show the $\log_{10}(x + 1)$-transformed real data in *black lines* (*Signal*) and the model fit in*red lines* (*Deseasonalised*). Middle panels show a linear trend on the model fit with 95 % confidence intervals of the trend. Lower panels show the seasonal (within the year) decomposition. Where applicable,*vertical dotted lines* indicate variance change points in the data and different linear trends were fitted for each period—see methods for details. *Wa* Wales, *W* West English Region, *Sc* Scotland, *N* North English Region, *E* East English Region

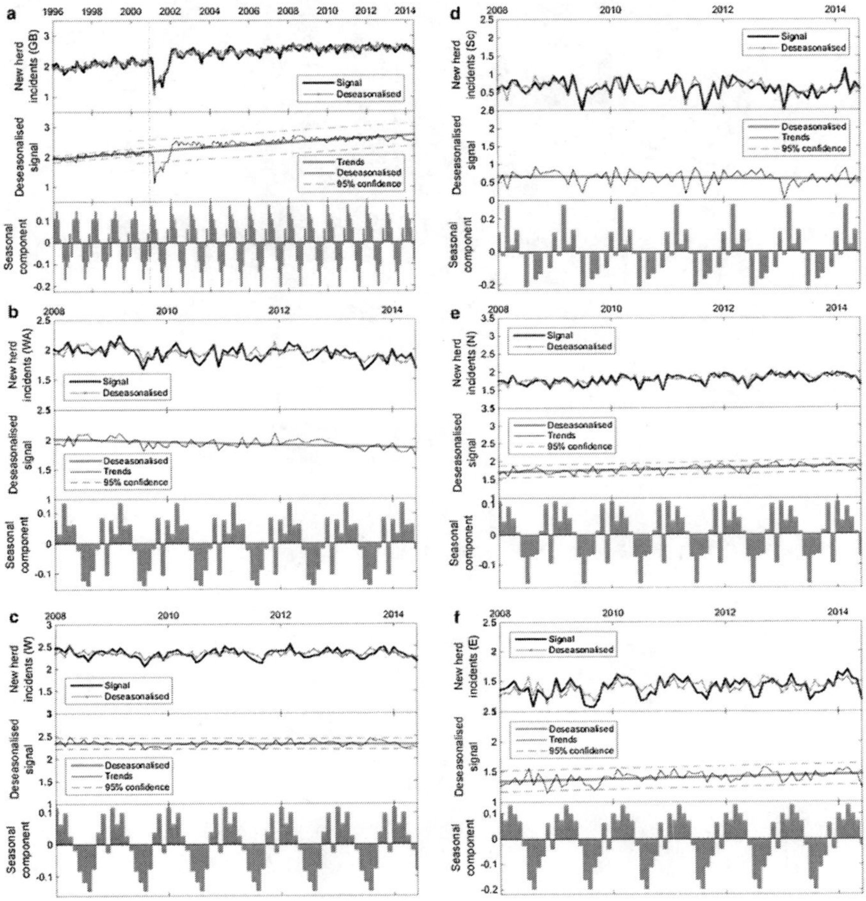

Figure 4. Time series decomposition of new herd incidents (log-transformed data) in GB (data from Jan 1996 to June 2014) and in different regions in GB (data from Jan 2008 to June 2014) on a monthly time step. Upper panels of each graph show the $\log_{10}(x + 1)$-transformed real data in black lines (*Signal*) and the model fit in *red lines*(*Deseasonalised*). Middle panels show a linear trend on the model fit with 95 % confidence intervals of the trend. Lower panels show the seasonal (within the year) decomposition. Where applicable, *vertical dotted lines* indicate variance change points in the data and different linear trends were fitted for each period—see methods for details. *Wa* Wales, *W* West English Region, *Sc* Scotland, *N* North English Region, *E* East English Region

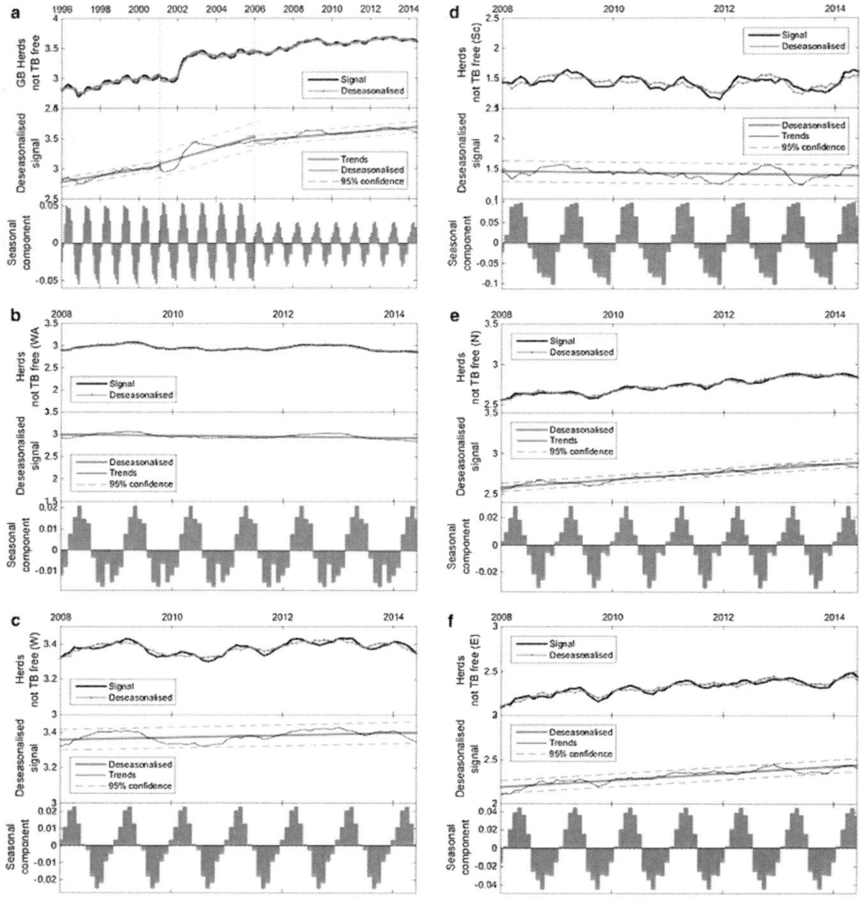

Figure 5. Time series decomposition of herds not TB free (log-transformed data) in GB (data from Jan 1996 to June 2014) and in different regions in GB (data from Jan 2008 to June 2014) on a monthly time step. Upper panels of each graph show the $\log_{10}(x + 1)$-transformed real data in *black lines* (*Signal*) and the model fit in *red lines*(*Deseasonalised*). Middle panels show a linear trend on the model fit with 95 % confidence intervals of the trend. Lower panels show the seasonal (within the year) decomposition. Where applicable, *vertical dotted lines* indicate variance change points in the data and different linear trends were fitted for each period—see methods for details. *Wa* Wales, *W* West English Region, *Sc* Scotland, *N* North English Region, *E* East English Region

3.3 The relationship between total tests on herds and total cattle slaughtered

Total cattle slaughtered were positively correlated with total tests conducted on herds in the North and East English regions, with high slopes of regression (Fig. 6). Total cattle slaughtered were also positively correlated with tests in Scotland and in West England, although in West England the slope was considerably lower (Fig. 6). There was no correlation between total cattle slaughtered and total tests on herds in Wales (Fig. 6). For further details regarding regression analysis between total cattle slaughtered and total tests on herds per region and figures at a higher resolution please refer to Supp. 1 'Regression analysis'. These results are also held when correlating the total cattle tests against the total cattle slaughtered (results not shown here).

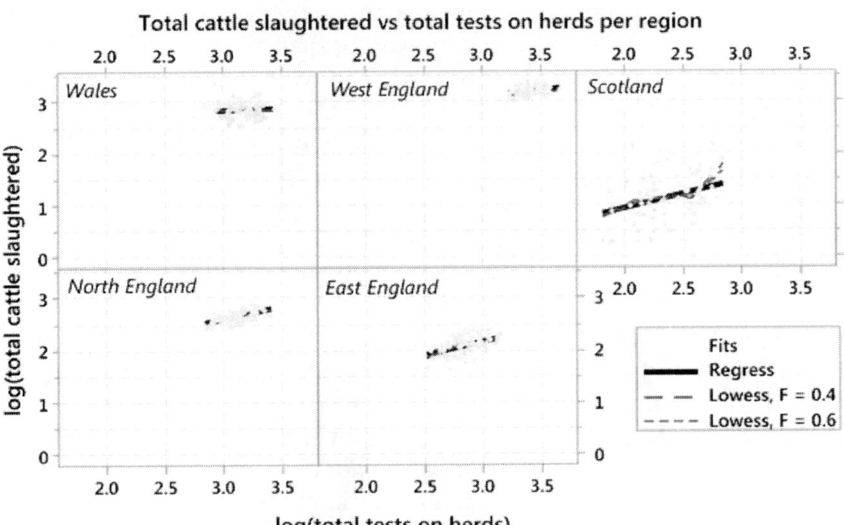

Figure 6. Linear regressions and locally-weighted smoother lines between total cattle slaughtered on the vertical axis and total tests on herds on the horizontal axis between regions in the GB. All panels are on the same scale. Regressions are plotted within the range of the data for each region. All data were $\log_{10}(x + 1)$-transformed. For a higher resolution of each regression fit including confidence intervals please see 'Regression Analysis' in Supp. 1

4. DISCUSSION

There are major spatial differences between the temporal trends of the dynamics of TB in cattle in the GB and thus spatio-temporal analysis provides useful insights into linking those differences with the underlying control strategies (Christakos and Hristopulos 1998; Christakos et al. 2014; Yu et al. 2013). In order to account for both TB prevalence and incidence in cattle we have analysed both herds not TB free, and new herd incidents. Incidence is rate of occurrence of new cases, and so new herd incidents were used to asses this. Prevalence is the proportion of the total number of cases to the total population and is more a measure of the burden of the disease on society with no regard to time, and herds not TB free (restricted) were used as a proxy of prevalence.

The majority of TB infected herds and cattle in GB are located in England (West, East, and North English regions), while both TB incidence and prevalence are decreasing in Wales and Scotland. In the West English region new herd incidents are marginally decreasing but herds not TB free are increasing indicating that once a herd is infected, the TB infection remains in the herd. The overall increase pattern of both TB incidence and prevalence in cattle in GB is thus driven by the English regions. Scotland and Wales both have a declining number of new herd incidents as well as herds not TB free and thus the current programme applied, all else being equal, appears to be leading to eradication or control of the disease. In Wales in particular for the period of January 2008 to December 2009, prior to applying annual or more frequent testing, the mean of new herd incidents was 99.33 and the mean of herds not TB free was 979.41, while for the period of January 2010 to June 2014 when annual or more frequent testing was applied the mean of new herd incidents was 83.57 and the mean of herds not TB free 863.31 (Fig. 7). All regional data analysed here could be approximated with a single linear trend, while the same data for GB were not; linear trends were fitted for total cattle slaughtered and new herd incidents (break points in Sept 2001, and Nov 2001 respectively), while herds not TB free had two break points in Jan 2002 and Dec 2006. The first break points are temporally close to the TB testing interruption due to the foot-and-mouth disease outbreak in GB in 2001. Both TB incidence and prevalence was affected by interrupted cattle testing: new herd incidents have a larger within year variance after the break point of Nov 2001, while herds not TB free had the highest rate of increase during Jan 2002–Dec 2006. This may serve as an inverse proof of the effects of testing on TB dynamics in cattle because when testing was removed TB infections rose so substantially.

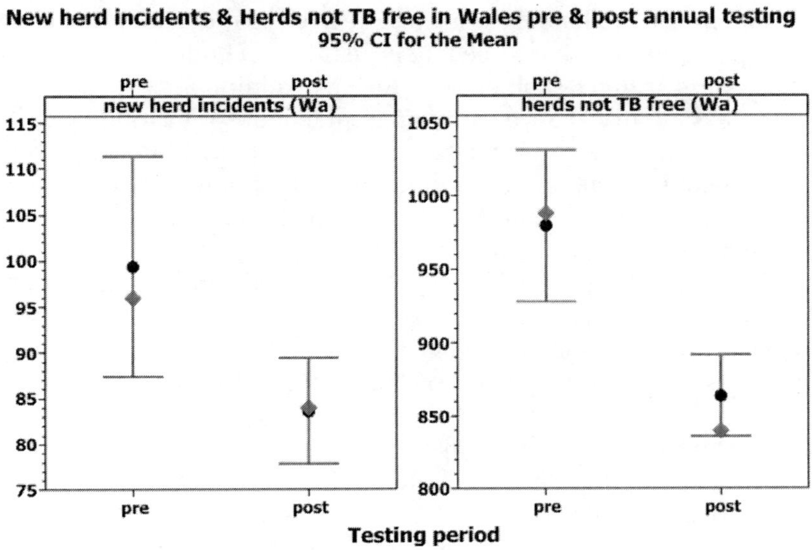

Figure 7 Confidence intervals of 95 % for the mean value (in *black circles*) of New herd incidents (*left panel*) and Herds not TB free (*right panel*) in Wales. The median values are also listed in *red squares*. Pre and Post testing periods refer at the period before annual testing was applied in Wales (*Pre*: Jan 2008–Dec 2009) and after introducing at least annual testing (*Post*: Jan 2010–June 2014). The original data are plotted here (no log-transformations were applied)

Scotland applies a risk based surveillance policy (Oidtmann et al. 2013) testing more often at high risk areas (DEFRA 2014b). This policy appears to be facilitating TB eradications as there are diminishing new herd incidents as well as number of herds not TB free. In addition Scotland is the only region in the GB where the total number of tests on herds has been decreasing through the period under consideration, while in all other regions it is increasing either due to the increasing number of new herd incidents (North and West English regions) and/or due to the introduction of an annual testing policy (Wales). Thus the risk based testing policy not only is leading to TB eradication but it is also cost effective as decreased numbers of TB infected herds are achieved via a lower total number of tests (Bessell et al. 2013). However, in Scotland the total number of cattle per hectare is low (Yang et al. 2014) and it remains unknown whether such testing policy would lead to TB eradication when the total number of cattle and subsequently the total cattle tests is high.

Cases in human epidemiology have concluded that in all but the lowest-risk populations, routine screening is justified on both clinical and cost-effectiveness grounds (Paltiel et al. 2005). In addition, risk based versus routine based surveillance in human epidemiology has indicated that routine based testing provides a better alternative (Beckwith et al. 2005; Jenkins et al. 2006). While there certainly are fundamental differences between human and animal epidemiology due to behavioural and social structure differences, a positive albeit weak correlation between total cattle slaughtered and total tests on herds/cattle was recorded in Scotland. This provides statistical inference for a potential for higher detection of TB infected herds/cattle if more tests were employed (Guo et al. 2006; Weisberg 2014).

Both the number of new herd incidents and number of herds not TB free are declining in Wales and the number of the total cattle slaughtered is not correlated to the total tests on herds or total cattle tested. Thus, more cattle tests would not have resulted in detecting more infected herds/cattle (Weisberg 2014). Interestingly, Wales is the only region in the GB that has adopted annual testing policy in all herds since Jan 2010, and in some counties in Wales cattle testing is exercised even every 6 months (DEFRA 2014b). Indeed there is a significant difference between pre and post Jan 2010, when at least annual testing was applied, of both new herd incidents and herds not TB free in Wales (Fig. 7).

4.1 Synthesis and management implications

4.1.1 Testing frequency

The main result derived here from statistical analysis of publicly available data from the British Government show that increased cattle testing leads to TB control or possibly eradication as exemplified by the results in Wales. This conclusion fully supports the outputs of a computational model suggesting that all eradication scenarios included cattle testing frequencies of annual or even more frequent testing (Moustakas and Evans 2015). This is the cattle testing frequency exercised in Wales since 2010, with annual or even more frequent testing applied (DEFRA 2014b).

4.1.2 Within year effects & winter housing

The pattern of new herd incidents is clearly peaking during winter months (Fig. 2d and de-seasonalised effects in Fig. 4). Herds not TB free are also peaking during late winter months (Fig. 2c and de-seasonalised effects in Fig. 5). However total tests on herds are also peaking during winter months (Fig. 2a). The fact that new herd incidents peak during winter months may support three hypotheses that merit further investigation: (i) TB prevalence in cattle is increased during winter months when winter housing is practised—i.e., all cattle within a farm are summoned in a winter housing facility with high densities of cattle within a small surface area. This has been suggested to increase cattle-to-cattle TB spread from Moustakas and Evans (2015). (ii) Herds not TB free and new herd incidents are peaking late in the winter simply because more tests are conducted during these months. Then this implies that the more one tests, the more infected cattle is likely to detect and thus should test more often. (iii) Lower testing during summer months when cattle are out in the field may create buffers of infection between cattle; cattle that get infected during summer months will more likely be detected during winter months as more tests are conducted then. During that interval they may infect other cattle as they will not be removed from the herd and the herd will not be under movement restrictions. Interestingly new herd incidents as well as total tests on herds are lowest during summer months, when cattle are out in the field, the period that interactions with badgers are maximised. If badgers are the agent of cattle infection it is logical to test cattle during summer months.

The issue of winter housing while it was only recently proposed by (Moustakas and Evans 2015) in animal TB epidemiology is not new in human TB epidemiology. Indeed several studies have found seasonality in TB and attributed it mainly to indoor winter crowding and lower vitamin D levels with lower levels of sunshine (Koh et al. 2013; Parrinello et al. 2012; Wingfield et al. 2014). There are also cases where winter crowding was ruled out (Top et al. 2013) but in any case this merits serious investigation: winter crowding was known to be a main factor of seasonal pick in infections and the fact that it was found not to be significant was a publishable result (Top et al. 2013). In order to quantify seasonal effects in cattle infections the analysis can include social networks of cattle (Corner et al. 2003; Croft et al. 2011) during winter housing and during summer months.

4.1.3 The importance of public data

We would like to highlight the importance of pubic data. In order to predict and mitigate disease spread informed decisions are needed. These decisions need to be taken based on data analysis and predictive models calibrated with data (Evans et al. 2013, 2014; Lonergan 2014). Our view is that making regional TB data available so that potential differences and underlying management decisions is a very good step forward. We argue that making publicly available the data regarding badger culling experiments as agents of TB infecting cattle will greatly facilitate their analysis and to informed decisions regarding TB control in GB.

ACKNOWLEDGMENTS

We thank Yannis Daliakopoulos and Manolis Grillakis for their help during time series decomposition. Their code was developed under the auspices of the CASCADE Project (Tsanis and Daliakopoulos 2015). Comments from two anonymous reviewers considerably improved an earlier manuscript draft.

REFERENCES

1. Abaurrea J, Asín J, Cebrián AC (2015) Modeling and projecting the occurrence of bivariate extreme heat events using a non-homogeneous common Poisson shock process. Stoch Environ Res Risk Assess 29:309–322

2. Abernethy D, Upton P, Higgins I, McGrath G, Goodchild A, Rolfe S, Broughan J, Downs S, Clifton-Hadley R, Menzies F (2013) Bovine tuberculosis trends in the UK and the Republic of Ireland, 1995–2010. Vet Rec 172:312

3. AHVLA (2013) Pre-movement and post-movement testing of cattle in Great Britain. Animal Health and Veterinary Laboratory Agencies. Publication number AG-TBYHE-04

4. Armstrong JS, Overton TS (1977) Estimating nonresponse bias in mail surveys. J Mark Res 396–402

5. Barucci A, Macaluso G, Mecatti D, Noferini L, Fanelli D, Facchini A, Materassi M, Pieraccini M, Atzeni C (2010) Universal fluctuations in tropospheric radar measurements. EPL (Europhysics Letters) 89:20006

6. Beckwith CG, Flanigan TP, del Rio C, Simmons E, Wing EJ, Carpenter CCJ, Bartlett JG (2005) It is time to implement routine, not risk-based, HIV testing. Clin Infect Dis 40:1037–1040

7. Bessell PR, Orton R, O'Hare A, Mellor DJ, Logue D, Kao RR (2013) Developing a framework for risk-based surveillance of tuberculosis in cattle: a case study of its application in Scotland. Epidemiol Infect 141:314–323

8. Booth GL, Hux JE, Fang J, Chan BT (2005) Time trends and geographic disparities in acute complications of diabetes in Ontario, Canada. Diabetes Care 28:1045–1050

9. Brockwell PJ, Davis RA (2002) Introduction to time series and forecasting

10. Caron A, Miguel E, Gomo C, Makaya P, Pfukenyi DM, Foggin C, Hove T, De Garine-Wichatitsky M (2013) Relationship between burden of infection in ungulate populations and wildlife/livestock interfaces. Epidemiol Infect 141:1522–1535

11. Carstensen B (2005) Demography and epidemiology: age-Period-Cohort models in the computer age. Department of Biostatistics, University of Copenhagen, Copenhagen

12. Christakos G, Hristopulos DT (1998) Spatiotemporal environmental health modelling: a tractatus stochasticus. Springer, New York

13. Christakos G, Olea RA, Serre ML, Yu HL, Wang LL (2006) Interdisciplinary public health reasoning and epidemic modelling: the case of black death: the case of black death. Springer, New York

14. Christakos G, Wang J-F, Wu J (2014) Stochastic medical reasoning and environmental health exposure. World Scientific, Singapore

15. Claridge J, Diggle P, McCann CM, Mulcahy G, Flynn R, McNair J, Strain S, Welsh M, Baylis M, Williams DJL (2012) Fasciola hepatica is associated with the failure to detect bovine tuberculosis in dairy cattle. Nat Commun 3:853

16. Cleveland WS (1979) Robust locally weighted regression and smoothing scatterplots. J Am Stat Assoc 74:829–836

17. Corner LAL, Pfeiffer DU, Morris RS (2003) Social-network analysis of *Mycobacterium bovis* transmission among captive brushtail possums (*Trichosurus vulpecula*). Prev Vet Med 59:147–167

18. Croft DP, Madden JR, Franks DW, James R (2011) Hypothesis testing in animal social networks. Trends Ecol Evol 26:502–507

19. DEFRA (2009) Bovine tuberculosis in England: towards eradication. Final Report of the Bovine TB Advisory Group

20. DEFRA (2014a) Annex. Background and methodologies to the National Statistics on the Incidence of Tuberculosis (TB) in cattle in Great Britain. Latest statistics on the incidence of tuberculosis (TB) in cattle in Great Britain

21. DEFRA (2014b) Monthly publication of National Statistics on the Incidence of Tuberculosis (TB) in Cattle to end June 2014 for Great Britain

22. Dorjee S, Revie C, Poljak Z, McNab W, Sanchez J (2013) Network analysis of swine shipments in Ontario, Canada, to support disease spread modelling and risk-based disease management. Prev vet Med 112:118–127

23. Evans MR, Bithell M, Cornell SJ, Dall SRX, Díaz S, Emmott S, Ernande B, Grimm V, Hodgson DJ, Lewis SL, Mace GM, Morecroft M, Moustakas A, Murphy E, Newbold T, Norris KJ, Petchey O, Smith M, Travis JMJ, Benton TG (2013) Predictive systems ecology. Proc R Soc B: Biol Sci 280:20131452

24. Evans MR, Benton TG, Grimm V, Lessells CM, O'Malley MA, Moustakas A, Weisberg M (2014) Data availability and model complexity, generality, and utility: a reply to Lonergan. Trends Ecol Evol 29:302–303

25. García-Portugués E, Barros AG, Crujeiras R, González-Manteiga W, Pereira J (2014) A test for directional-linear independence, with applications to wildfire orientation and size. Stoch Environ Res Risk Assess 28:1261–1275

26. Gilbert M, Mitchell A, Bourn D, Mawdsley J, Clifton-Hadley RS, Wint W (2005) Cattle movements and bovine tuberculosis in Great Britain. Nature 435:491–496

27. Gocheva-Ilieva SG, Ivanov AV, Voynikova DS, Boyadzhiev DT (2014) Time series analysis and forecasting for air pollution in small urban area: an SARIMA and factor analysis approach. Stoch Environ Res Risk Assess 28:1045–1060

28. Godfray HCJ, Donnelly CA, Kao RR, Macdonald DW, McDonald RA, Petrokofsky G, Wood JLN, Woodroffe R, Young DB, McLean AR (2013) A restatement of the natural science evidence base relevant to the control of bovine tuberculosis in Great Britain†. Proc R Soc B: Biol Sci 280:20131634

29. Guo Z, Kan Wan S, August DA, Ying J, Dunn SM, Semmlow JL (2006) Optical imaging of breast tumor through temporal log-slope difference mappings. Comput Biol Med 36:209–223

30. Hethcote HW (2000) The mathematics of infectious diseases. SIAM Rev 42:599–653

31. Jenkins TC, Gardner EM, Thrun MW, Cohn DL, Burman WJ (2006) Risk-based human immunodeficiency virus (HIV) testing fails to detect the majority of HIV-infected persons in medical care settings. Sex Transm Dis 33(329–333):3.

32. Koh GCKW, Hawthorne G, Turner AM, Kunst H, Dedicoat M (2013) Tuberculosis incidence correlates with sunshine: an ecological 28-year time series study. PLoS One 8:e57752

33. Koutsoyiannis D (2005) Uncertainty, entropy, scaling and hydrological stochastics. 2. Time dependence of hydrological processes and time scaling. Hydrol Sci J 50:405–426

34. Koutsoyiannis D (2011) Hurst–Kolmogorov dynamics and uncertainty1. JAWRA J Am Water Resour Assoc 47:481–495

35. Krebs JR, Anderson R, Clutton-Brock T, Morrison I, Young D, Donnelly CA, Frost S, Woodroffe R (1997) Bovine tuberculosis in cattle and badgers. London, p 191

36. Lavielle M (1999) Detection of multiple changes in a sequence of dependent variables. Stoch Process Appl 83:79–102

37. Lavielle M (2005) Using penalized contrasts for the change-point problem. Sig Process 85:1501–1510

38. Lijffijt J, Papapetrou P, Puolamäki K (2014) A statistical significance testing approach to mining the most informative set of patterns. Data Min Knowl Disc 28:238–263

39. Lin C-H, Wen T-H, Teng H-J, Chang N-T (2015) The spatio-temporal characteristics of potential dengue risk assessed by Aedes aegypti and Aedes albopictus in high-epidemic areas. Stoch Environ Res Risk Assess.

40. Lonergan M (2014) Data availability constrains model complexity, generality, and utility: a response to Evans et al. Trends Ecol Evol 29:301–302

41. Markonis Y, Koutsoyiannis D (2013) Climatic variability over time scales spanning nine orders of magnitude: connecting Milankovitch cycles with Hurst–Kolmogorov dynamics. Surv Geophys 34:181–207

42. Marsden T, Morley A (2014) Sustainable food systems: building a new paradigm. Routledge, London

43. McCallan L, McNair J, Skuce R, Branch B (2014) A review of the potential role of cattle slurry in the spread of bovine tuberculosis. Agri-Food and Biosciences Institute: Belfast. Available online at: http://www.dardni.gov.uk/tb-slurry-lit-review.pdf. Accessed May 2015

44. Moustakas A, Evans M (2015) Coupling models of cattle and farms with models of badgers for predicting the dynamics of bovine tuberculosis (TB). Stoch Environ Res Risk Assess 29:623–635

45. Oidtmann B, Peeler E, Lyngstad T, Brun E, Bang Jensen B, Stärk KDC (2013) Risk-based methods for fish and terrestrial animal disease surveillance. Prev Vet Med 112:13–26

46. Paltiel AD, Weinstein MC, Kimmel AD, Seage GR, Losina E, Zhang H, Freedberg KA, Walensky RP (2005) Expanded screening for HIV in the United States—an analysis of cost-effectiveness. N Engl J Med 352:586–595

47. Parrinello CM, Crossa A, Harris TG (2012) Seasonality of tuberculosis in New York City, 1990–2007. Int J Tuberc Lung Dis 16:32–37

48. Ruiz-Medina MD, Espejo RM, Ugarte MD, Militino AF (2014) Functional time series analysis of spatio–temporal epidemiological data. Stoch Environ Res Risk Assess 28:943–954

49. Sur M, Skidmore AK, Exo K-M, Wang T, Ens BJ, Toxopeus A (2014) Change detection in animal movement using discrete wavelet analysis. Ecol Inform 20:47–57

50. Thrusfield M (2013) Veterinary epidemiology. Wiley, Hoboken

51. Tomlinson I (2013) Doubling food production to feed the 9 billion: a critical perspective on a key discourse of food security in the UK. J Rural Stud 29:81–90

52. Top R, Boshuizen H, Dekkers A, Korthals Altes H (2013) Similar seasonal peak in clustered and unique extra-pulmonary tuberculosis notifications: winter crowding hypothesis ruled out? Int J Tuberc Lung Dis 17:1466–1471

53. Tsanis IK, Daliakopoulos IN (2015) D2.2 Drivers of change in the CASCADE Study Sites. Public Deliverable of the CASCADE Project: CAtastrophic Shifts in drylands: how CAn we prevent ecosystem DEgradation? (FP7/2007-2013 grant agreement no 283068)

54. Weisberg S (2014) Applied linear regression. Wiley, Hoboken

55. Wingfield T, Schumacher SG, Sandhu G, Tovar MA, Zevallos K, Baldwin MR, Montoya R, Ramos ES, Jongkaewwattana C, Lewis JJ, Gilman RH, Friedland JS, Evans CA (2014) The seasonality of tuberculosis, sunlight, vitamin D, and household crowding. J Infect Dis 210:774–783

56. Yang AL, Rounsevell MD, Wilson RM, Haggett C (2014) Spatial analysis of agri-environmental policy uptake and expenditure in Scotland. J Environ Manag 133:104–115

57. Ye T, Nie J, Wang J, Shi P, Wang Z (2015) Performance of detrending models of crop yield risk assessment: evaluation on real and hypothetical yield data. Stoch Environ Res Risk Assess 29:109–117

58. Yu H, Alonso WJ, Feng L, Tan Y, Shu Y, Yang W, Viboud C (2013) Characterization of regional influenza seasonality patterns in China and implications for vaccination strategies: spatio-temporal modeling of surveillance data. PLoS Med 10:e1001552

5

IMPORTANCE SAMPLING FOR EFFICIENT MODELLING OF HYDRAULIC LOADS IN THE RHINE–MEUSE DELTA

F. L. M. Diermanse · K. M. De Bruijn · J. V. L. Beckers · N. L. Kramer

Deltares, P.O. Box 177, 2600 MH Delft, The Netherlands

ABSTRACT

During flood events, breaching of flood defences along a river system can have a significant reducing effect on downstream water levels and flood risks. This paper presents a Monte Carlo based flood risk framework for policy decision making, which takes this retention effect into account. The framework is developed to estimate societal flood risk in terms of potential numbers of fatalities and associated probabilities. It is tested on the Rhine–Meuse delta system in the Netherlands, where floods can be caused by high flows in the Rhine and Meuse rivers and/or high sea water levels in the North Sea. Importance sampling is applied in the Monte Carlo procedure to increase computational efficiency of the flood risk computations. This paper focuses on the development and testing of efficient importance sampling strategies for the framework. The development of an efficient importance sampling strategy for river deltas is more challenging than for non-tidal rivers where only discharges are relevant, because the relative influence of river discharge and sea water level on flood levels differs from location to location. As a consequence, sampling methods that are efficient and accurate for one location may be inefficient for other locations or, worse, may introduce errors in computed design water levels. Nevertheless, in the case study described in this paper the required simulation time was reduced by a factor 100 after the introduction of an efficient importance sampling method in the Monte Carlo framework, while at the same time the accuracy of the Monte Carlo estimates were improved.

KEYWORDS

Flood risk, Monte Carlo simulation, Importance sampling, Rhine–Meuse delta

1. INTRODUCTION

An extensive system of flood defences has been constructed in the Netherlands to prevent regular floods from the sea, major rivers and lakes. In this system, 53 dike ring areas are distinguished, which are protected by a connected system of dikes, dunes and hydraulic structures. For the design and safety assessment of the flood defences, protection standards per dike ring area are defined in terms of "allowable flood frequencies". The present protection standards vary from 10^{-3} to 10^{-4} per year. The foundation of the flood protection standards was laid by Van Dantzig (1956) and formally established in Delta Committee (1958). The present protection standards are being reconsidered, following the advice of a newly established Delta Committee (2008). The proposed protection standards for safety assessments from 2017 onwards will be based on an advanced cost-benefit analysis (Kind 2013) and on flood fatality risk assessments. Fatality risks are considered from the viewpoint of individuals and from the perspective of society. The framework presented in this paper is developed to assess *societal* flood fatality risks. However, the concept can also be used to quantify economic flood risks.

Societal flood fatality risk in the Netherlands is defined as the probability of exceedance of a large number of flood fatalities occurring in a single year. This risk is typically quantified with FN-curves (Beckers et al. 2012; De Bruijn et al. 2010; Evans and Verlander 1997; Vrijling et al. 1995), where N is the number of fatalities and$F$ is the associated frequency of exceedance. The FN-curve allows for a differentiated evaluation of events with high probabilities and low numbers of fatalities versus events with low probabilities and high numbers of fatalities. The new protection standards for the Netherlands will be based on the criterion that events with 10*N fatalities are 100 times less tolerable than events resulting in N fatalities. The FN-curve provides the required information to evaluate such criteria.

For the analysis of societal flood risk, the potential influence of a dike breach on the flood risk of downstream areas needs to be taken into account. Otherwise, scenarios in which (almost) all dike ring areas are flooded may have a significant effect on the derived FN-curve, even though in reality these scenarios cannot occur. So far, this retention effect has not been taken into account in formal flood risk assessments in the Netherlands. This paper describes the set-up of the Monte Carlo based probabilistic framework for societal flood risk analysis, which takes the retention effects of dike breaches into account. The full framework will only be described briefly, as the main focus of the paper will be on a specific component: the Monte Carlo sampling method. Computation results of the framework and consequences for societal flood risk are discussed in a separate paper (De Bruijn et al. 2014). The main objective of the current paper is to construct a sampling strategy that is efficient in terms of computation times and at the same time provides sufficiently accurate results. The method is applied on the Rhine–Meuse delta in the Netherlands, but can be applied on other river deltas as well.

2. PROBABILISTIC RISK MODELING FRAMEWORK

2.1 Required capabilities

The modeling framework is developed with the aim to derive societal flood risks for large river deltas. Furthermore, it is required that mitigating measures can be evaluated and also that areas that contribute most to the societal flood risk can be identified. This leads to the following set of requirements:

1. The framework has to be able to quantify the (societal) flood risk of the delta as a whole and the individual polders and floodplains;

2. The main components of the risk chain need to be modeled explicitly, i.e. hydraulic loads, resistance and breaching of flood defences, flooding of polders and flood plains and evacuation response;

3. All relevant uncertainties need to be taken into account;

4. The method has to be able to deal with the combined influence of sea water level and river discharge;

5. The reduction of downstream water levels due to breaching flood defences has to be quantified;

6. The method has to be applicable for systems with a large number (>100) of dike sections and potential breach locations;

7. For practical purposes the runtime should preferably be less than 24 h on a standard personal computer.

2.2 Existing flood risk models in literature

The advantages of probabilistic flood risk methods over more traditional deterministic methods are widely recognized. The increase in computation power has given an impulse to the development of various flood risk models in which the entire 'chain' (sources, pathways, receptors) is explicitly modelled, including all relevant uncertainties. This section discusses a number of risk models from literature that are most relevant to our study.

The subject of downstream flood risk reduction due to upstream breaching in the Netherlands was explored in a research project as described by Van Mierlo et al. (2007) and Courage et al. (2013). Courage et al. (2013) successfully quantified the potential effects of dike breaches on downstream water levels and flood risks for a number of dike rings in the Rhine–Meuse delta. In their simulations, flood defences could breach due to the failure mechanisms 'piping' and 'erosion of the inner slope due to wave overtopping'. In case of a dike breach, a detailed 2D flood simulation model was used to compute inundation depths in the protected polders. As a result, their approach was computationally time-consuming: a single flood simulation took about 2–6 days on a standard issue 2 GHz Linux PC.

Apel et al. (2004) and (2009) also quantified the retention effects of breaching on downstream flood frequency curves for stretches of the Lower Rhine River in Germany. They used a Monte Carlo based approach in which upstream discharges were sampled from the derived distribution function and subsequently routed through the river stretch. In their model, flood defences could only breach due to the failure mechanism 'wave overtopping'. In both studies the retention effects were demonstrated to have significant influences on the flood frequency curves of downstream locations, especially for events with high return periods. Apel et al. (2009) therefore concluded that their approach provides more realistic results than the traditional flood frequency approach in which the retention effect is not taken into account. Vorogushyn et al. (2012) extended the work of Apel et al. (2009) by quantifying additional flood intensity indicators and introducing 'piping and heave' and 'micro-instability' as additional failure mechanisms in their framework. Their method was applied on a stretch of the Elbe River in Germany. Due to the relatively small storage of the Elbe floodplains, dike breaching was not as influential on downstream flood levels as it was for the lower Rhine in the studies of Apel et al. (2004) and (2009).

Lamb et al. (2010) developed a statistical conditional exceedance model to describe the joint probability of extreme river flows or sea levels at multiple locations. The joint probability method was developed by Keef et al. (2009), based on the earlier works of Heffernan and Tawn (2004). One of the main advantages of this model is that it offers the possibility to handle dependencies of a large set of locations with mutually correlated hydraulic loads. Wyncoll and Gouldby (2013) linked the load model of Lamb et al. (2010) to the consequence model of Gouldby et al. (2008) to develop a fully risk based approach. Gouldby et al. (2008) used fragility curves to describe the resistance of the flood defences, a rapid flood model to derive flood depths and standard depth-damage functions to quantify economic losses. In the models of Lamb et al. (2010) and Wyncoll and Gouldby (2013) the hydraulic loading conditions are assumed to be fully dependent in terms of recurrence interval (return period). Retention effects of breaching flood defences were not taken into account.

Recently, Zhong et al. (2013) have implemented a hydraulic load model for the Rhine–Meuse delta in a Monte Carlo framework. They applied importance sampling functions for the most relevant input variables to speed up the convergence process. The importance sampling procedure used normal distribution functions which were centred around the values that lead to critical water levels for design and safety assessment. They did not report on the efficiency of this sampling strategy compared to crude Monte Carlo, which is the focal point of our study. Dawson and Hal (2006) developed a Monte Carlo based flood risk method in which adaptive importance sampling was applied to speed up the convergence of the sampling process. Fragility curves were used to describe the reliability of the flood defences of a combined coastal/fluvial system in Towyn, North Wales. Five different failure modes, including dune erosion, were taken into account and represented by fragilty curves. A simplified two-dimensional inundation model and standard damages curves were used to compute flood damages for each simulated event. Retention effects were not taken into account. The output of their study consisted of flood risk maps of the study area.

The papers described in this section provide valuable flood risk modeling concepts that can be used for our modeling purposes. However, to our best knowledge there is no model available that has all modeling capabilities as desired for our study (Sect. 2.1). Especially a system that takes the retention effects of dike breaches into account for a tidal river system with combined influences of sea water levels and river discharges does not seem to be available. Therefore a new modeling framework was required that takes these issues into account.

2.3 Framework components

The objective of the probabilistic framework is to quantify FN-curves for river deltas. The computation of FN curves involves dealing with multiple sources of uncertainty. In our framework, uncertainties in the hydraulic loads, the resistance of the flood defences, the evacuation response and the resulting number of flood fatalities from a breach are all taken into account. The different sources of uncertainty are described with probability distribution functions, which are input for the framework. The framework consists of the following components:

[i] Generation of synthetic events characterized by load, strength and response variables;

[ii] Hydrodynamic modeling of the synthetic events; and

[iii] Post-processing for deriving FN-curves.

The generation of synthetic events (component [i]) starts with sampling of hydraulic load variables like river discharge and sea water level from derived distribution functions. Subsequently, the resistance of the flood defences is sampled from fragility curves for potential breach locations. Breaching of flood defences can potentially occur anywhere in the system. The framework, however, requires a finite set of potential breach locations. These breach locations are selected in such a way that all relevant potential flood scenarios are captured. For this purpose, the system of flood defences is subdivided into several stretches, based on the criterion that flood consequences are approximately the same for breaches at any location within a single stretch. For each stretch, a single representative potential breach location was selected. For each potential breach location, fragility curves are derived for all relevant failure mechanisms based on characteristics of the flood defences. Three failure mechanisms are considered that are known to be dominant for riverine flood defences: "erosion of the inner slope", "piping" and "slope instability". This means for each simulated event and each location, three breaching levels are sampled from the fragility curves, representing the water levels at which the flood defence will breach due to the corresponding failure mechanism. The lowest of the three breaching levels is the water level at which the flood defence will breach. The final step in component [i] consists of sampling a success rate of the evacuation response, which is used later on to determine the expected number of casualties in a flood event.

Each event is simulated with a hydrodynamic model (component [ii]). The sampled hydraulic load variables like river discharge and sea water level serve as boundary conditions for this model. The sampled breaching levels from the fragility curves are also input to the hydrodynamic model. At each simulation time step, locations are identified for which the water level exceeds the breaching level. For these locations the formation of a breach is simulated and water is abstracted from the river. This leads to a reduction of downstream water levels as well as inundation of the protected polders. As such, the framework is able to take the reducing effect of breaching on downstream water levels and flood probabilities into account. More details on the hydrodynamic model simulation are provided in Sect. 4.3.

In the post-processing procedure (component [iii]), the expected number of fatalities in the simulated events is derived from indicators like the flooded area, the number of inhabitants and the evacuation success rate. These numbers are used to derive FN-curves that show frequencies of exceedance of (large) numbers of flood fatalities. More details on the computation of the FN curves are provided in the following section.

2.4 Probabilistic computation method

To estimate the probability of failure of complex systems, probabilistic computation methods are required. In such computations, generally the probability is computed that the load (S) of the systems exceeds the strength, or resistance (R). For food defences, the load typically consists of a combination of water levels and waves and in some cases currents. The resistance depends on the geometry of the flood defence and soil characteristics. For convenience, a limit state function Z is defined as follows:

$$Z = R - S \tag{1}$$

With this definition, $Z < 0$ refers to "failure" and $Z \geq 0$ to the opposite, i.e. "no failure". This means the failure probability is equal to: $P[Z < 0]$. This definition of function Z can also be used to quantify *exceedance*probabilities of e.g. a threshold water level w^*. In that case the "load", S, is taken equal to the actual water level, w, and the "resistance", R, is taken equal to w^*. A value of $Z < 0$ then corresponds to the situation that$w$ exceeds w^*, so again $P[Z < 0]$ is the probability that needs to be quantified.

The limit state function, Z, is a function of a number of random variables representing both load and resistance variables. This means Z is a random variable as well. The probability of failure can be written as follows:

$$P_f = P[Z < 0] = \int_{Z(\mathbf{x}) < 0} f_{\mathbf{X}}(\mathbf{x})\, d\mathbf{x} \tag{2}$$

where $\mathbf{x}$ is the vector of variables: $\mathbf{x} = (x_1, \ldots x_n)$ and $f_{\mathbf{X}}$ is the joint probability density function of $\mathbf{x}$. To evaluate $Z(\mathbf{x})$, often numerical simulation models are required. This means Eq. (2) is generally too complex to evaluate in an exact, analytical way. Therefore, probabilistic techniques are required to provide an estimate of the failure probability. There are different probabilistic methods available, each with its advantages and disadvantages. Extensive overviews are given in Ditlevsen and Madsen (1996) and Melchers (2002). The "best" choice of the probabilistic method depends on the problem under consideration. Grooteman (2011) identifies the following three selection criteria on which the choice of a computational method should be based: accuracy, efficiency (i.e. computation time) and robustness.

For the existing statutory safety assessment procedure of flood defences in the Rhine–Meuse delta, numerical integration is used to determine probabilities of hydraulic loads (Geerse, 2005). Numerical integration scores high on robustness and accuracy, if grid cells are chosen sufficiently small. However, numerical integration becomes very time-consuming if more than just a few random variables are involved. For the probabilistic model of the statutory safety assessment this poses no problem as it only considers hydraulic loads in order to determine design water levels. The number of random variables is therefore limited. However, in our study the resistance of the flood defences is also considered and the number of random variables is well over a hundred. This makes the application of numerical integration infeasible.

As a potential alternative, FORM (Rackwitz, and Fiessler, 1978) is known to be computationally efficient. This is the reason why this is the preferred method in the computational software that is being developed for the subsequent statutory safety assessment of flood defences in the Netherlands (Den Heijer and Diermanse, 2012). The disadvantage of FORM is that it relies on an iterative procedure which sometimes does not converge. Furthermore, FORM relies on linearization of the *Z*-function, which means errors are introduced if the actual *Z*-function is highly non-linear. For application in the framework of the current study, FORM has a further disadvantage that it computes the failure probability (or exceedance probability) at a single location. Since the framework needs to be applicable for systems with large numbers of potential breach locations, the efficiency gain of FORM is easily lost in this case.

Monte Carlo methods do not have the disadvantages of FORM and numerical integration as described above. The required computation time does not increase with increasing number of variables and the failure analysis of multiple locations can be efficiently combined. For this reason it was decided to implement a Monte Carlo based method in the framework. For Monte Carlo methods, the main challenge is provide accurate estimates for cases in which the probability of failure is very small. In those cases, crude Monte Carlo sampling requires a large number of *Z*-function evaluations for an accurate estimate of the failure probability. If *Z*-function evaluations are very time-consuming, the number of evaluations will have to be limited from a practical point of view. This will automatically be at the expense of the accuracy of the estimate. Fortunately, the efficiency of Monte Carlo simulation can be enhanced through application of advanced sampling techniques like Latin hypercube sampling (Georgiou 2009; HSU et al. 2011; Olsson et al.2003; Owen 1994; Ye 1998), directional sampling (Bjerager 1988; Ditlevsen et al. 1990; Grooteman, 2011; Melchers 2002), stratified sampling (May et al. 2010; Keskintürk and Er 2007; Christofides 2003) or importance sampling (Engelund and Rackwitz 1993; Koopman et al. 2009; Sezer 2009; Yuan and Druzdzel2006). A further efficiency gain may be achieved if these methods are combined with adaptive response surface techniques (Liu et al. 2010; Steenackers et al. 2009 Allaix and Carbone, 2011) which help explore the failure space, $Z(\mathbf{x}) < 0$, at a low computational cost.

For most practical problems, the advanced sampling techniques reduce the required number of *Z*-function evaluations in comparison with the crude Monte Carlo approach. The efficiency gain of Latin Hypercube sampling is relatively small compared to the other methods in cases where extreme events are relevant, and therefore not appropriate for our purpose. The efficiency of directional sampling decreases if a large number of random variables are involved (Waarts 2000). Furthermore, directional sampling is less efficient in case of a large number of objective functions that need to be evaluated individually. The large number of potential breach locations considered in the current study, brings with it an even larger set of random variables and objective functions in the hydraulic load model, which is why directional sampling is not the preferred option. Stratified sampling and importance sampling are anticipated to be the most efficient Monte Carlo techniques for our framework. We chose to apply importance sampling as it was considered the most practical of the two methods to implement and apply in the test procedure as described in Sect. 4.4

Application of importance sampling means the actual multivariate distribution function, F, of the set, $\mathbf{x}$, of random variables is replaced in the sampling procedure by an alternative distribution function, H, in order to increase the probability of sampling events which are most relevant for the flood risk in the study area. The changes in sampling probabilities need to be corrected for in the Monte Carlo probability estimate by a factor that is equal to $c = f(\mathbf{x})/h(\mathbf{x})$, where f and h are the density functions corresponding to F and H. This correction factor c is determined for each sample of $\mathbf{x}$. The estimated probability of exceedance of a threshold number of N^* fatalities is then equal to:

$$P[N > N^*] = \frac{1}{n}\sum_{i=1}^{n} 1_{[N_i > N^*]} c_i \qquad (3)$$

In which n is the number of simulated events, P is the exceedance probability *per event*, N is the number of fatalities in an event, N^* is a possible realization of N, N_i is the number of fatalities in event i, $1_{[..]}$ is an indicator function which is equal to 1 if $N_i > N^*$ and 0 otherwise and c_i is the correction factor for importance sampling for event i. The FN-curve is derived through application of Eq. (3) for a range of values of N^*.

The Monte Carlo simulation procedure in our framework uses random variables to model the hydraulic load (river discharge, sea water level, possible failure of a barrier), the resistance of flood defences (fragility curves for potential breach locations) and the rate of success of the evacuation response. In the method, *n*years are simulated, where *n* can be selected by the user. The choice of *n* is generally based on the trade-off between computation time and desired accuracy. The simulated years do not represent a series of *n*subsequent years, but *n* possible realizations of a reference year. This is a relevant distinction, because in the first case, the sampling method for resistance variables needs to take the high correlations between samples of subsequent years into account, whereas in the second case the samples for each simulated year can be generated independently.

For each simulated year, two synthetic events are considered: (i) the event in which the annual maximum river discharge occurs and (ii) the event in which the annual maximum sea water level occurs. The motivation is that in a river delta both types of events can cause floods and they generally do not coincide. This means in total $2n$ synthetic events are generated and simulated according to steps [i] and [ii] of Sect.2.3. In this approach, the estimated annual probability of exceedance of a threshold number of N^* fatalities is equal to:

$$P[N > N*] = \frac{1}{n}\sum_{i=1}^{2n} 1_{[N_i > N^*]} c_i \tag{4}$$

In Eq. (4) P refers to the *annual* exceedance probability, whereas in Eq. (3) P referred to the exceedance probability per *event*. This explains the number '2' in Eq. (4)which is absent in Eq. (3): there are two simulated events per year.

3. CRITERIA FOR SELECTING IMPORTANCE SAMPLING STRATEGIES

The main purpose of this paper is to establish an efficient sampling strategy for the framework as described in the previous section. The optimal strategy usually depends on the case study area. In other words:

no sampling strategy can be expected to be efficient for all river deltas in the world. However, our proposed approach to derive the sampling strategy is generic. The method is to iteratively develop and apply various sampling strategies and to test the efficiency. Sampling strategies will be formulated based on test results from previous iterations.

One of the possible approaches to 'learn' from previous simulation results is to use adaptive sampling schemes (e.g. Steenackers et al. 2009; Allaix and Carbone 2011). Dawson and Hal (2006) applied such an approach in a flood risk analysis of a combined coastal/fluvial system in Towyn, North Wales. In their method, the sampling density function $h(\mathbf{x})$ is iteratively adapted in such a way that events that were found to contribute most to the flood risk in earlier iterations are given higher sampling densities in subsequent iterations. They demonstrated that such an approach can significantly speed up the convergence of the Monte Carlo simulation procedure. Nevertheless, such an approach was not adopted in our study. The main reason is that for our decision making process not only the flood risk of the entire delta area is required, but also the flood risk of all the individual polder areas that are protected by flood defences. An approach that optimizes the sampling scheme for the flood risk of the delta as a whole may not provide reliable estimates for each individual polder.

For this reason a 'learn-by-doing' approach was followed in which sampling schemes are judged based on their ability to provide reliable Monte Carlo estimates at each location in the delta within an acceptable simulation time. This means the sampling scheme needs to be efficient, accurate and robust for all locations. As stated before, importance sampling can increase the efficiency of a Monte Carlo simulation by sampling the vector, $\mathbf{x}$, of random variables from an alternative distribution function, $H(\mathbf{x})$, instead of the actual distribution function, $F(\mathbf{x})$. Function $H(\mathbf{x})$ should be chosen in such a way that the required number of samples to obtain accurate results is reduced as much as possible. The accuracy of the sampling procedure can be verified with the following two criteria for the Monte Carlo estimate:

1. the bias should be equal to 0;

2. the standard deviation should be lower than an acceptable threshold.

The first criterion implies that the Monte Carlo estimate (Eq. (4)) should converge to the correct result if the number of samples, n, goes to infinity. More formally, this criterion can be described as follows: The bias is equal to zero if for any combination of (small) positive values ε_1 and ε_2 there is a value n^* for which the following holds if the number of samples, n, is higher than n^*:

$$P[\varepsilon > \varepsilon_1] < \varepsilon_2 \tag{5}$$

In which ε is the error in the Monte Carlo estimate. A bias in the estimate will be introduced if, and only if, there are events that result in failure which are awarded a probability of 0 in the importance sampling procedure. The second criterion above can be verified by carrying out the Monte Carlo experiment multiple (M_s) times, and to subsequently verify if the standard deviation of the resulting M_s estimates is below the acceptable threshold. Note that the choice of the acceptable threshold is subjective. The standard deviation can always be kept as low as possible by increasing the number of samples, n. However, the purpose of importance sampling is to keep this number as low as possible. So, if we assume the number of n to be fixed, notorious contributors to high standard deviations are potential realisations, **x**, for which $Z(\mathbf{x}) < 0$ and$f(\mathbf{x})/h(\mathbf{x}) \gg 1$. These are events that contribute to failure, but have a relatively small probability of being sampled in the importance sampling procedure. However, *if* they are sampled, they have a relatively large contribution to the Monte Carlo estimator because the factor $c = f(\mathbf{x})/h(\mathbf{x})$ is large. Such realisations of **x** can cause relatively large differences in the successive Monte Carlo estimators and therefore contribute strongly to the standard deviation. This should therefore be avoided, which means the events that contribute to the flood risk in the area should never be awarded a sampling probability that is significantly lower than the actual probability.

The bias and standard deviation are used in the remainder of this paper as criteria to test the accuracy of various importance sampling strategies. Our main objective is to construct a sampling function, $H(\mathbf{x})$, that is efficient (i.e. low number of hydraulic model simulations) and sufficiently accurate (i.e. no bias and acceptably small standard deviations) for all considered locations in the area. Different sampling strategies are tested and mutually compared in terms of bias and standard deviation, in order to obtain an "optimal" sampling strategy.

4. CASE STUDY

4.1 Area

The Rhine–Meuse delta in the Netherlands (see Fig. 1) is a densely populated region that includes the cities of Rotterdam, Utrecht, Arnhem, Nijmegen and Dordrecht. The area is of high economic value, in particular the port of Rotterdam. The protection standards for this region are among the highest in the world: the flood defenses must be able to withstand hydraulic loads associated with a return period of 10,000 years. With respect to the hydraulic characteristics, the area can be subdivided into three regions: a *tidal region*, where high water levels are caused by high sea water levels; a *non-tidal region*, where high water levels are caused by high river discharges and a *transitional region*, where high water levels can be caused by both high sea water levels and high river discharges.

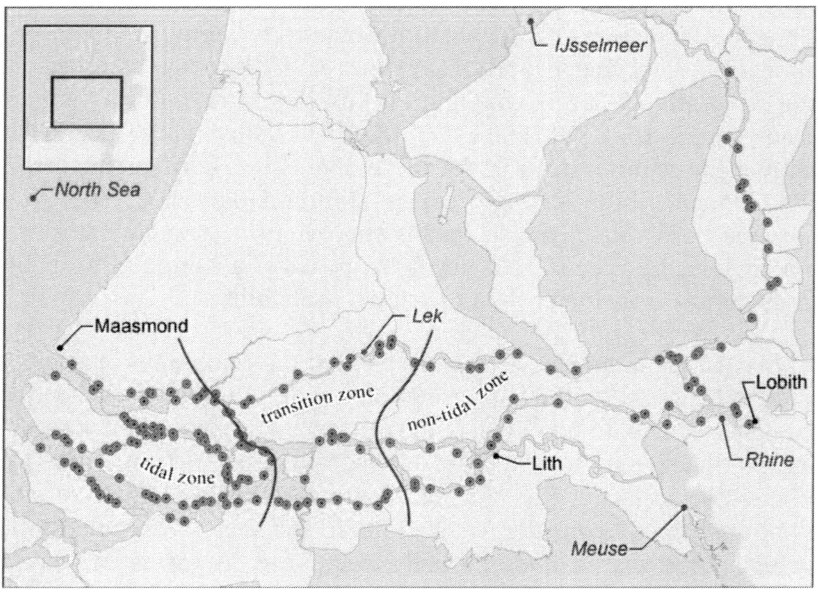

Figure 1. Model area and potential breach locations (*red dots*)

The tidal and transitional regions are protected from high sea water levels by the Maeslant storm surge barrier, near location Maasmond (see Fig. 1). This barrier closes when the water level at Rotterdam is expected to exceed the level of NAP + 3 m,[1] or if the water level at Dordrecht is expected to exceed a water level of NAP + 2.9 m. A closure request for the barrier due to high storm surges is expected to occur on average approximately once every 10 years. There is an estimated probability of 1 % that the barrier fails to close upon request.

Flood defence breaches can occur at any location in the area and consequences in terms of numbers of fatalities can vary strongly from location to location. The framework, however, requires a finite set of potential breach locations. The potential breach locations of the model are selected in such a way that all relevant potential flood scenarios are captured by the modelling framework. For this purpose, the system of flood defences is subdivided into several stretches, based on the criterion that flood consequences are approximately the same for breaches at any location within a single stretch. In total 171 different stretches were identified by De Bruijn and Van der Doef (2011), varying in length from 400 m to 34 km. Figure 1 shows the corresponding 171 representative breach locations.

4.2 Statistics of hydraulic loads

High water levels in the Rhine–Meuse delta are mainly determined by the discharge of the River Rhine at upstream boundary Lobith, the discharge of the river Meuse at upstream boundary Lith, the sea water level at downstream boundary Maasmond and the functioning of the barrier near Maasmond, which may fail to close upon request. As the focus of the study is on flood risk, statistics of high river discharges and high sea water levels are most relevant. Statistical distribution functions for high rivers discharges and sea water levels have been adopted as much as possible from the probabilistic model that was developed to derive design water levels for the formal safety assessment of flood defences in the Rhine–Meuse delta (see e.g. Geerse 2005 for a description of that model). Statistics of extreme river discharges and sea water levels have been derived by fitting extreme value distribution functions through observed annual maximum water levels and peaks-over-threshold series. Sea water level statistics are described by a conditional Weibull distribution:

$$P(M \leq m | m > \omega) = 1 - \exp\left[(\omega/\sigma)^{\xi} - (m/\sigma)^{\xi}\right], \tag{6}$$

where M is the annual maximum sea water level, relative to NAP, m a potential realization of M and ω, σ, and ξ are the location, scale and shape parameter respectively. Exceedance *frequencies* of high sea water levels can be derived by multiplying the probabilities that follow from equation (6) with the frequency of exceedance, λ, of threshold ω. The value of λ is determined by counting the number of peaks above the threshold and dividing by the number of years of record. For location Maasmond, $\omega = 1.97, \lambda = 7.24, \xi = 0.57$ and $\sigma = 0.0157$.

Probabilities of high river discharges in the Rhine and Meuse are described with a Gumbel distribution:

$$P(Q<q) = \exp\left(-\exp\left(-\frac{q-b}{a}\right)\right), \quad (7)$$

where Q is the annual maximum peak discharge, q is a potential realization of Q and a and b are the two parameters of the distribution function, with a = 1,316 m^3/s, b = 6,612 m^3/s for the Rhine River at Lobith and$a$ = 342 m^3/s, b = 1,190 m^3/s for the Meuse River at Lith.

These statistical distribution functions quantify probabilities for the high range of peak values. However, high river discharges are likely to occur jointly with 'normal' sea water level conditions and vice versa, so statistics of 'daily' conditions are also relevant. To describe probabilities of the whole range of conditions, histograms were derived from all observed daily discharges in a period of approximately 100 years of measurements. Similarly, histograms of tidal peaks were derived from all observed tidal peaks in a period of approximately 100 years. The histograms are used directly as input for the sampling procedure, so no curve fitting was applied. The probability distribution for the Maeslant barrier is binominal: there is a 1 % probability that the barrier fails to close upon request, and therefore a 99 % probability that the barrier closes upon request. High water events in the Rhine and Meuse rivers often occur simultaneously. This means discharges of these rivers are correlated and this is taken into account in the modeling framework. However, in the current paper this correlation is not relevant as the Rhine is the only river we focus on to determine our optimal importance sampling scheme.

The statistical distribution functions of discharges and sea water levels refer to peak values, while the hydrodynamic simulation model requires time series as input. To describe the temporal evolution of the river discharge, a normalised hydrograph is used, i.e. a dimensionless hydrograph with a peak value equal to 1. For each simulated event, the normalized hydrograph is multiplied by the sampled peak discharge to form a discharge time series, which is used as input of the hydrodynamic simulation model. The normalised hydrograph is based on average durations of threshold exceedances, as observed during high water events on the Rhine and Meuse rivers. The hydrograph of the sea water level is a combination of a standardised surge hydrograph and average tidal conditions. The standardised surge hydrograph is also based on averages of observed high storm surge events. The standardised hydrographs for storm surge and river discharge are applied in each simulated event. The following assumptions are made with regard to "timing":

- The peak of the Rhine discharge at Lobith occurs at the same time as the peak of the Meuse discharge at Lith.
- The peak of the sea water level occurs 2 days after the peak of the river discharge. This means the peaks arrive approximately at the same time in the transitional area, i.e. the area that is influenced by both river discharges and sea water level.

Events with high discharges in the Rhine and Meuse rivers may last several weeks. For long duration events, the probability that the sea water level exceeds a given high threshold at some stage during the event is higher than for short duration events. This is especially relevant for the transitional area, where water levels are influenced by both river discharges and sea water levels. In the probabilistic model for the statutory safety assessment of flood defences, the total duration of a river induced flood event in the Rhine river is assumed to be in the order of 30 days (Geerse 2005). This duration includes the rising and falling limb of the hydrograph, which means high river discharges only occur during a smaller sub-period. In our model, the duration of this sub-period is equal to τ tidal periods and starts at $\tau/2$ tidal periods before the peak discharge and ends at $\tau/2$ tidal periods after the peak discharge. The value of τ is chosen to be 12 tidal periods, i.e. about 6.5 days. The value of τ is applied in the sampling procedure of the (peak) sea water level. The probability distribution function, $F_M(m)$, for the peak sea water level is available for the tidal period:

$$F_M(m) = P[M < m] \tag{8}$$

This means the function $F_M(m)$ describes the probability that the maximum sea water level during a single tidal period is less than or equal to m. In the Monte Carlo procedure, the maximum sea water level of a period of τ tidal periods will be sampled. This value has the following distribution function:

$$F_M^*(m) = P[M < m]^\tau = \left(F_M(m)\right)^\tau \tag{9}$$

Note that this formula is based on the assumption of independence between subsequent tidal peaks. Equation (9) shows the assumed duration τ influences the distribution function from which the peak sea water level is sampled. An increase in the value of τ increases the probability of higher peak sea water levels being sampled. This is exactly the duration effect that needed to be incorporated in the approach: the longer the duration of a high discharge event, the higher the probability that the sea water level exceeds a given high threshold at some stage during this event.

4.3 Hydrodynamic simulations

The Sobek hydrodynamic model (see e.g. Stelling and Verwey 2005) was used to compute water levels at all potential breach locations of Fig. 1, for the selected combinations of river discharges, sea water levels and barrier states. The formation and consequences of breaches are also simulated in this hydrodynamic model. Breaches occur if the river water level at a location exceeds the maximum water level that a dike section can withstand, as sampled from the derived fragility curves. The process of breaching is not further considered in the remainder of this paper, even though it is highly relevant for the estimated societal flood risk. This is because the focus of the paper is on the development of efficient Monte Carlo importance sampling techniques and it is most practical to test these on the relatively simplified case in which it is assumed no breaches can occur. This will be further explained in Sect. 4.4. For more information on the hydrodynamic modeling of breaches and resulting societal flood risks, the interested reader is referred to De Bruijn et al. 2014.

4.4 Test set for selecting importance sampling strategies

Section 3 describes criteria for the selection of efficient sampling schemes. One of the criteria is the standard deviation of the Monte Carlo estimator. In order to quantify this standard deviation, the Monte Carlo simulation needs to be repeated multiple times with different (random) seeds. In order to prevent having to carry out millions of time-consuming model simulations, the tests for the selection of sampling strategies were only carried out for the hydraulic loads, i.e. not for breaches and flood consequences. To construct the test procedure, hydraulic simulations were carried out for combinations of seven sea water levels, 11 river discharges and 2 barrier situations, so in total 7 × 11 × 2 = 154 simulations. The simulated sea water levels and river discharges cover the complete range of events that are relevant for flood risk assessments. In the simulations, dike breaching was not modelled. For this particular test, the river discharges of the Rhine and Meuse were assumed to be fully correlated, to further simplify the test procedure. The simulated maximum water levels at the 171 potential breach locations, as obtained from the 154 simulations, served as a lookup table for the Monte Carlo test simulations. In this way, water levels at the 171 potential breach locations can be derived for all potential realisations of the random load variables with negligible computation time.

Another advantage of the relatively low number of random variables of the test case is that results of the Monte Carlo simulations can be compared with the "exact" results as computed with numerical integration. The computations with numerical integration were carried out on a very fine grid consisting of 400,000 combinations of river discharges and sea water levels. This grid was applied for both barrier states, i.e. 'functioning' and 'malfunctioning'. *T*-year water levels were derived for all potential breach locations for a range of values of *T*. The computed *T*-year water levels of the numerical integration procedure were compared with analytical results for locations in the non-tidal area, which showed that the error in water levels as estimated with the numerical integration procedure were less than two millimeters. This is considered small enough to serve as the reference for the detection of errors (bias) in the Monte Carlo simulations.

5. RESULTS AND ANALYSIS

5.1 Sampling strategies for the non-tidal area

In total six different sampling strategies were tested. The strategies are labeled ISS1... ISS6 and the associated distribution functions and parameters are summarized in Table 1. The first tests for importance sampling strategies were only carried out for locations in the non-tidal area. For these locations, the river discharge is the only random variable of interest, i.e. the influence of the sea water level and barrier state is negligible. This means the most efficient sampling strategies for these locations only need to focus on high river discharges. The following two importance sampling strategies were tested:

- ISS1: Sampling from the highest quantiles of river discharges only, with sampling probability densities proportional to actual probability densities;
- ISS2: Uniform sampling of river discharges.

In formula, this means the following sampling density functions are applied on the variable discharge:

$$h_1(x) = \begin{cases} 0 & ; x \leq x_T \\ \dfrac{f(x)}{1 - F(x_T)} & ; x > x_T \end{cases} \tag{10}$$

$$h_2(x) = \begin{cases} 0 & ; x \notin [x_L, x_U] \\ \dfrac{1}{x_U - x_L} & ; x \in [x_L, x_U] \end{cases} \tag{11}$$

In which x_T is the threshold discharge above which discharges are sampled in strategy 1 (ISS1), x_L and $x_U$are the lower and upper bounds of the interval from which discharges are sampled in strategy 2 (ISS2) and$f$(x) is the actual density function of the discharge. The generic variable name '*x*' is used in the equations above because these sampling functions will be applied on other random variables later on as well. In strategy 1, the sampling density for discharges above threshold x_T is proportional to the original density function $f(x)$, whereas in strategy 2 the probability density is uniform for all discharges in the interval $[x_L, x_U]$. Threshold x_T was taken equal to the 100-year discharge (≈12.700 m^3/s for the Rhine at Lobith), x_L and $x_U$are taken equal to 10,000 m^3/s and 24,000 m^3/s respectively. These bounds were carefully chosen make sure that [a] the interval is not too large and hence the sampling method inefficient and [b] the interval is large enough to have all discharges included that are relevant for flood risk in the non-tidal area.

Table 1. Overview of sampling strategies for the annual maximum discharge of the Rhine river at Lobith (QR_AM), the daily maximum discharge of the Rhine river at Lobith (QR_D), the annual maximum sea water level at Maasmond (SWL_AM), the tidal peak of the sea water level at Maasmond (SWL_T) and the functioning of the barrier at Maasmond (B). Functions f and $h_1,\ldots,h_4$ are sampling density functions as defined in the bottom part of the Table, the values between brackets are the corresponding function parameters

No.	QR_AM	QR_D	SWL_AM	SWL_T	B
1	h_1 (12,700)	h_1 (7,500)	f	f	f
2	h_2 (10,000; 24,000)	h_2 (10,000; 24,000)	f	f	f
3	h_2 (10,000; 24,000)	h_2 (10,000; 24,000)	h_2 (2; 7)	h_2 (1.2; 6)	f
4	h_2 (6,000; 24,000)	h_2 (500; 14,000)	h_2 (2; 7)	h_2 (1.2; 6)	f
5	h_2 (6,000; 24,000)	h_2 (500; 14,000)	h_2 (2; 7)	h_3 (1.2;2.5;6)	f
6	h_2 (6,000; 24,000)	h_2 (500; 14,000)	h_2 (2; 7)	h_3 (1.2;2.5;6)	h_4 (0.1)

Importance sampling functions: f, original density function (no importance sampling); h_1, scaling factor in probability; parameter x_T, see Eq. 10; h_2, uniform sampling; parameters x_L and x_U, see Eq. 11; h_3, split distribution function; parameters x_1, x_2, x_3, see Eq. 12; h_4, binomial distribution function; parameter p_1, see Eq. 13

Monte Carlo simulations were carried out 100 times for both strategies to obtain the standard deviation of computed T-year water levels. Furthermore, the mean value of the T-year water level over the 100 simulations was compared with the "exact" results from numerical integration to quantify the (potential) bias that may be introduced by importance sampling methods. Figure 2 shows the resulting bias and standard deviation of the 100, 1,000 and 10,000-year water level of the following sampling strategies: [a] crude Monte Carlo with 100,000 simulated years, [b] crude Monte Carlo with 1,000 simulated years [c] ISS1 with 1,000 simulated years and [d] ISS2 with 1,000 simulated years. The subplots on the left show the bias and the subplots on the right show the standard deviation for estimated water levels at 7 locations in the non-tidal area with return periods of 100 years (top panel), 1,000 years (centre panel) and 10,000 years (lower panel). These return periods cover the range that is most relevant for flood risk estimates in the area.

It is no surprise that the crude Monte Carlo results for n = 100,000 are more accurate than the crude Monte Carlo results for n = 1,000. For n = 1,000 the absolute value of the bias is larger and the standard deviation is substantially larger. Furthermore, the 10,000 year water level could not be obtained with n = 1,000 samples, as it requires at least 10,000 samples to quantify this water level without the (undesired) use of extrapolation techniques. The positive value of the bias as observed in the left panel subplots may require some further explanation. Formally, the bias as defined in Sect. 3 is equal to 0 for crude Monte Carlo sampling. In other words: the error in the Monte Carlo estimate reduces to 0 if n goes to infinity. The bias in Fig. 2 for the crude Monte Carlo methods is therefore caused by the fact that a limited set of samples is used. Increasing the number of samples will decrease the bias, which is demonstrated by the fact that the bias for n = 100,000 is much smaller than for n = 1,000.

Since errors of 0.1 m or more in estimated water levels are considered unacceptable for flood risk analysis in this area, n = 1,000 simulated years is not sufficient for crude Monte Carlo simulation, whereas n = 100,000 leads to an acceptable bias and standard deviation. However, it would be unpractical to carry out 100,000 hydraulic model simulations whereas n = 1,000 would be acceptable. This is the reason why importance sampling is required in our probabilistic framework. Figure 2 shows that both importance sampling strategies lead to acceptable results, i.e. errors well below 0.1 m, for n = 1,000 simulated years. Results of strategy 1 are nearly the same as the crude Monte Carlo results with n = 100,000 years. The reason is that for the analysis of events with return periods > = 100 years the two methods are essentially the same, since strategy 1 [a] only samples from discharges with return period > 100 years and [b] was applied with a factor 100 lower number of samples. So with ISS1, similar results can be obtained as crude Monte Carlo with a factor 100 lower computation time. Note that this is only the case for return values of 100 years and higher, for lower return periods this importance sampling strategy will not provide results. This is no problem if only the extreme events are relevant, but this is not always the case as will be demonstrated in the next section.

Figure 2 shows that strategy 2 provides even more accurate results than strategy 1 and the crude Monte Carlo simulations, especially for the 1,000-year and 10,000-year water level. The reason is that in strategy 2 the extremely high discharges up to 24,000 m^3/s have a significantly higher probability of being sampled in comparison with strategy 1 and crude Monte Carlo sampling. This gives the method the potential to provide more reliable estimates for particularly the high return periods.

5.2 Sampling strategies for all locations

So far, the analysis has focused on locations for which the river discharge is the only relevant variable. For these locations, sampling strategy 2 turned out to be very efficient. How bbbbbbbbever, this success is partly explained by the fact that this sampling strategy only focuses on river discharges. For sea water dominated locations this strategy is less efficient, as can be seen from Fig. 3 (red open circles). This Figure shows results for all 171 potential breach locations. Locations are ordered based on the longitudinal coordinate, which means sea-dominated locations (tidal area) are on the left and river dominated locations (non-tidal area) are on the right. Clearly, the bias with sampling strategy 2 is unacceptably large for locations in the tidal area.

In order to develop a sampling strategy that provides reliable results for all locations, using a limited set of 1,000 simulated years, importance sampling on sea water levels is required as well. Therefore, a uniform sampling strategy (ISS3) is adopted for the sea water level, with bounds 2 m + NAP and 7 m + NAP. This is done for event type (ii), i.e. for events with the annual maximum sea water level. For event type (i) (annual maximum discharge with co-inciding sea water level) the bounds 1.2 m + NAP and 6 m + NAP are adopted (see Table 1 for the details of the sampling strategies). For river discharges, strategy 3 uses the same sampling distributions as strategy 2.

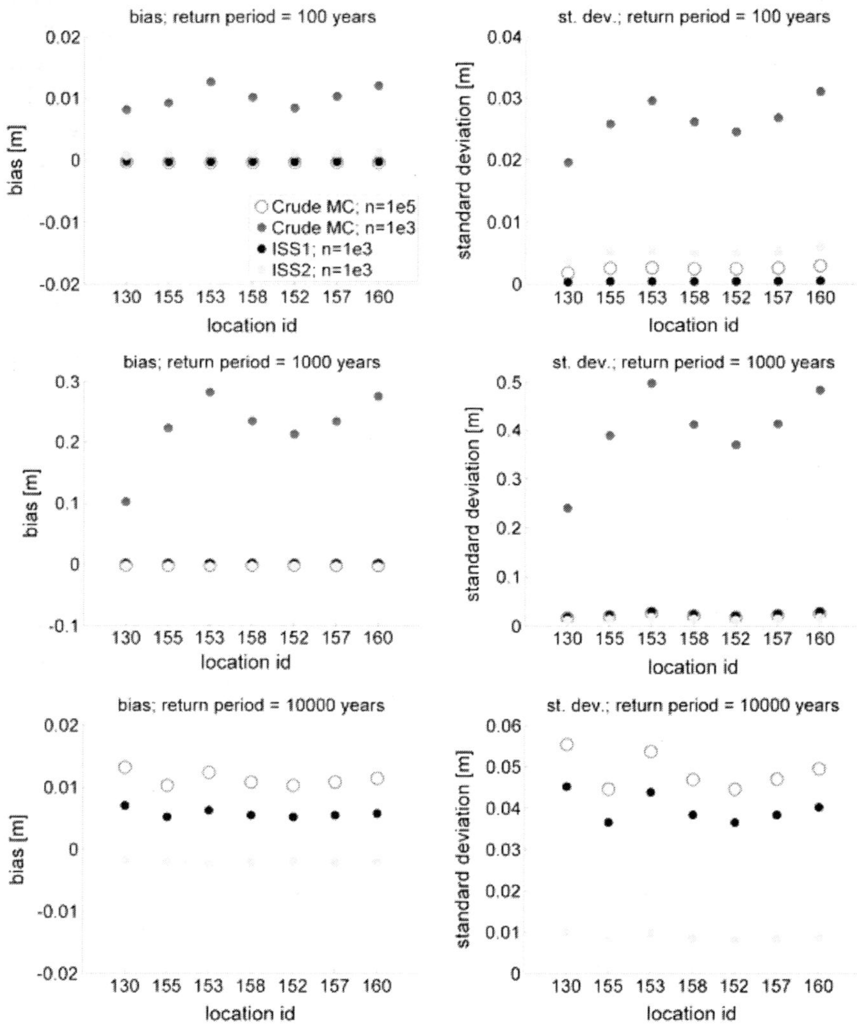

Figure 2. Bias (*left*) and standard deviation (*right*) for estimated water levels with return periods of 100 years (*top panel*), 1,000 years (*centre panel*) and 10,000 years (*lower panel*); comparison of results of importance sampling strategies 1 and 2 with 1,000 simulated years and crude MC with 1,000 and 100,000 simulated years. All locations are discharge dominated

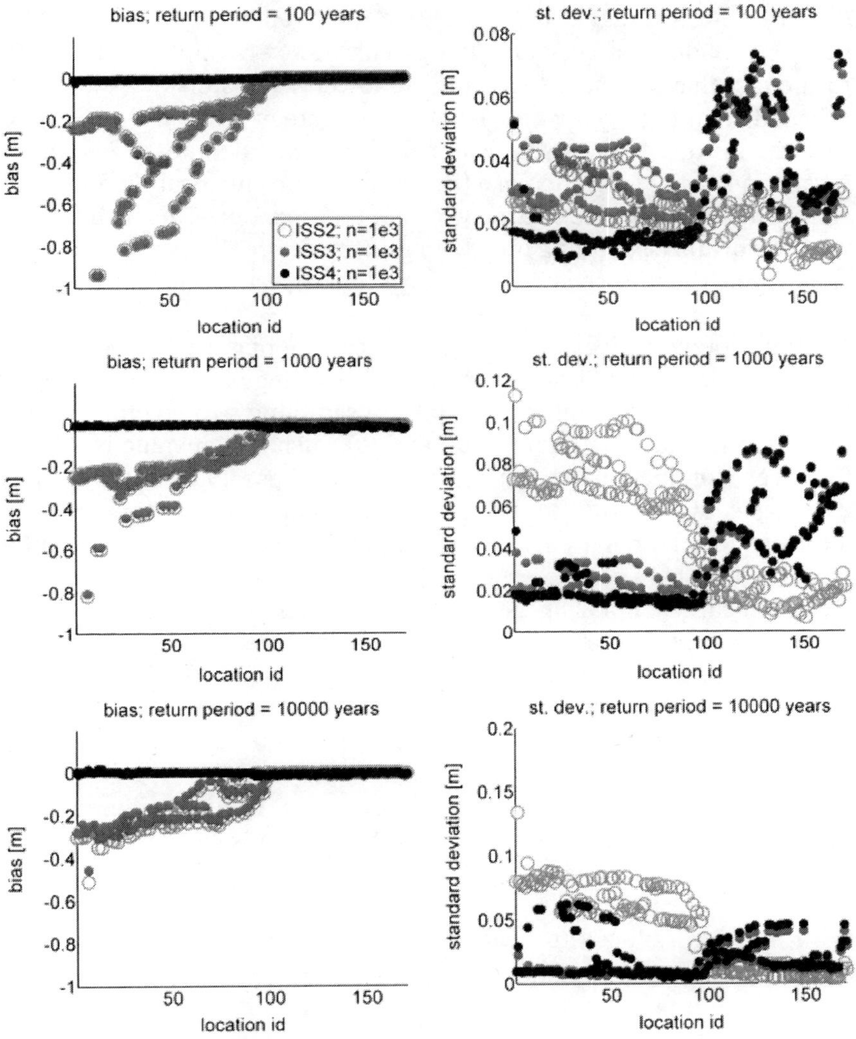

Figure 3. Bias (*left*) and standard deviation (*right*) for estimated water levels with return periods of 100 years (top panel), 1,000 years (centre panel) and 10,000 years (lower panel); comparison of results of importance sampling strategies 2, 3 and 4 with 1,000 simulated years

Figure 3 compares the results for strategies 2 and 3. In general, strategy 3 results in a reduction of standard deviations in comparison with strategy 2 for locations in the tidal area (location id's 1–100), but the bias for these locations is still large. Furthermore, it can be seen that for locations in the non-tidal area (location id's 100–171) the standard deviation for strategy 3 is higher than for strategy 2, even though the same sampling strategies for river discharges were applied. This is due to the fact that for these locations the lower sea water levels are more relevant than high sea water levels and strategy 3 shifts the sampling density to the higher sea water levels. This demonstrates that a sampling strategy can increase the accuracy for one location and at the same time decrease the accuracy for another location. This is the reason why importance sampling in a delta like the Rhine–Meuse delta is more challenging than for rivers where only discharges are relevant.

Another noteworthy aspect of Fig. 3 is that the bias for locations in the tidal area (id < 100) is large for sampling strategy 3. This is due to the fact that the lower bound of the uniform sampling strategy for the river discharge is relatively high (10,000 m^3/s). Such a high value is efficient for river dominated locations because for these locations only high discharges will result in relevant high water levels. However, for locations in the tidal area, low and moderate discharges are relevant as well, because high sea water levels usually coincide with low/moderate river discharges. Ignoring these events in the sampling scheme clearly results in an underestimation of design water levels for these locations. Therefore, in a subsequent sampling strategy 4 (ISS4) the lower bounds of the uniform sampling distributions for river discharges where reduced to 6,000 m^3/s for event type (i), i.e. the events with annual maximum river discharges, and to 500 m^3/s for event type (ii), i.e. the events with annual maximum sea water levels. Figure 3 demonstrates that this has the desired effect on the bias, which is close to zero for all locations for sampling strategy 4.

With the bias reduced to near zero, the next objective is to adapt the sampling strategy in such a way that the standard deviation is further reduced, without simultaneous increase of the bias. The first point of consideration is the set of river dominated locations (id's > 100), for which the standard deviation of the 10,000-year water level is close to 0.1 m. Figure 3 shows that these relatively large numbers arose when sampling strategy 3 was introduced, in sampling strategy 2 these numbers were much smaller. In sampling strategy 3, importance sampling for sea water levels was introduced, which resulted in a significant reduction of standard deviations for locations in the tidal area. However, this was at the expense of the standard deviations for discharge dominated locations. For discharge dominated locations, events with high discharges in combination with moderate to low sea water levels are relevant. The introduction of importance sampling for high sea water levels caused a decrease in the number of samples of moderate/low sea water levels, i.e. a decrease in the number of samples that are relevant for discharge dominated locations. This gave rise to the increase in the standard deviation of the Monte Carlo estimate for river dominated locations. In order to decrease the standard deviation of these locations, the obvious way is to increase the probability of sampling lower sea water levels. The problem is that this will be at the expense of the accuracy for sea water level dominated locations, which is the reason why sampling strategy 3 was introduced in the first place.

In order to obtain accurate probability estimates for both discharge dominated and sea water level dominated locations, the sampling strategy for sea water level therefore has to be a compromise between the importance sampling function on one hand and the original density function (i.e. no importance sampling) on the other hand. This compromise is reached by dividing the relevant range of sea water levels into two intervals $[x_1, x_2]$ and $[x_2, x_3]$, with $x_1 < x_2 < x_3$. The first interval represents the low/moderate sea water levels and the second interval represents the high/extreme sea water levels. Since both intervals are relevant for the flood risk in the area, we decided to give equal probability weights to both. In other words: each individual sea water level sample has a 0.5 probability of being low/moderate (interval 1) and also a 0.5 probability of being high/extreme (interval 2). In the first interval, the sampling density is taken proportional to the original density function, $f(x)$. In the second interval, the sampling density is taken uniform. In formula:

$$h_3(x) = \begin{cases} \dfrac{0.5f(x)}{F(x_2) - F(x_1)} & ; x \in [x_1, x_2] \\ \dfrac{0.5}{x_3 - x_2} & ; x \in [x_2, x_3] \end{cases} \tag{12}$$

This sampling strategy for sea water levels is only applied for event type (i), i.e. events that represent annual maximum discharges in combination with a coinciding sea water level. Figure 4 shows the results for sampling strategy 5, which uses function h_3. For river dominated locations (id > 100) the standard deviations are now all below 0.05 m. This is a significant reduction in comparison with sampling strategy 4, while the bias remains in the same order of magnitude.

Unfortunately, the new sampling strategy caused an increase in the estimate of the 10,000 year water level for some locations in the tidal area (location id < 100 in the lower right subplot of Fig. 4. The estimated 10,000 year water level with sampling strategy 5 is around 0.1 m for these locations. For these locations, extremely high water levels are caused by a combination of high sea water levels and a malfunctioning barrier. The probability of failure to close upon request for this barrier is 1 in 100, so a malfunctioning barrier only occurs in 1 % of the samples on average. The fact that in the Monte Carlo simulations there are only a few samples of this type of event is the main reason why the standard deviation of the Monte Carlo estimate is relatively high for sea water level dominated locations: one or two samples more (or less) may lead to significant changes in the estimated *T*-year water levels. The obvious way to reduce this effect is to increase the probability of sampling events in which the barrier fails to close upon request. Since the random variable 'barrier' has only two states (functioning or malfunctioning), the importance sampling distribution is binomial:

$$h_4(x) = \begin{cases} p_1 & \text{; malfunctioning barrier} \\ 1 - p_1 & \text{; functioning barrier} \end{cases} \quad (13)$$

In sampling strategy 6 (ISS6), the probability of a malfunctioning barrier, p_1, was chosen to be 10 %, whereas the actual probability is estimated to be 1 %. Of course, this increase with a factor 10 needs to be corrected for in the usual manner in the Monte Carlo estimator. For samples in which the barrier malfunctions, the correction factor is equal to 1/10, because the sampling probability of a malfunctioning barrier was increased with a factor 10. For samples in which the barrier closes upon request, the correction factor is equal to 99/90, because the sampling probability of a functioning barrier was decreased from 99 to 90 %.

Figure 4 compares the results for sampling strategies 5 and 6. The objective of reducing the remaining "high" standard deviations has clearly been successful. With sampling strategy 6, the standard deviations in the estimated 100, 1,000 and 10,000 year water levels are all less than 0.06 m. Further fine-tuning of the sampling strategy did not lead to significant further improvements since an improvement for one location is almost automatically at the expense of the quality of the results for other locations. Sampling strategy 6 is therefore chosen as the preferred option in the framework for societal flood risk analysis. Figure 4 also compares the results of sampling strategy 6 for 1,000 simulated years with the crude Monte Carlo results with 100,000 simulated years. It can be seen that for the 100-year water level the crude Monte Carlo method performs better due to its abundance in the applied number of samples. However, for the 10,000-year water level the importance sampling strategy clearly outperforms the crude Monte Carlo approach, because the bias and standard deviation are lower for all locations. For many locations, the crude Monte Carlo method results in a standard deviation of more than 0.1 m, whereas with importance sampling this standard deviation is less than 0.06 m for all locations and all considered *T*-year water levels. So in spite of the fact that for the Crude Monte Carlo method 100 times more samples were used, the importance sampling method provides overall more accurate estimates *T*-year water levels. This clearly shows the added value of importance sampling: with 100 times lower computation time, more accurate results can be obtained.

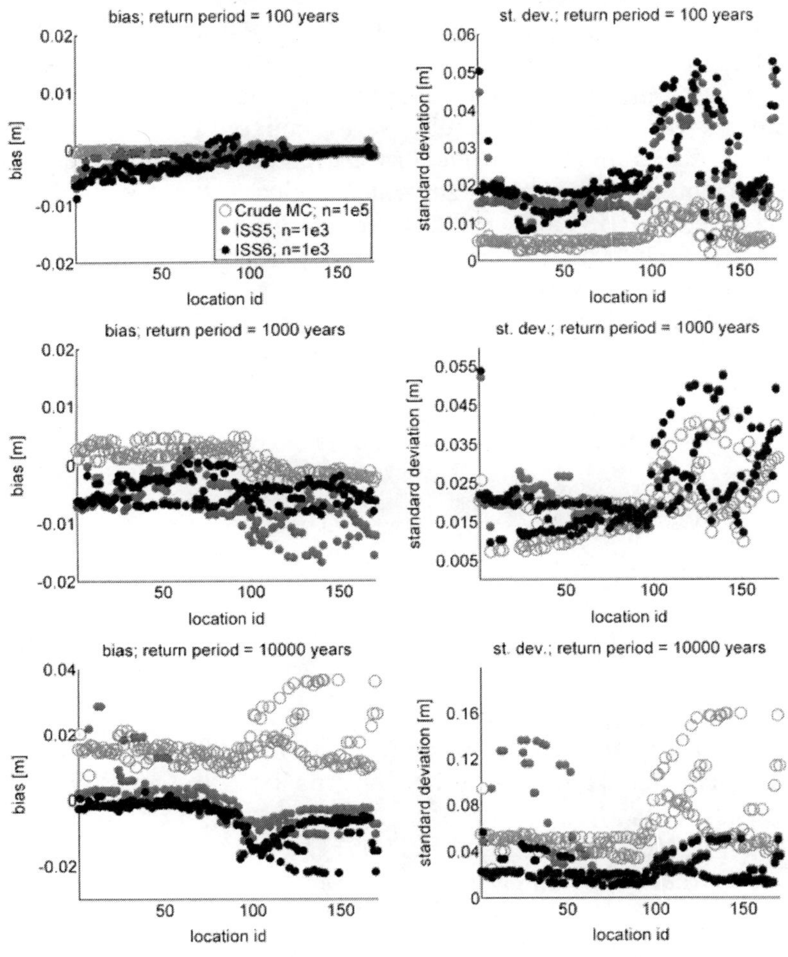

Figure 4. Bias (*left*) and standard deviation (*right*) for estimated water levels with return periods of 100 years (*top panel*), 1,000 years (*centre panel*) and 10,000 years (*lower panel*); comparison of results of importance sampling strategies 5 and 6 with 1,000 simulated years and crude MC with 100,000 simulated years

6. CONCLUSIONS

This paper described the benefits of the application of importance sampling in a probabilistic framework for societal flood risk analysis in the Rhine–Meuse delta in the Netherlands. The choice of efficient importance sampling techniques in a delta like the Rhine–Meuse delta is more challenging than for non-tidal rivers where only discharges are relevant, because the relative influence of the forcing factors like river discharge and sea water level differ from location to location. As a consequence, sampling methods that are efficient and accurate for one location may be very inefficient for other locations or, worse, may introduce errors in computed design water levels. Several sampling strategies were tested and results were compared in terms of bias and standard deviation in the probability estimate. The analysis resulted in an efficient sampling strategy which reduces the required model simulation time by a factor 100 compared to crude Monte Carlo simulation, while at the same time the probability estimates of the relevant extreme water levels are more accurate for all locations considered in the area. This is a very valuable result, as it reduces the required computation times of the probabilistic framework to acceptable quantities, while at the same time the output is more accurate.

REFERENCES

1. Allaix DL, Carbone VI (2011) An improvement of the response surface method. Struct Saf 33:165–172

2. Apel H, Thieken A, Merz B, Bloschl G (2004) Flood risk assessment and associated uncertainty. Nat Hazards Earth Syst Sci 4:295–308

3. Apel H, Merz B, Thieken AH (2009) Influence of dike breaches on flood frequency estimation. Comput Geosci 35(2009):907–923

4. Beckers J, De Bruijn KM, Riedstra D (2012) Life safety criteria for flood protection standards. In: Chavoshian A, Takeuchi T (eds.) (2012). Floods: from risk to opportunity (IAHS Publ. 357). ISBN: 0144-7815, pp 21–26

5. Bjerager P (1988) On computation methods for structural reliability analysis. In: Frangopol DM (ed) New directions, in structural system reliability. University of Colorado, Boulder, pp 52–67

6. Christofides TC (2003) Randomized response in stratified sampling. J Stat Plan Inference 128(2005):303–310

7. Courage W, Vrouwenvelder T, Van Mierlo T, Schweckendiek T (2013) System behaviour in flood risk calculations. Georisk 7(2):62–76

8. Dawson R, Hal J (2006) Adaptive importance sampling for risk analysis of complex infrastructure systems. Proc R Soc A 462(2075):3343–3362

9. De Bruijn KM, Van der Doef M (2011) Gevolgen van overstromingen—Informatie ten behoeve van het project Waterveiligheid in de 21e eeuw. Project 1204144.004, Deltares, Delft (In Dutch)

10. De Bruijn KM, Beckers J, van der Most H (2010) Casualty risks in the discussion on new flood protection standards in the Netherlands. In: Brebbia C (ed) Second International Conference on Flood recovery, innovation and response. Wessex Institute of Technology, Ashurst, p 73

11. De Bruijn KM, Diermanse FLM, Beckers JVL (2014) An advanced method for flood risk analysis in river deltas, applied to assess societal flood fatality risks in the Netherlands. Nat Hazards Earth Syst Sci 2:1637–1670

12. Delta Committee (1958) Report of the Delta Committee, part 1, Final report. 's Gravenhage, Staatsdrukkerij en uitgeversbedrijf

13. Delta Committee (2008) Samen werken met water. Bevindingen van de Deltacommissie 2008. Sept 2008, ISBN/EAN 978-90-9023484-7 (in Dutch)

14. Den Heijer F, Diermanse FLM (2012) Towards risk-based assessment of flood defences in the Netherlands: an operational framework, In: Comprehensive flood risk management: research for policy and practice. Proceedings of the flood risk 2012 conference in Rotterdam, Nov 2012, CRC Press, ISBN 9780415621441

15. Ditlevsen O, Madsen HO (1996) Structural reliability analysis. Wiley, Chichester

16. Ditlevsen O, Melchers R, Gluver H (1990) General multi-dimensional probability integration by directional simulation. Comput Struct 36(2):355–368

17. Engelund S, Rackwitz R (1993) A benchmark study on importance sampling techniques in structural reliability. Struct Saf 12(1993): 255–276

18. Evans AW, Verlander NQ (1997) What is wrong with criterion fn-lines for judging the tolerability of risk? Risk Anal 17(2):157–168

19. Geerse CPM (2005) Probabilistic model to assess dike heights in part of the Netherlands. International Symposium on Stochastic Hydraulics, Nijmegen

20. Georgiou SD (2009) Orthogonal Latin hypercube designs from generalized orthogonal designs. J Stat Plan Inference 139(2009): 1530–1540

21. Gouldby B, Sayers P, Mulet-Marti J, Hassan MAAM, Benwell D (2008) A methodology for regional-scale flood risk assessment. Proc ICE Water Manag 161(WM3):169–182

22. Grooteman F (2011) An adaptive directional importance sampling method for structural reliability. Probab Eng Mech 26(2011):134–141

23. Heffernan JE, Tawn JA (2004) A conditional approach for multivariate extreme values (with discussion). J R Stat Soc B 66(3):497–546

24. Hsu Y-C, Tung Y-K, Kuo J-T (2011) Evaluation of dam overtopping probability induced by flood and wind. Stoch Environ Res Risk Assess 25(1):35–49

25. Keef C, Tawn J, Svensson C (2009) Spatial risk assessment for extreme river flows. Appl Stat 58(5):601–618

26. Keskintürk T, Er S (2007) A genetic algorithm approach to determine stratum boundaries and sample sizes of each stratum in stratified sampling. Comput Stat Data Anal 52(2007):53–67

27. Kind JM (2013) Economically efficient flood protection standards for the Netherlands. J Flood Risk Manag 7:103–117

28. Koopman SJ, Shephard N, Creal D (2009) Testing the assumptions behind importance sampling. J Econom 149(2009):2–11

29. Lamb R, Keef C, Tawn J, Laeger S, Meadowcroft I, Surendran S, Dunning P, Batstone C (2010) A new method to assess the risk of local and widespread flooding on rivers and coasts. J Flood Risk Manage 3(4):323–336

30. Liu X, Cardiff MA, Kitanidis PK (2010) Parameter estimation in nonlinear environmental problems. Stoch Environ Res Risk Assess 24(7):1003–1022

31. May RJ, Maier HR, Dandy GC (2010) Data splitting for artificial neural networks using SOM-based stratified sampling. Neural Netw 23(2010):283–294

32. Melchers (2002) Structural reliability analysis and prediction, Robert E. Melchers, Wiley, New York. ISBN0471983241

33. Olsson A, Sandberg G, Dahlblom O (2003) On Latin hypercube sampling for structural reliability analysis. Struct Saf 25(2003):47–68

34. Owen A (1994) Controlling correlations in Latin hypercube samples. J Am Stat Assoc 89(428):1517–1522

35. Rackwitz R, Fiessler B (1978) Structural reliability under combined random load sequences. Comput Struct 9:489–494

36. Sezer AD (2009) Importance sampling for a Markov modulated queuing network. Stoch Proc Appl 119:491–517

37. Steenackers GF, Presezniak F, Guillaume P (2009) Development of an adaptive response surface method for optimization of computation-intensive models. Comput Ind Eng 57:847–855

38. Stelling GS, Verwey A (2005) Numerical flood simulation. In: Anderson MG (ed) Encyclopedia of hydrological sciences. Wiley, New York, pp 257–270

39. Van Dantzig D (1956) Economic decision problems for flood prevention. Econometrica 24:276–287

40. Van Mierlo MCLM, Vrouwenvelder ACWM, Calle EOF, Vrijling JK, Jonkman SN, De Bruijn KM, Weerts AH (2007) Assessment of floods risk accounting for river system behaviour. J River Basin Manag 5(2):93–104

41. Vorogushyn, S, Merz B, and Apel H (2012) Analysis of a detention basin impact on dike failure probabilities an 54 flood risk for a channel-dike-floodplain system along the river Elbe, Germany. J Hydrol 55:436–437, 120–131

42. Vrijling JK, Van Hengel W, Houben RJ (1995) A framework for risk evaluation. J Hazard Mater 43(1995):245–261

43. Waarts PH (2000) Structural reliability using finite component analysis; an appraisal of DARS: Directional Adaptive Response Surface Sampling, PH. D. Thesis, Delft University of Technology

44. Wyncoll D, Gouldby B (2013) Application of a multivariate extreme value approach to system flood risk analysis. J Flood Risk Manag 1:3–10

45. Ye KQ (1998) Orthogonal column Latin hypercubes and their application in computer experiments. J Am Stat Assoc 93:1430–1439

46. Yuan C, Druzdzel MJ (2006) Importance sampling algorithms for Bayesian networks: Principles and performance. Math Comp Model 43(2006):1189–1207

47. Zhong H, Van Overloop P-J, Van Gelder PHAJM (2013) A joint probability approach using a 1D hydrodynamic model for estimating high water level frequencies in the Lower Rhine Delta. Nat Hazards Earth Syst Sci 13:1841–1852

6

BAYESIAN TSUNAMI FRAGILITY MODELING CONSIDERING INPUT DATA UNCERTAINTY

Raffaele De Risi[1], Katsuichiro Goda[1], Nobuhito Mori[2], Tomohiro Yasuda[2]

[1]Department of Civil Engineering, Queen's Building University Walk, University of Bristol, Bristol BS8 1TR, UK

[2] Disaster Prevention Research Institute, Kyoto University, Kyoto 611-0011, Japan

ABSTRACT

Empirical tsunami fragility curves are developed based on a Bayesian framework by accounting for uncertainty of input tsunami hazard data in a systematic and comprehensive manner. Three fragility modeling approaches, i.e. lognormal method, binomial logistic method, and multinomial logistic method, are considered, and are applied to extensive tsunami damage data for the 2011 Tohoku earthquake. A unique aspect of this study is that uncertainty of tsunami inundation data (i.e. input hazard data in fragility modeling) is quantified by comparing two tsunami inundation/run-up datasets (one by the Ministry of Land, Infrastructure, and Transportation of the Japanese Government and the other by the Tohoku Tsunami Joint Survey group) and is then propagated through Bayesian statistical methods to assess the effects on the tsunami fragility models. The systematic implementation of the data and methods facilitates the quantitative comparison of tsunami fragility models under different assumptions. Such comparison shows that the binomial logistic method with un-binned data is preferred among the considered models; nevertheless, further investigations related to multinomial logistic regression with un-binned data are required. Finally, the developed tsunami fragility functions are integrated with building damage-loss models to investigate the influences of different tsunami fragility curves on tsunami loss estimation. Numerical results indicate that the uncertainty of input tsunami data is not negligible (coefficient of variation of 0.25) and that neglecting the input data uncertainty leads to overestimation of the model uncertainty.

KEYWORDS

Bayesian regression, Tsunami fragility, Logistic regression, Multinomial regression, Markov Chain Monte Carlo simulation, 2011 Tohoku earthquake

1. INTRODUCTION

In the last two decades, tsunamis triggered by earthquakes were responsible for 33 % of total deaths and 35 % of total economic losses globally (Guha-Sapir et al. 2015). A reliable quantification of tsunami risk becomes increasingly important for emergency officers to manage critical infrastructures and for insurance companies to quantify the expected economic losses (Goda 2015). An accurate risk analysis encompasses reliable assessments of hazard, exposure, and vulnerability. The tsunami hazard assessment is often given in a form of inundation maps, reporting inundation depths at various locations, for different tsunami rupture scenarios (Goda et al. 2014; Fukutani et al. 2015; Goda and Song 2016). Exposure assessment identifies the elements at risk, including human, built, and natural environments in coastal areas. Finally, vulnerability is represented by fragility curves, i.e. probability of reaching or exceeding specific damage states for a given hazard intensity (Porter et al. 2007). Different methods for deriving fragility curves can be found in the literature (Rossetto et al. 2014). They can be classified into four approaches: (a) empirical methods based on statistical analysis of observed damage data; (b) judgmental methods based on expert elicitation; (c) analytical methods based on evaluation of the performance response through structural analysis; and (d) hybrid techniques by combining the preceding methods. This work focuses on the empirical methods for tsunami fragility modeling.

Empirical tsunami fragility modeling requires numerous pairs of tsunami damage observations (e.g. the number of buildings reaching or exceeding a specific damage state) and explanatory variables related to hazard and exposure. Tsunami inundation depth is widely adopted as tsunami intensity measure in developing tsunami fragility curves (Koshimura et al. 2009; Reese et al. 2011; Suppasri et al. 2011, 2013; Charvet et al. 2014a). It is important to recognize that observed tsunami intensity measures are subjected to errors. Measurement errors are always present, influenced by survey techniques, equipment, and conditions. When direct measurements are not available and inundation needs to be assessed over a vast area quickly, interpolation of measured tsunami depths at nearby locations may be considered, which inevitably introduces additional errors (both random and systematic) in the gathered tsunami data. In the context of empirical fragility modeling, uncertainty associated with input hazard data should be treated adequately, because neglecting this kind of uncertainties results in potential overestimation of model uncertainty associated with developed fragility curves (Cetin et al. 2002; Der Kiureghian 2002; Straub and Der Kiureghian2008). In the literature, problems of implementing different sources of uncertainties in fragility functions have been tackled mainly in the field of earthquake engineering (e.g. Der Kiureghian 2002; Porter et al. 2007; Bradley 2010; Baker 2015; Jalayer et al. 2015; Lallemant et al. 2015). The general frameworks for developing robust fragility models (e.g. Bayesian statistics) help implement both inherent and epistemic uncertainties in the assessment of the statistics describing fragilities (Gardoni et al. 2007). In tsunami fragility modeling, incorporation of input data errors and uncertainty has not been explored rigorously (Tarbotton et al. 2015).

The aims of this paper are twofold. Firstly, uncertainty of input tsunami hazard parameters is evaluated by consulting with two extensive tsunami inundation/run-up datasets by the Ministry of Land, Infrastructure, and Transportation (MLIT 2014) and by the Tohoku Tsunami Joint Survey (TTJS) group (Mori et al. 2011), which were compiled after the 11th March 2011 Tohoku earthquake (Fraser et al. 2013). Secondly, the effects of propagating the input data uncertainty on tsunami fragility functions are investigated by adopting Bayesian regression methods. It is noteworthy that, in general, the empirical assessment of input inundation data uncertainty is very limited due to the lack of observed data. The 2011 Tohoku tsunami offers a unique opportunity to evaluate the accuracy and consistency of the tsunami inundation data from extensive post-event field surveys and tsunami damage inspections. The MLIT database contains more than 200,000 buildings (each data entry includes building type, location, tsunami damage level, and inundation depth) and is particularly useful for developing empirical tsunami fragility curves (Suppasri et al. 2013; Charvet et al.2014a). The TTJS database contains more than 5000 surveyed inundation and run-up heights along the Tohoku coastline and is useful for examining the tsunami inundation/run-up characteristics at both regional and local levels. The uncertainty of tsunami inundation data is evaluated by comparing the MLIT and TTJS data, noting that this comparison is not straightforward because spatial distributions of the MLIT and TTJS data differ and conversions of height to depth data are necessary by adopting a suitable digital elevation model (DEM). A potential log-normal distribution of the observations' error is found; such result is in line with the findings of Kim et al. (2014) given the extension of the analyzed coastline.

Among procedures for developing tsunami empirical fragility curves in the literature (Tarbotton et al. 2015), three statistical approaches are considered: (a) lognormal fragility model (Peiris 2006; Porter et al. 2007), (b) binomial logistic regression (Reese et al. 2011), and (c) multinomial logistic regression (Charvet et al. 2014a; Yazdi et al. 2015). For each method, two sets of tsunami fragility functions are developed by neglecting and considering the input data uncertainty. Innovative aspects of this work in implementing the preceding three methods are that they are based on a Bayesian framework and thus the uncertainty of input data is adequately propagated in conducting the point estimation of fragility parameters. As the complexity of the fitted models increases from the lognormal to multinomial logistic models, a Markov Chain Monte Carlo (MCMC) simulation (Cheung and Beck 2010) has been conducted for the parameter estimation. To authors' knowledge, Bayesian regression methods are applied in this paper for the first time in assessing tsunami empirical fragilities. The systematic implementation of the data and methods facilitates the quantitative comparison of tsunami fragility assessments under different assumptions.

Moreover, developed tsunami fragility models (three methods with/without input data uncertainty) are implemented in tsunami loss estimation to investigate the impact of different tsunami fragility models on risk assessment. For this, a portfolio of wooden buildings in the Tohoku region is considered. The loss assessment results are calculated using a procedure similar to Yu et al. (2013) and are presented in a form of total economic loss as well as economic loss as a function of distance from the shoreline to evaluate the spatial variation of the impact of different fragility functions.

The paper is organized as follows. Section 2 presents the mathematical formulations of the three tsunami fragility models based on the Bayesian framework. In Sect. 3, empirical uncertainty of tsunami inundation data is evaluated based on the MLIT and TTJS databases for the 2011 Tohoku earthquake. Section 4presents the development and systematic comparison of tsunami fragility models by considering the three statistical methods with/without input data uncertainty. The tsunami loss estimation results based on different fragility models are discussed in Sect. 5, and key conclusions are drawn in Sect. 6.

2. BAYESIAN FRAGILITY MODELS

The three fragility models are conventionally developed through the least squares fitting procedure for the lognormal model and through the maximum likelihood method for the binomial/multinomial logistic models. In this work, these fragility models are developed through the Bayesian method by incorporating the input data uncertainty. The Bayesian approach is problem-specific and requires a good understanding of physical nature of the problem and observations (Der Kiureghian 2002). After a brief review of the Bayesian updating formula, methods of point estimation, neglecting and considering input data uncertainty are presented.

2.1 Bayesian estimation

Let $\boldsymbol{\theta}$ represent fragility parameters that are to be estimated based on the observed data $\mathbf{D}$. $\boldsymbol{\theta}$ can be treated as a set of random variables characterized by probability distribution functions. According to the Bayesian paradigm, the distribution of $\boldsymbol{\theta}$ can be updated as new observational information becomes available (Box and Tiao 1992):

$$f(\boldsymbol{\theta}|\mathbf{D}) = c^{-1} \cdot L(\mathbf{D}|\boldsymbol{\theta}) \cdot f(\boldsymbol{\theta}) \tag{1}$$

where $f(\boldsymbol{\theta})$ is the prior distribution of fragility parameters and represents the information available on $\boldsymbol{\theta}$ prior to the estimation; $L(\mathbf{D}|\boldsymbol{\theta})$ is the likelihood function and represents the information contained in the observation;$f(\boldsymbol{\theta}|\mathbf{D})$ is the posterior distribution describing the updated estimate of $\boldsymbol{\theta}$; and c is a normalizing factor:

$$c = \int L(\mathbf{D}|\boldsymbol{\theta}) \cdot f(\boldsymbol{\theta}) \cdot d\boldsymbol{\theta} \quad (2)$$

The likelihood function depends on the adopted type of regression and is proportional to the conditional probability of the data **D** given the parameter **θ**; therefore, assuming that observations are independent, the likelihood is given by:

$$L(\mathbf{D}|\boldsymbol{\theta}) = \prod_{i=1}^{n} f(\boldsymbol{D}_i|\boldsymbol{\theta}) \quad (3)$$

where n is the number of observations, $\boldsymbol{D}_i$ is the ith observation composed by the intensity measure value and the associated damage outcomes for damage states ($\boldsymbol{D}_i = [im_i, n_{i,\mathrm{DS1}},\ldots,n_{i,\mathrm{DSk}}]$); and $f(\boldsymbol{D}_i|\boldsymbol{\theta})$ is the value of the likelihood for the ith observation given **θ**.

The likelihood function represents the key factor for the propagation of the data uncertainty in the regression. According to the total probability theorem (Jaynes 2003), for the ith observation, the likelihood function can be written as:

$$f(\boldsymbol{D}_i|\boldsymbol{\theta}) = \int_{-\infty}^{+\infty} f(\boldsymbol{D}_i|\varepsilon,\boldsymbol{\theta}) \cdot f_i(\varepsilon) \cdot d\varepsilon \quad (4)$$

where $f_i(\varepsilon)$ is the probability density function (pdf) of the error associated with the ith observation, modeled as a Gaussian distribution with zero mean and standard deviation σ_ε. The subscript i indicates that σ_ε can be different for each observation. $f(\boldsymbol{D}_i|\varepsilon, \boldsymbol{\theta})$ is the value of likelihood for the ith observation given **θ** and the error associated with observation ε (i.e. the likelihood for the sum of the logarithm of im_i and ε). The parameters that maximize the posterior represent the solutions of the problem. In the following, this is referred to as Bayesian maximum likelihood.

2.2 Lognormal method

In the lognormal method, exceedance probabilities for damage states are calculated and median values are plotted against a range of equally spaced bins of tsunami hazard parameter (e.g. inundation depth interval of 0.5 m). The probability of occurrence of damage is:

$$P(DS \geq ds|h) = \Phi\left(\frac{\ln h - \ln \eta}{\beta}\right) \tag{5}$$

where Φ is the standard normal distribution function, h is the inundation depth, η is the median, and β is the logarithmic standard deviation (or dispersion parameter). The two parameters η and β are obtained by plotting the logarithm of inundation depth versus the inverse cumulative distribution function of the exceedance probability and by performing a linear regression analysis according to the following relation:

$$\ln h = \ln \eta + \beta \cdot \Phi^{-1}[P(DS \geq ds|h)] + \varepsilon_R \tag{6}$$

where ε_R is the term representing the regression error, which is normally distributed with zero mean and unknown standard deviation σ_R. In Eq. (6), $P(DS \geq ds|h)$ can be obtained based on the data analysis of damage outcomes for each inundation depth bin. It follows that ln$h$ is normally distributed with mean function of unknown parameters η and β, and unknown standard deviation equal to σ_R.

The linear regression is generally carried out through a least squares fitting procedure by minimizing the square of residuals between empirical data and calculated values. By adopting the Bayesian regression procedure, the fundamental equation is Eq. (1) where in the specific case **θ** = [η, β, σ_R]. The prior distribution can be decomposed into three marginal distributions of independent variables. In particular, as priors, uniform distributions are used for η, β, and σ_R in absence of other information. The likelihood function is:

$$L(\eta, \beta, \sigma_R|\mathbf{D}) = \prod_{i=1}^{n} \frac{1}{\sqrt{2\pi} \cdot \sigma_R} \cdot \exp\left\{-\frac{1}{2 \cdot \sigma_R^2} \cdot \left[\ln h_i - \ln \eta - \beta \cdot \Phi^{-1}(P(DS \geq ds|h_i))\right]^2\right\} \tag{7}$$

The triplet of the parameters is determined by maximizing the posterior numerically over the model parameter space.

When the observations are uncertain or not free from error, additional uncertainty should be considered in the problem. In this case, Eq. (6) becomes:

$$\ln h + \varepsilon_{\ln h} = \ln \eta + \beta \cdot \Phi^{-1}[P(DS \geq ds|h)] + \varepsilon_R \tag{8}$$

where $\varepsilon_{\ln h}$ is the error of observational data. According to Eq. (4), the ith term of the product in Eq. (7) becomes:

$$\int_{-\infty}^{+\infty} \frac{1}{\sqrt{2\pi} \cdot \sigma_R} \cdot \exp\left\{-\frac{1}{2 \cdot \sigma_R^2} \cdot \left[\ln h_i + \epsilon_{\ln h} - \ln \eta - \beta \cdot \Phi^{-1}(P(DS \geq ds|h_i))\right]^2\right\} \cdot f(\varepsilon_{\ln h}) \cdot d\varepsilon_{\ln h} \tag{9}$$

where $f(\varepsilon_{\ln h})$ is the pdf of the error. The error is normally distributed with zero mean and standard deviation $\sigma_{\ln h}$. The parameter $\sigma_{\ln h}$ needs to be estimated from the analysis of the error associated with observations.

Assuming independent uniform priors, likelihood, and error distribution, the formulation can be simplified further. In fact, the likelihood function remains the same as defined in Eq. (7) where the term σ_R is replaced by the square root of the sum of the two variances corresponding to the regression error and data error, respectively:

$$\sigma_{TOT} = \sqrt{\sigma_R^2 + \sigma_{\ln h}^2} \tag{10}$$

This simplification facilitates the efficient estimation of the model parameters, since it avoids the integration related to Eq. (9).

2.3 Binomial logistic method

Logistic regression is a special case of a generalized linear model and can be used for developing empirical fragility functions based on binomial data. For each damage state, individual building damage survey results provide with a binary indication of whether the considered damage state is exceeded or not and the maximum water depth at the building site. Unlike the lognormal model, the data are not organized in bins.

Let π_i denote the probability that the ith observed building is diagnosed as attaining a specific damage state*ds*. The probability that all observed buildings are classified with *ds* is:

$$\prod_{i=1}^{n} \binom{1}{y_i} \cdot \pi_i^{y_i} \cdot (1 - \pi_i)^{1-y_i} \tag{11}$$

where y_i is equal to 1 if the ith observation falls in the examined damage state and it is zero otherwise. Therefore, Eq. (11) is the likelihood function representing the probability of observed data. The term π may assume different forms, such as probit, logit, and loglog (Hosmer et al. 2013). In this study, the logit function is considered:

$$\pi_i = \frac{\exp(b_1 + b_2 \cdot \ln h_i)}{1 + \exp(b_1 + b_2 \cdot \ln h_i)} \tag{12}$$

where b_1 and b_2 are the model parameters. In a non-Bayesian framework, the point estimation is carried out through the maximum likelihood procedure.

When a Bayesian regression is carried out, the posterior of the model parameters can be formulated as Eq. (1), where $\boldsymbol{\theta} = [b_1, b_2]$. The prior distribution can be decomposed into two marginal distributions of independent variables. As priors, uniform distributions are used for b_1 and b_2 in absence of other information. The likelihood function is defined in Eq. (11). By maximizing the posterior, the model parameters can be determined numerically. An advantage of the Bayesian regression is that uncertainty of the data can be incorporated in the parameter estimation. More specifically, the ith term of the likelihood function becomes [see Eq. (4)]:

$$\int_{-\infty}^{+\infty} \binom{1}{y_i} \cdot \left[\frac{\exp(b_1 + b_2 \cdot (\ln h_i + \varepsilon_{\ln h}))}{1 + \exp(b_1 + b_2 \cdot (\ln h_i + \varepsilon_{\ln h}))}\right]^{y_i} \cdot \left[1 - \frac{\exp(b_1 + b_2 \cdot (\ln h_i + \varepsilon_{\ln h}))}{1 + \exp(b_1 + b_2 \cdot (\ln h_i + \varepsilon_{\ln h}))}\right]^{1-y_i} \cdot f(\varepsilon_{\ln h}) \cdot d\varepsilon_{\ln h} \tag{13}$$

A pair of parameters that maximize the posterior represents the solution of the problem.

2.4 Multinomial logistic method

Multinomial regression is a generalized linear model that allows considering more than two outcomes at the same time and can be employed to develop empirical fragility models for multiple damage states. Charvet et al. (2014a) applied the procedure by considering binned data. More specifically, denoting the probability that structures corresponding to the ith observation data bin fall in the jth damage state ds_j by π_{ij}, the probability that all buildings of the ith bin fall in the respective damage state class is given by the multinomial probability distribution:

$$\frac{m_i!}{\prod_{j=1}^{k} y_{ij}!} \prod_{j=1}^{k} \pi_{ij}^{y_{ij}} \tag{14}$$

where m_i is the total number of structures composing the ith observation bin, k is the number of damage states, and y_{ij} is the number of structures for the ith observation bin attaining the specific damage state ds_j. In Eq. (14), π_{ij} may take different functions (e.g. probit, logit, and log–log) and may assume different features (e.g. nominal, ordinal, and hierarchical), as described in detail in McCullagh and Nelder (1989). In the following, the logit function is considered, and the hierarchical partially-ordered approach is adopted for the regression. In the hierarchical method, π_{ij} assumes the following form:

$$\pi_{ij} = \frac{\exp\left(b_{1,j} + b_{2,j} \cdot \ln h_i\right)}{1 + \exp\left(b_{1,j} + b_{2,j} \cdot \ln h_i\right)} \cdot \left(1 - \sum_{l=1}^{j-1} \pi_{il}\right) \tag{15}$$

where $b_{1,j}$ and $b_{2,j}$ are the model parameters (i.e. the intercept and the slope of the generalized linear model, respectively); the approach is partially-ordered because the model parameters are different for all the considered damage states. Considering k damage states, it is possible to write k-1 sets of Eq. (15). The first equation presents only the fraction (i.e. no term in parenthesis) and Eq. (15) for the kth damage state is equal to one. In a non-Bayesian method, point estimates of the model parameters are calculated, in accordance with the maximum likelihood approach, by computing the first and second derivatives of the likelihood (or log-likelihood) function that is expressed as follows:

$$\prod_{i=1}^{n}\prod_{j=1}^{k}\pi_{ij}^{y_{ij}} \tag{16}$$

where n is the number of bins.

The preceding problem can be evaluated within a Bayesian framework, based on Eq. (1) with $\boldsymbol{\theta} = [b_{1,1}, b_{2,1},\ldots,b_{1,k\text{-}1}, b_{2,k-1}]$. The prior distribution can be composed of 2 × (k-1) marginal uniform distributions of independent variables. The likelihood function is given in Eq. (16). The Bayesian parameter estimation can be carried out by maximizing the posterior. Given a high number of model parameters, the parameter estimation is numerically achieved through a MCMC simulation. The Bayesian regression facilitates the inclusion of the data uncertainty. In such cases, the the ith term of the likelihood function becomes:

$$\prod_{j=1}^{k}\int_{-\infty}^{+\infty}\frac{\exp\left(b_{1,j}+b_{2,j}\cdot(\ln h+\varepsilon_{\ln h})\right)}{1+\exp\left(b_{1,j}+b_{2,j}\cdot(\ln h+\varepsilon_{\ln h})\right)}\cdot\left(1-\sum_{l=1}^{j-1}\pi_{il}\right)\cdot f(\varepsilon_{\ln h})\cdot d\varepsilon_{\ln h} \tag{17}$$

3. UNCERTAINTY OF TSUNAMI INUNDATION DATA

3.1 MLIT database

The input data for developing tsunami fragility models are obtained from the MLIT damage database. Buildings located in the inundated areas during the 2011 Tohoku earthquake are included in the database and are characterized by various attributes, such as geographical location, structural material, story number, tsunami inundation depth, and damage level. More than 200,000 structures are located in Iwate, Miyagi and Fukushima prefectures (from North to South). Among them, 83.9 % of the structures (176,215 buildings) have information about structural material, inundation depth, and damage state (which is necessary for developing fragility functions). The majority of the surveyed buildings (83.8 %) are wooden structures, followed by masonry structures (8.8 %), steel structures (5.0 %), and reinforced concrete (RC) structures (2.4 %). In this study, only wooden structures are considered for tsunami fragility modeling due to statistical stability.

The original MLIT database adopts the tsunami damage scale with seven discrete states, namely no damage (DS1), minor damage (DS2), moderate damage (DS3), major damage (DS4), complete damage (DS5), collapse (DS6), and wash-away (DS7). Figure 1a shows the locations of the surveyed wooden buildings along about 300 km coastline near the epicenter, depicting the geographical distribution of damage states. In terms of sustained damage levels, only 1.5 % of the surveyed buildings did not suffer any damage, while 40.4 % of the buildings were washed away. The statistics of the damage levels are presented in Fig. 1b. Figure 1c shows a histogram of inundation depths that were experienced by the surveyed buildings ranging between 0.1 m and 27.0 m. Prior to tsunami fragility modeling, the original damage data are modified by combining DS6 and DS7 data, noting that these two damage states are two different descriptions of a collapse mode. Therefore, as also suggested in Charvet et al. (2014b), the seven damage states (DS1-DS7) are reduced to six (DS1-DS6/7) in the tsunami fragility analysis in the following.

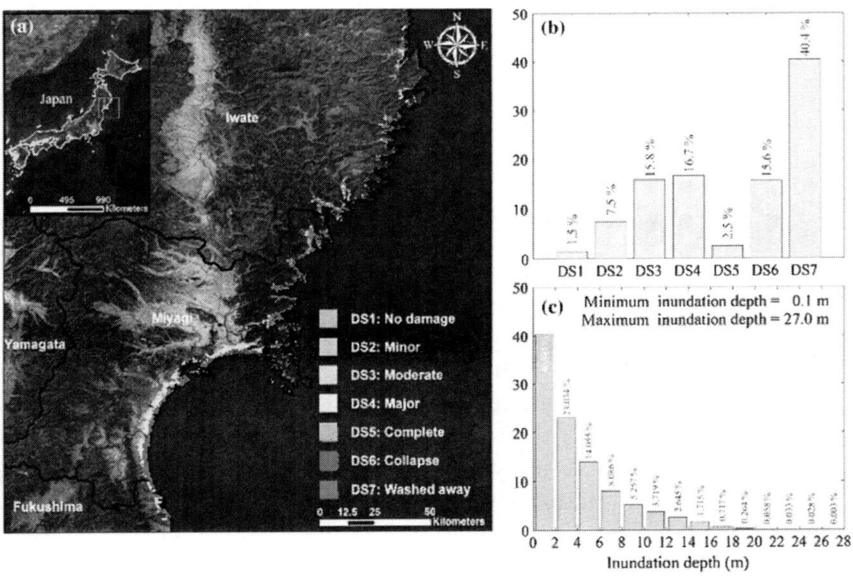

Figure 1. a Spatial distribution of surveyed wooden buildings having different damage states, **b** histogram of the tsunami damage states, and **c** histogram of the inundation depths

The inundation depth data of the MLIT damage database are assigned based on the MLIT 100-m mesh inundation data (a separate database developed by the MLIT); in the MLIT damage database, local variations of elevation at sub-mesh levels are taken into account based on the DEM data that are developed by the Geospatial Information Authority of Japan (GSI). The GSI-DEM data are obtained by airborne laser scanner surveys and have 5-m resolution; the vertical accuracy is plus/minus 0.3 m in terms of standard deviation. The MLIT 100-m data contain information of locations, elevation, inundation depth, and inundation height. The original inundation depth data were obtained by conducting surveys of tsunami marks and interviews to local residents and from existing reports/data. When direct observations were not available, interpolation of inundation depths/heights at nearby locations was carried out. In other words, the MLIT inundation data, which are used as input tsunami hazard parameters in the fragility analysis (e.g. Suppasri et al. 2013; Charvet et al. 2014a), are subjected to errors and uncertainty.

3.2 TTJS database

The TTJS database, which was developed independently from the MLIT database, contains post-event surveyed tsunami inundation/run-up heights along the Tohoku coast. Heights of watermarks on buildings, trees, and walls were measured using a laser range finder, a level survey, a real-time kinematic global positioning system (RTK-GPS) receiver with a cellular transmitter, and total stations (Mori et al. 2011). Generally, the accuracy of the measurements is within a few centimeters vertically. The database includes information of location (latitude and longitude as well as address), measurement date/time (used for tidal level corrections), tsunami heights, run-up distance from shoreline, tidal levels, reliability of measurements, and target objects/marks. The TTJS inundation height data can be adopted as a benchmark in assessing the errors/uncertainty associated with the MLIT inundation data as they may be considered to be more accurate (or controlled); only TTJS data with high reliability (Rank A) are used in this study. Based on the TTJS height data, the corresponding depth data can be obtained by using the GSI-DEM data (note: the correction potentially introduces systematic errors).

3.3 Uncertainty of tsunami inundation data

The MLIT inundation depth data (as in the MLIT damage database) are subjected to two sources of uncertainty: (i) they are based on the MLIT 100-m data (i.e. errors due to interpolation/smoothing) and (ii) elevation data at the building sites are not available (thus the MLIT depth data cannot be converted to height deterministically). In other words, it is not straightforward to evaluate the accuracy of the MLIT 100-m data (both depth and height are available) quantitatively, because the data are only available at 100-m resolution (coarse for detailed assessment) and the locations corresponding to the representative values for the meshes are unknown. By taking into account the characteristics of the available data, uncertainty associated with the tsunami inundation data is assessed by comparing the MLIT versus TTJS inundation height data.

To relate the MLIT data points with the TTJS data points (which have different spatial distributions and coverage), several distance radii between 5 and 50 m are considered (i.e. 5, 10, 20, and 50 m). A schematic representation of the MLIT and TTJS observation points is given in Fig. 2. The radius of 5 m is suitable for a lower bound because the baseline GSI-DEM data for the MLIT database are at this resolution. The radius of 50 m (a half of 100 m) is considered as an upper bound. With the increase of the radius, the number of data pairs increases because more points in the TTJS and MLIT databases can be associated each other (although the similarity of the selected data decreases with the separation distance due to the changes of the elevation and the local variability of the tsunami waves). Figure 3a shows the scatter plots of the MLIT and TTJS height data for the radii of 5 and 50 m. Figure 3b presents the differences between the TTJS and MLIT height data as a function of GSI elevation, while Fig. 3c displays the ratios between the TTJS and MLIT height data as a function of GSI elevation.

Table 1 lists inundation height statistics for different values of radii between 5 and 50 m; the Pearson's linear correlation coefficient (ρ) between the TTJS and MLIT 100-m data, the statistics of the differences between the TTJS and MLIT height data, and the statistics of the ratios between the TTJS and MLIT height data (in logarithm) are included. When the TTJS and MLIT data points are close (e.g. less than 5 m), the TTJS and the MLIT inundation height data are consistent. With the increase of the separation distance between the TTJS and MLIT data points, the consistency of the TTJS and the MLIT data decreases and the data points start to scatter more widely. This can be inspected from the decreasing trends of the linear correlation coefficient and the increasing trends of the standard deviation of the differences and ratios. Another important observation is that when the radius for interpolation becomes large (e.g. 50 m), the consistency of the inundation metrics is deteriorated significantly. Therefore, a caution is necessary in carrying out such interpolation. Taking the TTJS data as benchmark, the standard deviation of the differences of the height data can be assigned as 1.5–2.0 m and the corresponding logarithmic standard deviation of the ratios of the height data is 23 % (Table 1). Moreover, Fig. 4 shows the empirical distributions of the height differences and the height ratios for the TTJS and the MLIT data. It indicates that the height ratios are well fitted by a normal distribution (i.e. the ratios are log-normally distributed), while the height differences are not. Based on the preceding results, the inundation data are assumed to be log-normally distributed and the data uncertainty equal to 25 % is considered in the Bayesian tsunami fragility analysis (Sect. 4).

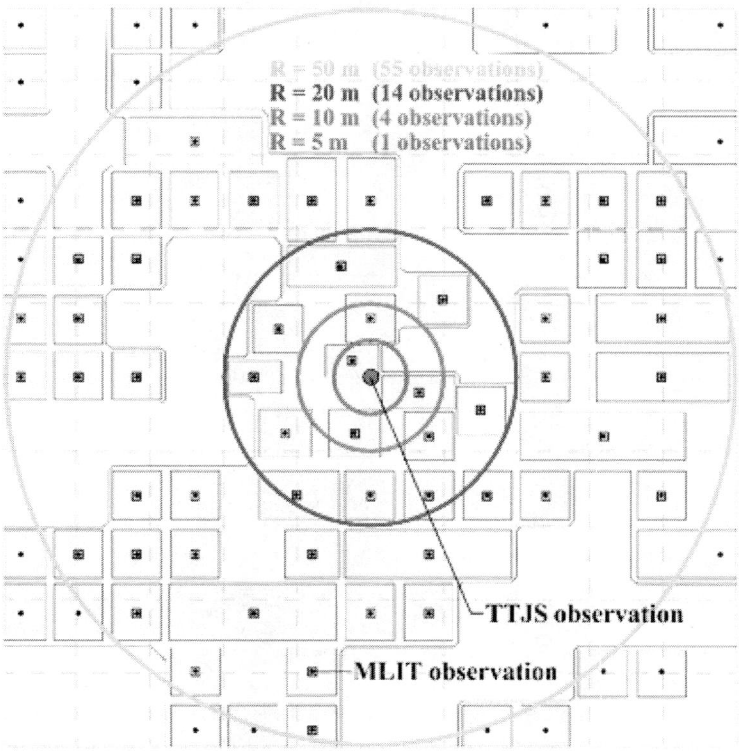

Figure 2. Schematic representation of the MLIT and TTJS observation points

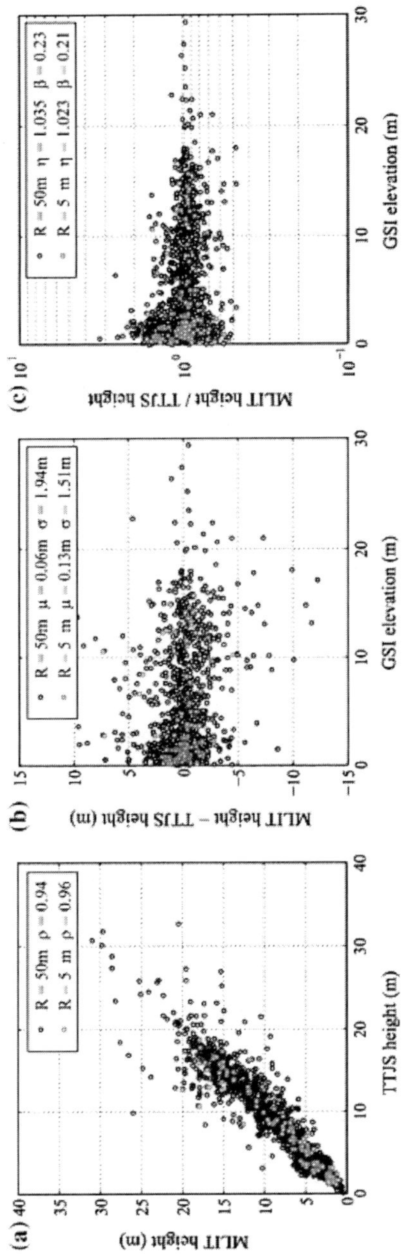

Figure 3. Comparison of the TTJS and MLIT inundation height data for radii of 5 and 50 m: **a** scatter plot of the TTJS and MLIT inundation height data, **b** difference between the TTJS and MLIT inundation height data with respect to the GSI-elevation, and **c** ratio between the TTJS and MLIT inundation height data with respect to the GSI elevation

Table 1. Comparison of the TTJS and MLIT inundation height data

Radius (m)	ρ	$\mu_{h\mathrm{MLIT}-h\mathrm{TTJS}}$ (m)	$\sigma_{h\mathrm{MLIT}-h\mathrm{TTJS}}$ (m)	$\eta_{\log(h\mathrm{MLIT}/h\mathrm{TTJS})}$	$\beta_{\log(h\mathrm{MLIT}/h\mathrm{TTJS})}$
5	0.96	0.13	1.50	1.023	0.21
10	0.94	0.21	1.83	1.035	0.21
20	0.95	0.24	1.65	1.053	0.22
50	0.94	0.06	1.94	1.035	0.23

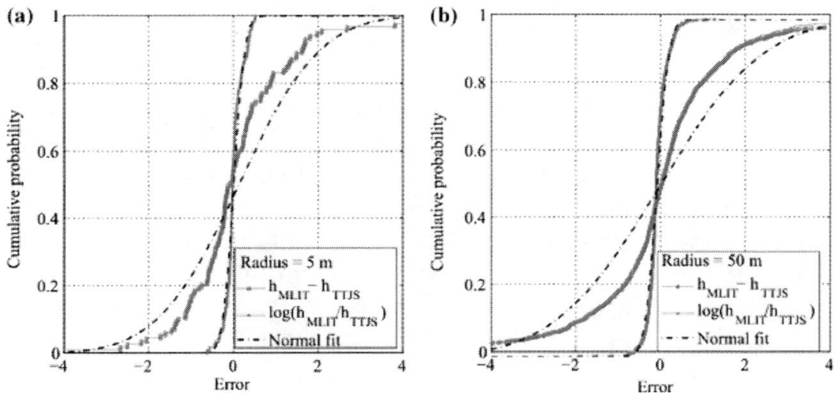

Figure 4. Empirical distributions of the differences (*blue dotted line*) and of the logarithmic ratios (*purple line*) between the MLIT and TTJS data in comparison with the normal fit (*dashed dotted black line*): **a** radius of 5 m and **b**radius of 50 m

4. TSUNAMI FRAGILITY ASSESSMENT

This paper focuses on wooden buildings damaged due to tsunami triggered by the 11th March 2011 Tohoku earthquake. In the following, empirical fragility curves are obtained with the different regression methods considering and neglecting the input data uncertainty.

4.1 Results for lognormal method

For the lognormal method, the inundation depth data are binned with equal intervals of 0.50 m, as suggested by Suppasri et al. (2013). Figure

5 shows the posterior joint distributions of the regression parameters (i.e. η, β, and σ_R) without (black contours with solid line) and with (red contours with dashed line) input data errors for the collapse damage state. Figure 6 shows the corresponding marginal distributions of the three parameters for the collapse damage state. Neglecting the input data error, the Bayesian maximum likelihood estimations (i.e. red points) and the least squares results (hollow black circles) are identical. On the other hand, considering the input data error, σ_R changes significantly, showing a reduction from 0.29 to 0.14. This reduction is due to the fact that the consideration of the input data error reduces the regression error; this is expected because the term $\varepsilon_{\ln h}$ is moved from the left-hand side to the right-hand side in Eq. (8). The reduction in σ_R translates into a smaller confidence interval around the central estimate of the fragility function. For example, Fig. 7a shows the central estimate of the collapse (DS6/7) fragility function for wooden structures and the 90 % confidence intervals without and with input data error. Figure 7b shows the resulting fragility curves for all damage states, and Table 2 lists the parameters of the fragility curves obtained without and with input data error. The numerical results indicate that medians and logarithmic standard deviations are equal for the two considered cases, whereas the regression confidence intervals are lower when the input data error is taken into account, with a reduction of about 50 % according to available data.

It is noteworthy that the data need to be binned (grouped) to apply this method. From a practical point of view, when the lognormal method is used by taking into account the input data error, the influence of considering the uncertainty is revealed as the narrower fragility confidence interval, whereas the central fragility curve does not change. In other words, the model dispersion is not affected.

4.2 Results for binomial logistic method

For the binomial logistic method, individual records are used as unbinned data, as suggested by Reese et al. (2011). Figure 8a, b show the joint posterior distributions for DS6/7 neglecting and considering input data error, respectively. Figure 8a shows that the Bayesian maximum likelihood (red point) coincides with the maximum likelihood approach (hollow black circle). Figure 8b shows that the parameters obtained with the Bayesian and non-Bayesian procedures are different. This difference can be observed in terms of resulting fragility functions. Figure 9 shows the fragility functions for all damage states neglecting (solid line) and considering (dashed line) the input data error. Table 3 lists the numerical values of the estimated fragility parameters. The medians remain the same and the logarithmic standard deviations are decreased by about 10 % (e.g. DS5 and DS6/7).

For the Bayesian binomial logistic method using ungrouped data, the effect of considering the input data error shows up in the reduced logarithmic standard deviation of a fragility curve. This means that the dispersion of the fragility model is decreased and thus more confidence on the central estimate of the tsunami fragility is achieved. It is noted that the preceding result is generally applicable to binned data (as in the multinomial logistic method in Sect. 4.3).

4.3 Results for multinomial logistic method

For the multinomial logistic method, the inundation depth data are binned with non-uniform intervals in order to obtain equally populated bins (Charvet et al. 2014a). In the estimation, the reference damage state (i.e. the damage state with probability equal to 1) is set to DS1 (Sect. 2.4); thus the order of damage states in Eq. (15) should be reversed. Figure 10 shows the results of the MCMC simulations for the Bayesian approach, neglecting and considering the input data error, represented by continuous line and dashed line, respectively. The blue and red lines are the intercept (b_1) and slope (b_2) of the generalized linear model, respectively. The dashed black line represents the value obtained through the maximum likelihood point estimate procedure. From the plot, the effects of the initial values of parameters take a while to disappear before the process begins to look stationary (1000 simulation steps are sufficient to achieve the convergence).

Neglecting the input data error, the final values (blue and red light numbers), obtained after 1000 simulations, converge to the value obtained with the maximum likelihood point estimate procedure (black numbers). Considering the input data error, the model parameters (blue and red bold numbers) are different with respect to the preceding case. This difference can be observed in terms of resulting empirical fragility functions, as presented in Fig. 11 neglecting (continuous lines) and considering (dashed lines) the input data error. Moreover, Table 4 lists the parameters of the obtained fragility curves for all damage states. The results show that the medians remain the same and the logarithmic standard deviations related to the case with input data errors are decreased by as large as 16 %.

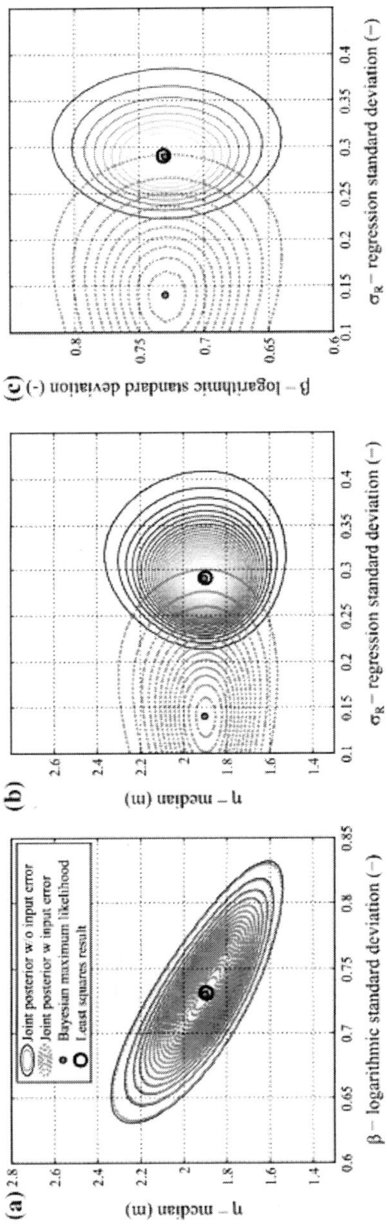

Figure 5. Posterior distributions of the regression parameters neglecting (*black contours*) and considering (*red contours*) the input data error for the collapse damage state (DS6/7) based on the lognormal method: **a** η–β joint distribution, **b** η–σ_R joint distribution, and **c** β–σ_R joint distribution

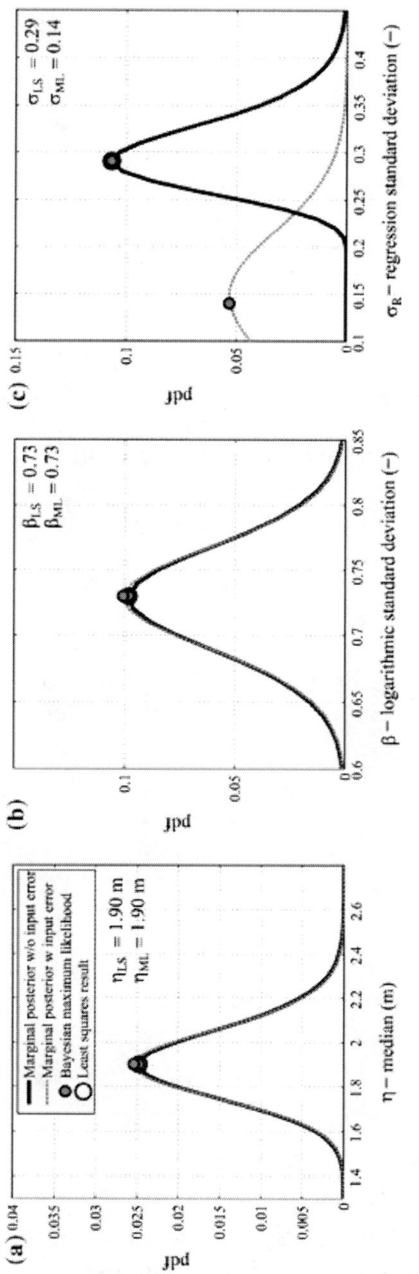

Figure. 6 Posterior distributions of the regression parameters neglecting (*black lines*) and considering (*red lines*) the input data uncertainty for the collapse damage state (DS6/7) based on the lognormal method: **a** η marginal distribution, **b** β marginal distribution, and **c** σ_R marginal distribution

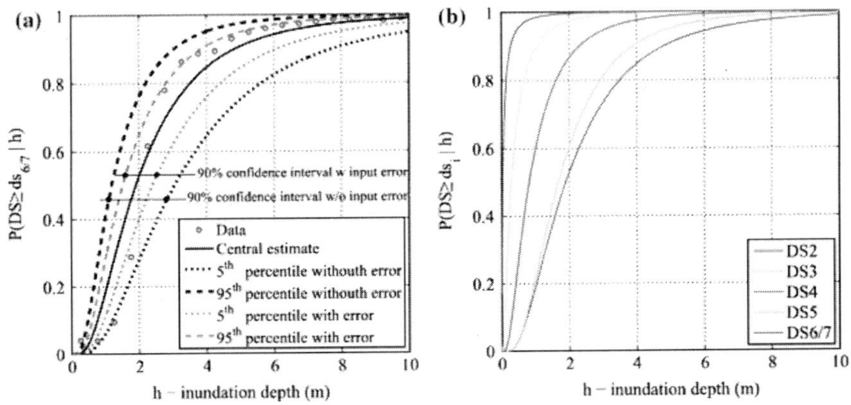

Figure 7. Tsunami fragility curves based on the lognormal method: **a** collapse fragility curves and 90 % confidence interval neglecting and considering input data error, and **b** tsunami fragility curves for all damage states

Table 2. Parameters of fragility curves neglecting and considering the input data error based on the lognormal method

Damage state	Without input data error			With input data error		
	η (m)	β (−)	σ_R (−)	η (m)	β (−)	σ_R (−)
DS2	0.04	1.49	0.40	0.04	1.49	0.31
DS3	0.29	1.00	0.33	0.29	1.00	0.22
DS4	0.82	0.81	0.25	0.82	0.81	0.01
DS5	1.68	0.66	0.32	1.68	0.66	0.20
DS6/7	1.90	0.73	0.29	1.90	0.73	0.14

Similarly to the binomial logistic method, the incorporation of input data errors into the tsunami fragility modeling only affects the model dispersion. The preceding general results are applicable to the case when un-binned data, rather than binned data, are used. This is because the multinomial logistic regression, as extension of the binomial logistic regression, can be performed for both types of data.

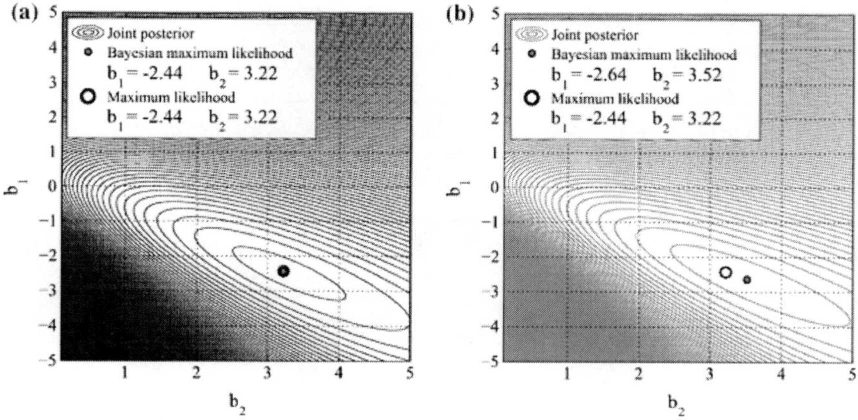

Figure 8 Posterior distribution of regression parameters b_1 and b_2 for **a** without and **b** with input data error for the collapse damage state (DS6/7) based on the binomial logistic method.

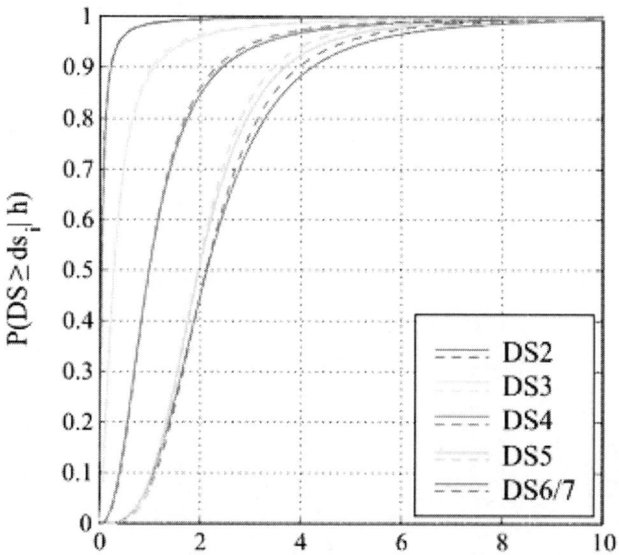

Figure 9. Tsunami fragility curves neglecting (*continuous line*) and considering (*dashed line*) input data error based on the binomial logistic method

Table 3. Parameters of fragility curves neglecting and considering the input data error based on the binomial logistic method

Damage state	Without input data error		With input data error	
	η (m)	β (−)	η (m)	β (−)
DS2	0.04	1.23	0.04	1.23
DS3	0.29	1.02	0.28	0.99
DS4	0.99	0.68	0.99	0.65
DS5	1.97	0.47	1.96	0.43
DS6/7	2.13	0.52	2.12	0.47

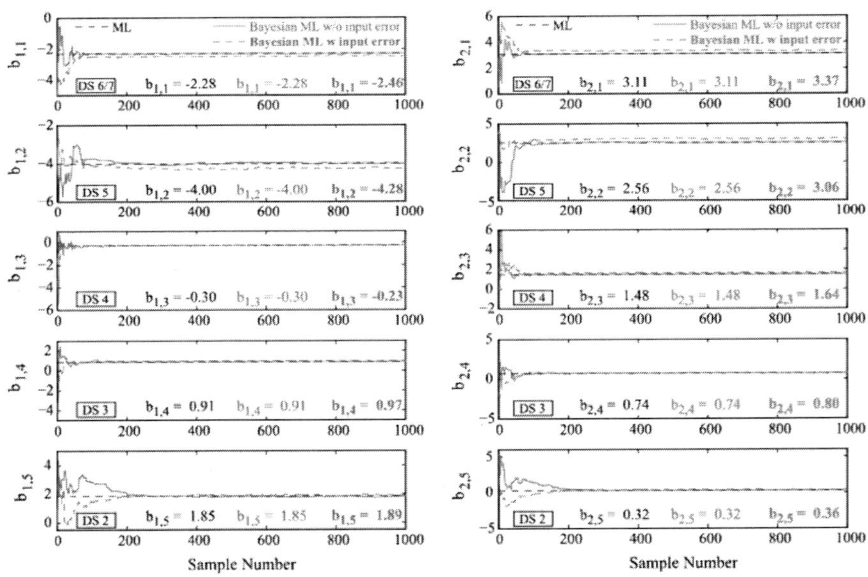

Figure 10. Convergence of parameter estimation based on Markov Chain Monte Carlo simulations neglecting and considering the input data error (*continuous* and *dashed line* respectively)

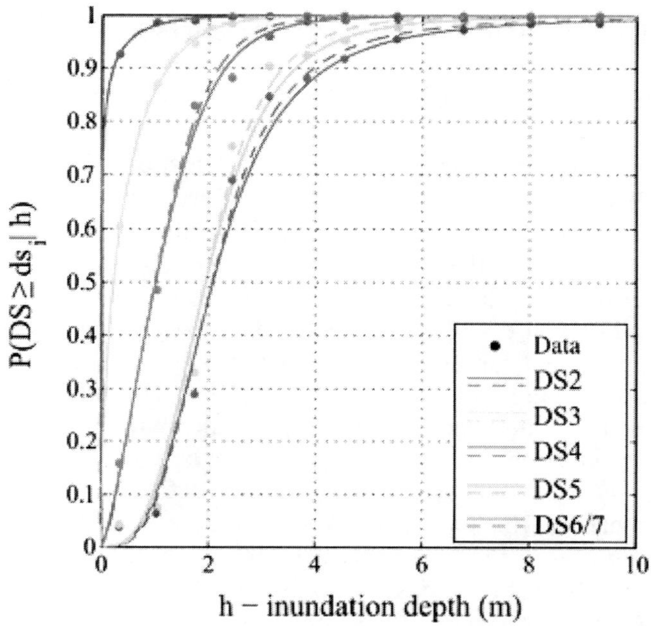

Figure 11. Tsunami fragility curves neglecting (*continuous line*) and considering (*dashed line*) input data error based on the multinomial logistic method

Table 4. Parameters of fragility curves neglecting and considering the input data error based on the multinomial logistic method

Damage state	Without input data error		With input data error	
	η (m)	β (−)	η (m)	β (−)
DS2	0.01	2.50	0.01	2.38
DS3	0.23	1.92	0.24	1.61
DS4	1.02	0.83	1.01	0.77
DS5	1.95	0.51	1.95	0.45
DS6/7	2.08	0.54	2.08	0.49

4.4 Comparison of tsunami fragility models

The systematic applications of three Bayesian regression methods facilitate the meaningful comparisons of the tsunami fragility curves developed under different assumptions. To inspect the differences of the fragility curves visually, Fig. 12 compares the tsunami fragility curves for DS4, DS5, and DS6/7 based on the lognormal, binomial logistic, and multinomial logistic methods without and with the input data uncertainty (note: for the lognormal model, the inclusion of the input data error in the analysis has no effect on the developed fragility curve). It can be observed from Fig. 12 that medians for the lognormal method are smaller than those for the binomial/multinomial logistic method, while model dispersions for the lognormal method are greater than those for the binomial/multinomial logistic method (note: these observations are applicable to both cases neglecting and considering the input data error). This means that at low inundation depths, the fragility curves based on the lognormal method predict higher damage probabilities than those based on the binomial/multinomial logistic method. On the other hand, the opposite tendency is applicable at high inundation depths. The differences can be attributed to the forms of the base functions (lognormal versus logistic) and partly to different grouping schemes used for the three methods. The latter has relatively minor effects because the binomial and multinomial methods produce similar results although they adopt different grouping procedures.

Regarding the impact of incorporating the uncertainty of inundation depth data in the fragility modeling, the lognormal method and the binomial/multinomial logistic method affect the developed fragility models differently. When the lognormal method is adopted, only a reduction of the confidence interval can be observed, which is consistent with Cetin et al. (2002) for liquefaction triggering analysis. In contrast, when the binomial/multinomial logistic method is used, a reduction of the model dispersion is observed; the result is in accord with Der Kiureghian (2002) for seismic fragility analysis. These differences of the tsunami fragility curves caused by the different modeling approaches, have direct influence on the tsunami damage assessment (and the seismic loss estimation). In fact, for the logistic regression methods, the central estimate fragility curve obtained considering the input data error returns a greater damage probability for high values of inundation depth and a smaller damage probability for low inundation depths, in comparison with the case in which the input data error is neglected (note: the consideration of the input data uncertainty does not affect the median and thus the changing point of the increased/decreased damage probability correspond to the median inundation depth).

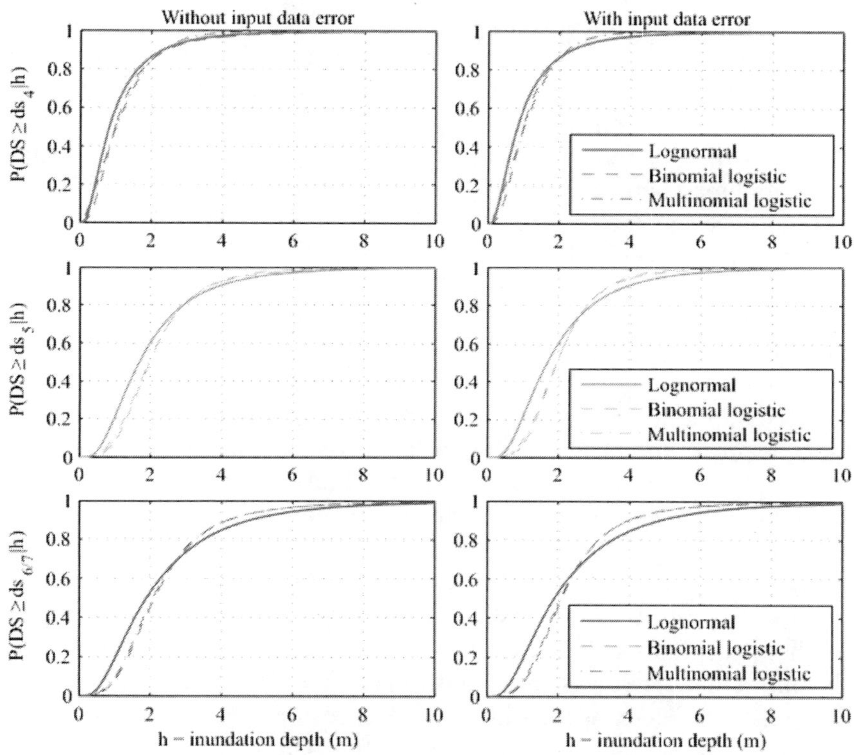

Figure 12. Comparison of tsunami fragility curves based on the lognormal, binomial logistic, and multinomial logistic methods for DS4, DS5, and DS6/7

The differences between fragilities with and without input data uncertainty are larger for severer damage states (DS4, DS5, and DS6/7). These differences reflect the not uniform number of observations across damage states. When more observations are available, the model dispersion without input data uncertainty becomes smaller, and consequently, the effect of the input data uncertainty becomes more relevant. This is because the number of observations governs both the likelihood function and the model dispersion.

Further insights can be obtained by examining regression residuals between observed data and fitted fragility curves. Figure 13 shows the residuals (i.e. the difference between the observed data and the fitted fragility curve) obtained for each regression method; the results are presented for all damage states, neglecting and considering the input data error. To concentrate on the interval where the greatest differences occur, the residual range is focused on inundation depth between 0.1 m and 4.0 m. To discuss the differences of the observations and model predictions quantitatively, the square root of sum of squares (SRSS) of the residuals for inundation depth between 0.1 and 27.0 m is evaluated for the three methods and the calculated values are presented in the figure (note: the percentages in parenthesis show the differences between the cases without and with input data error). The results shown in Fig. 13 suggest that the profiles of the residual variations in terms of inundation depth for the three methods are similar although the magnitudes of the deviations from zero residual line for the binomial/multinomial logistic methods are smaller than those for the lognormal method (these trends can be also inspected by comparing the SRSS values). In terms of the inclusion/exclusion of the input data uncertainty in the regression analysis, (*i*) no changes are observed for the lognormal method (as expected); (*ii*) the binomial logistic method produces residuals quite similar to the case in which the input data error is neglected for DS2, DS3, and DS4, whereas it produces smaller residuals for DS5 and DS6/7; and (*iii*) for the multinomial logistic method, residuals are consistently smaller for all damage states when the input data error is taken into account in the fragility modeling. Furthermore, focusing on the SRSS values for the structural damage states (i.e. DS3, DS4, DS5, and DS6/7), the binomial logistic regression presents the smaller residuals, followed in order by the multinomial logistic regression and the lognormal regression.

It can be concluded that for the tsunami fragility data for wooden houses in the Tohoku region, using the binomial logistic method with un-binned data leads to smaller overall residuals between observed data and fitted models, and therefore is preferred with respect to the other two methods. Nevertheless, further investigations related to the multinomial logistic method are required by considering un-binned data, as this is not directly investigated in this paper.

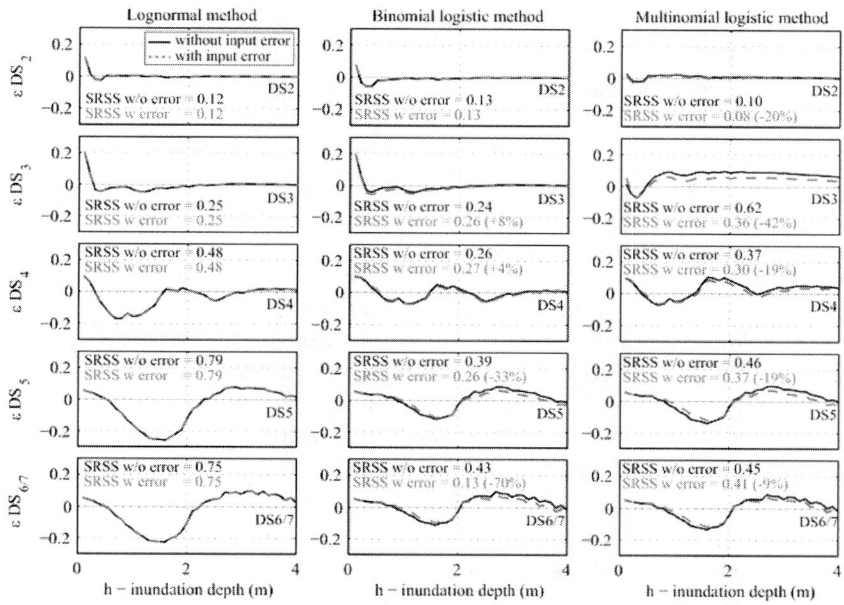

Figure 13. Residuals (observation minus prediction) based on the lognormal, binomial logistic, and multinomial logistic methods for all damage states neglecting (*black continuous line*) and considering (*red dashed line*) input data error

5. TSUNAMI LOSS ASSESSMENT

This section presents the tsunami loss assessment according to a methodology that is similar to a performance-based earthquake engineering (PBEE) framework (Cornell et al. 2002; Goda and Song 2016). The PBEE aims to quantify the extent of damage and consequences probabilistically and is useful for assessing financial and socioeconomic impact of earthquake-related hazards (e.g. Goda 2015). Therefore, the presented results can be viewed from a decision support system perspective, which are valuable to stakeholders and decision makers.

5.1 Set up

Empirical tsunami fragility curves based on the three different methods, considering and neglecting the input data error, can be used to carry out tsunami loss estimation. In this way, the effect induced by the input data error on the fragility curves can be further propagated to tsunami loss. The expected economic loss $E[L]$ for a single building can be calculated as:

$$E[L] = \sum_{j=1}^{k} R_j \cdot \left[P(DS \geq ds_j) - P(DS \geq ds_{j+1})\right] \tag{18}$$

where k is the number of damage states; R_j is the repair cost associated with the damage state j; and$P(DS \geq ds_j)$ is the exceedance probability for the damage state j. The buildings considered for the loss estimation are the same low-rise wooden structures (1 or 2 stories) that are contained in the MLIT database and used for the fragility modeling. Observed tsunami inundation depths at building sites are used as tsunami hazard parameter for the loss assessment. In particular, the inundation depth values are considered to be log-normally distributed with the central values corresponding to the observed ones and logarithmic standard deviation equal to 25 % (i.e. consistent with the results presented in Sect. 3.3).

Repair/replacement costs associated with damage states are computed on the basis of damage ratios 0, 5, 20, 40, 60, and 100 % for DS1, DS2, DS3, DS4, DS5, and DS6/7 (MLIT 2014), respectively. The assumed damage ratio and the replacement cost are multiplied to obtain the values of R. According to the MLIT (2015) and to the Japanese Construction Research Institute (CRI 2011), the mean unit construction cost is equal to 1600 $/m^2 and the coefficient of variation is 32 %. These are the values adopted for residential low-rise wooden buildings. The unit construction cost is multiplied by the footprint area and the number of stories (information available in the MLIT database) to obtain the total construction cost for each building. The total economic loss is the sum of expected loss for all the buildings in the portfolio. The total economic loss for wooden houses is computed considering uncertainties related to the inundation depth and those related to the construction cost through a Monte Carlo simulation. In particular, 1000 simulations are performed by taking into account variability of the construction costs and the tsunami inundation depths for individual buildings.

5.2 Total tsunami loss and risk disaggregation

Table 5 lists the central estimates of the total tsunami loss based on the three regression methods neglecting and considering the input data error. The lognormal method results in the greatest total expected loss (by about 4 %) when compared with the binomial and multinomial logistic methods. Whereas the logistic methods return similar values of total expected loss with a mean difference of 0.5 %. The major difference of the total expected loss is a consequence of the differences in the tsunami fragility curves at smaller inundation depths (Fig. 12).

The consideration of the input data uncertainty does not affect the expected total loss for the lognormal method, whereas there are slight increases of the expected tsunami loss by 0.68 and 0.43 % for the binomial logistic method and the multinomial logistic method, respectively. These numbers are small in relative terms, but they result in significant differences as actual values (i.e. hundreds of millions of dollars). They can be explained by the changes of the tsunami fragility curves (i.e. same median with reduced model dispersion). In fact, the greatest part of the exposure is located in the first 1500 m from the shoreline, where very large tsunami inundation depths were observed at many coastal cities and towns during the 2011 Tohoku tsunami (Fraser et al. 2013; Goda et al. 2014). Therefore, increased damage probabilities for collapse and complete damage, combined with greater damage ratios, result in greater tsunami loss estimates.

To further investigate the spatial distribution of the tsunami loss, a risk disaggregation is shown in Fig. 14 in terms of distance from the shoreline. The disaggregation is calculated by summing up the tsunami loss with an interval of 5 m. Generally, the tsunami loss profiles for the three methods are similar; the peaks are observed at 100 m from the coastline (where a large number of buildings are located and they experience high inundation depths). The results obtained with the lognormal method (Fig. 14a) indicate that the central estimate does not change depending on how the input data error is treated in the fragility modeling (as expected from Sect. 4.1); while the 90 % confidence interval becomes narrower around the central estimate when the input data uncertainty is taken into account. The reduction of the confidence interval is approximately equal to 13 %. Figure 14b shows the results obtained with the binomial and multinomial logistic methods. Tsunami loss profiles for the two cases are similar. Finally, Fig. 14c presents the ratio between expected tsunami loss profiles obtained considering and neglecting the input data error for the different regression methods. In Fig. 14c, the result for the lognormal method is constant with a ratio of 1.0. For the logistic regression approaches, the expected tsunami loss obtained considering the input data error is greater by about 1 % with respect to the case in which the input data error is ignored. This trend is observed up to about 1500 m from the shoreline, where 90 % of the total loss exposure is concentrated.

Table 5. Expected losses with and without input data error for the three regression methods

Regression method	$E[L]$ expected losses (\$)	
	Without input data error	With input data error
Lognormal method	1.534×10^{10}	1.534×10^{10}
Binomial logistic method	1.465×10^{10}	1.475×10^{10}
Multinomial logistic method	1.475×10^{10}	1.481×10^{10}

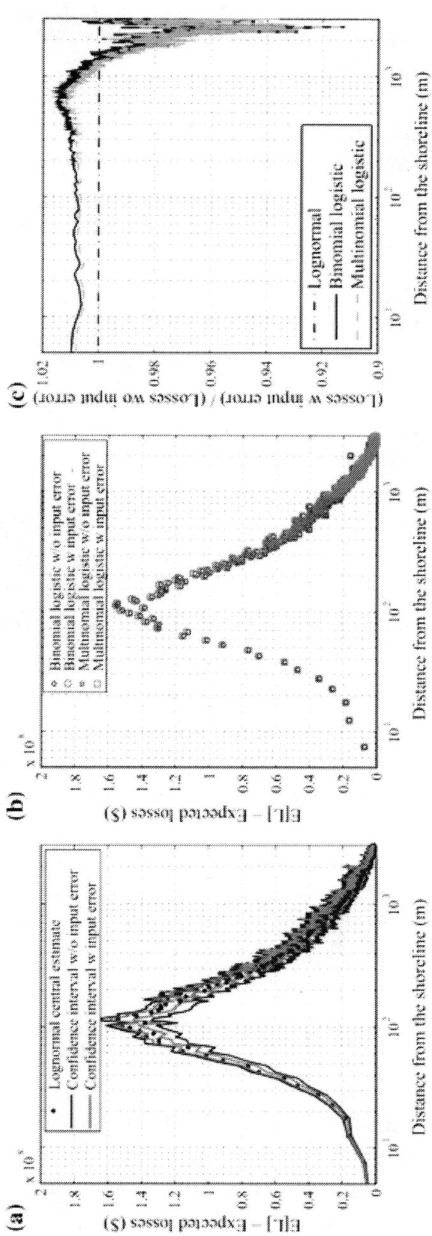

Figure 14. Expected tsunami losses as a function of distance from the shoreline based on the lognormal method (**a**) and the binomial and multinomial logistic methods (**b**), and **c** ratios of expected tsunami losses neglecting and considering input data error based on the three regression methods

6. SUMMARY AND CONCLUSIONS

This work presented a Bayesian statistical framework to consider the input data uncertainty in the empirical tsunami fragility modeling and investigated the effects of this kind of uncertainty on risk assessment for tsunamis. Three most common methods for tsunami fragility modeling were considered: the lognormal model based on linear regression, the binomial logistic regression, and the multinomial logistic regression. The developed Bayesian approaches were applied to the extensive MLIT tsunami damage data for residential wooden structures obtained from the 11th March 2011 Tohoku earthquake by neglecting and considering the input data error. To evaluate the uncertainty of tsunami inundation (i.e. input data for tsunami fragility modeling) from empirical perspectives, inundation heights available in the MLIT and TTJS databases were used. The data analysis indicated that a potential input error of 20–25 % in terms of the coefficient of variation is suitable for the Tohoku tsunami inundation data. This magnitude of the input data uncertainty was then propagated in the tsunami fragility modeling as well as tsunami loss estimation using the three regression procedures.

The systematic assessment and comparison of the tsunami fragility curves developed using different regression approaches within the Bayesian statistical framework indicated that considering the input data error leads to a reduction of the confidence interval for the lognormal method, whereas the incorporation of the input data uncertainty results in decreased model dispersion. The tsunami fragility models based on the lognormal method generally have smaller medians with greater dispersions, in comparison with those based on the binomial/ multinomial logistic method. It is important to highlight that the differences in the developed fragility curves have consequential influence on the estimated tsunami loss (e.g. reduced confidence interval of the central loss estimate for the lognormal regression and increased expected tsunami loss for the logistic regressions). It was also observed that considering the input data error through the proposed Bayesian framework leads to a reduction of the residuals between observed data and fitted fragility models. Overall, the binomial logistic method may be preferred among the three regression methods, since it achieved the smaller residuals.

It is important to underline that in developing empirical tsunami fragility curves, accounting for the input data uncertainty makes their use in damage and loss assessment more reliable. In fact, such fragilities are suitable for a more general use, since the restriction of application with respect to the areas for which they are calibrated is removed (in a sense of input data uncertainty; regional dependence of the fragility models on the characteristics of local structures still exists). Further extensions and developments of this work can be considered in three aspects. One is to consider the entire posterior distributions of regression parameters in fragility modeling, in order to obtain a more robust estimation of the fragility functions. The second is to consider binned data for the binomial logistic regression and un-binned data for the multinomial logistic regression in order to facilitate the more complete and systematic comparisons of the tsunami fragility methods, going beyond the scope that the current literature concerns. The last aspect is to consider models accounting for inundation flow velocity, fluid momentum or the other relevant fluid dynamic information as explanatory hazard variable in addition to flow depth, such as the models developed by Charvet et al. (2015) for inundation flow velocity. For the latter case, the quantification of uncertainty is a real challenge given the limited amount of observed data in terms of tsunami flow velocity.

ACKNOWLEDGMENTS

This work is funded by the Engineering and Physical Sciences Research Council (EP/M001067/1).

REFERENCES

1. Baker JW (2015) Efficient analytical fragility function fitting using dynamic structural analysis. Earthq Spectra 31(1):579–599

2. Box GEP, Tiao GC (1992) Bayesian inference in statistical analysis. Wiley, New York

3. Bradley BA (2010) Epistemic uncertainties in component fragility functions. Earthq Spectra 26(1):41–62

4. Cetin KO, Der Kiureghian A, Seed RB (2002) Probabilistic models for the initiation of seismic soil liquefaction. Struct Saf 24(1):67–82

5. Charvet I, Ioannou I, Rossetto T, Suppasri A, Imamura F (2014a) Empirical fragility assessment of buildings affected by the 2011 Great East Japan tsunami using improved statistical models. Nat Hazards 73(2):951–973

6. Charvet I, Suppasri A, Imamura F (2014b) Empirical fragility analysis of building damage caused by the 2011 Great East Japan tsunami in Ishinomaki City using ordinal regression and influence of key geographical features. Stoch Environ Res Risk Assess 8(7):1853–1867

7. Charvet I, Suppasri A, Kimura H, Sugawara D, Imamura F (2015) Fragility estimations for Kesennuma City following the 2011 Great East Japan tsunami based on maximum flow depths, velocities and debris impact, with evaluation of the ordinal model's predictive accuracy. Nat Hazards 79(3):2073–2099

8. Cheung SH, Beck JL (2010) Calculation of posterior probabilities for Bayesian model class assessment and averaging from posterior samples based on dynamic system data. Comput-Aided Civ Infrastruct Eng 25(5):304–321

9. Construction Research Institute (2011) Japan building cost information. Construction Research Institute, Tokyo, p 547

10. Cornell CA, Jalayer F, Hamburger RO, Foutch DA (2002) Probabilistic basis for 2000 SAC Federal Emergency Management Agency steel moment frame guidelines. J Struct Eng 129(4):526–533

11. Der Kiureghian A (2002) Bayesian methods for seismic fragility assessment of lifeline components. Acceptable risk processes: lifelines and natural hazards. In: Taylor C, VanMarcke E (eds) Council on disaster reduction and technical council on lifeline earthquake engineering, ASCE Monograph No. 21:61–77

12. Fraser S, Pomonis A, Raby A, Goda K, Chian SC, Macabuag J, Offord M, Saito K, Sammonds P (2013) Tsunami damage to coastal defences and buildings in the March 11th 2011 M_w9.0 Great East Japan earthquake and tsunami. Bull Earthq Eng 11(1):205–239

13. Fukutani Y, Suppasri A, Imamura F (2015) Stochastic analysis and uncertainty assessment of tsunami wave height using a random

source parameter model that targets a Tohoku-type earthquake fault. Stoch Environ Res Risk Assess 29(7):1763–1779

14. Gardoni P, Reinschmidt KF, Kumar R (2007) A probabilistic framework for Bayesian adaptive forecasting of project progress. Comput-Aided Civ Infrastruct Eng 22(3):182–196

15. Goda K (2015) Seismic risk management of insurance portfolio using catastrophe bonds. Comput-Aided Civ Infrastruct Eng 30(7):570–582

16. Goda K, Song J (2016) Uncertainty modeling and visualization for tsunami hazard and risk mapping: a case study for the 2011 Tohoku earthquake. Stoch Environ Res Risk Assess.

17. Goda K, Mai PM, Yasuda T, Mori N (2014) Sensitivity of tsunami wave profiles and inundation simulations to earthquake slip and fault geometry for the 2011 Tohoku earthquake. Earth Planets Space 66:105.

18. Guha-Sapir D, Below R, Hoyois PH (2015) EM-DAT: international disaster database. Available via www.emdat.be. Accessed 1 May 2015

19. Hosmer DW Jr, Lemeshow S, Sturdivant RX (2013) Applied logistic regression. Wiley, Hoboken, p 528

20. Jalayer F, De Risi R, Manfredi G (2015) Bayesian Cloud Analysis: efficient structural fragility assessment using linear regression. Bull Earthq Eng 13(4):1–21

21. Jaynes ET (2003) Probability theory: the logic of science. Cambridge University Press, Cambridge, p 729

22. Kim D, Kim BJ, Lee SO, Cho YS (2014) Best-fit distribution and log-normality for tsunami heights along coastal lines. Stoch Environ Res Risk Assess 28(4):881–893

23. Koshimura S, Oie T, Yanagisawa H, Imamura F (2009) Developing fragility functions for tsunami damage estimation using numerical model and post-tsunami data from Banda Aceh. Indones Coast Eng J 51(3):243–273

24. Lallemant D, Kiremidjian A, Burton H (2015) Statistical procedures for developing earthquake damage fragility curves. Earthq Eng Struct Dyn.

25. McCullagh P, Nelder JA (1989) Generalized linear models. CRC Press, Florida, p 532

26. Ministry of Land, Infrastructure, and Transportation (MLIT) (2014) survey of tsunami damage condition. Available via http://www.mlit.go.jp/toshi/toshi-hukkou-arkaibu.html. Accessed 1 July 2014

27. Ministry of Land, Infrastructure, and Transportation (MLIT) (2015) National statistics. Available via http://www.mlit.go.jp/statistics/details/index.html. Accessed 1 June 2015

28. Mori N, Takahashi T, Yasuda T, Yanagisawa H (2011) Survey of 2011 Tohoku earthquake tsunami inundation and run-up. Geophys Res Lett 38(7):L00G14.

29. Peiris N (2006) Vulnerability functions for tsunami loss estimation. In: Proceedings of the first European conference on earthquake engineering and seismology, Geneva, Paper 1121

30. Porter K, Kennedy R, Bachman R (2007) Creating fragility functions for performance-based earthquake engineering. Earthq Spectra 23(2):471–489

31. Reese S, Bradley BA, Bind J, Smart G, Power W, Sturman J (2011) Empirical building fragilities from observed damage in the 2009 South Pacific tsunami. Earth Sci Rev 107(1):156–173

32. Rossetto T, D'Ayala D, Ioannou I, Meslem A (2014) Evaluation of existing fragility curves. SYNER-G: typology definition and fragility functions for physical elements at seismic risk. Springer, Dordrecht, pp 47–93

33. Straub D, Der Kiureghian A (2008) Improved seismic fragility modeling from empirical data. Struct Saf 30(4):320–336

34. Suppasri A, Koshimura S, Imamura F (2011) Developing tsunami fragility curves based on the satellite remote sensing and the numerical modeling of the 2004 Indian Ocean tsunami in Thailand. Nat Hazards Earth Syst Sci 11(1):173–189

35. Suppasri A, Mas E, Charvet I, Gunasekera R, Imai K, Fukutani Y, Imamura F (2013) Building damage characteristics based on surveyed data and fragility curves of the 2011 Great East Japan tsunami. Nat Hazards 66(2):319–341

36. Tarbotton C, Dall'Osso f, Dominey-Howes D, Goff J (2015) The use of empirical vulnerability functions to assess the response of buildings to tsunami impact: comparative review and summary of best practice. Earth Sci Rev 142:120–134

37. Yazdi AJ, Haukaas T, Yang T, Gardoni P (2015) Multivariate fragility models for earthquake engineering. Earthq Spectra.

38. Yu JJ, Qin XS, Larsen O (2013) Joint Monte Carlo and possibilistic simulation for flood damage assessment. Stoch Environ Res Risk Assess 27(3):725–735

7

MODELING AND VISUALIZING UNCERTAINTIES OF FLOOD BOUNDARY DELINEATION: ALGORITHM FOR SLOPE AND DEM RESOLUTION DEPENDENCIES OF 1D HYDRAULIC MODELS

Sven Anders Brandt

Department of Industrial Development, IT and Land Management, University of Ga¨vle, 801 76 Ga¨vle, Sweden

ABSTRACT

As flood inundation risk maps have become a central piece of information for both urban and risk management planning, also a need to assess the accuracies and uncertainties of these maps has emerged. Most maps show the inundation boundaries as crisp lines on visually appealing maps, whereby many planners and decision makers, among others, automatically believe the boundaries are both accurate and reliable. However, as this study shows, probably all such maps, even those that are based on high-resolution digital elevation models (DEMs), have immanent uncertainties which can be directly related to both DEM resolution and the steepness of terrain slopes perpendicular to the river flow direction. Based on a number of degenerated DEMs, covering areas along the Eskilstuna River, Sweden, these uncertainties have been quantified into an empirically-derived disparity distance equation, yielding values of distance between true and modeled inundation boundary location. Using the inundation polygon, the DEM, a value representing the DEM resolution, and the desired level of confidence as inputs in a new-developed algorithm that utilizes the disparity distance equation, the slope and DEM dependent uncertainties can be directly visualized on a map. The implications of this strategy should benefit planning and help reduce high costs of floods where infrastructure, etc., have been placed in flood-prone areas without enough consideration of map uncertainties.

KEYWORDS

1D hydraulic modeling, River flood inundation, Uncertainty, Quantile regression, Geographical information systems (GIS), Digital elevation model (DEM)

1. INTRODUCTION

1.1 Background

Hydraulic modeling of river floods has received a significant boost during the last 10 years; not only thanks to improved computers and hydraulic modeling software, but also to the capabilities and user-friendliness of geographical information systems (GIS). During the same period, new legislation, such as EU's flood directive, demands that flood risks are incorporated into risk and management plans, and together, this has led to production of numerous flood risk maps. Although these maps may have been produced by professionals who are aware of the different inaccuracies and uncertainties underlying the maps, they are often used by people who have little or no experience of neither hydraulic nor digital elevation modeling. Furthermore, as these maps tend to form the basis for many decisions in spatial and physical planning of the built environment, there is a need for tools that can communicate the intrinsic uncertainties always present in the maps.

There are different types of uncertainties involved in flood risk mapping (see e.g. Pappenberger et al. 2008and Merwade et al. 2008, for general treatise on this subject). The most immediate is which model to be used (e.g. Wagener and Gupta 2005), but in practice, the most commonly treated uncertainty is which magnitude of flow to use for a certain flood return period. This can be handled by running the model with different water discharges and thereby get a range of flood inundation areas. Other ways of treating uncertainty is through Monte Carlo simulations, i.e., feeding the model with slightly varied input of all input parameters, and where the large number of output maps in turn can be related to flood prediction uncertainty on how accurate the modeled results are (e.g. Apel et al. 2008). Obviously, specific objects and parameters in the hydraulic model can also influence the accuracy of the produced results. For example Koivumäki et al. (2010) studied the effects of buildings in the model, Pappenberger et al. (2006) looked at the effects of boundary conditions and bridges, and Cook and Merwade (2009) and Castellarin et al. (2009) treated the effects of cross-section location and spacing. As the hydraulic modeling usually involves calibration against a previous flood event, the importance of roughness is also widely known. By varying river bed and floodplain roughness values in a range that theoretically can be expected, minimum and maximum extents of the flooded area can be modeled. Although modelers have acknowledged the implications of roughness for a long time, it is not until recent years that any efforts have been made to see how much this type of uncertainty affects the results (see e.g. Pappenberger et al. 2005; Werner et al. 2005; Casas et al. 2006; Schumann et al. 2007; Wilson and Atkinson 2007; Brandt 2009; Warmink et al. 2013; Wu in press). This is probably due to the type of uncertainty which earlier has been considered the main constraint for successful modeling, viz. the quality of the digital elevation model (DEM).

1.2 Previous research on delineation uncertainties related to DEMs

Before the advent of LiDAR, the results from hydraulic models, which could be based on detailed surveyed cross sections, were overlain on DEMs of poor resolution. In Sweden, e.g., up to only a couple of years ago, the only elevation database of national coverage has been Lantmäteriet's (the Swedish mapping, cadastral and land registration authority) with 50 m cell resolution (other countries have had similar resolutions). Very rarely, there have been DEMs of higher quality available. Due to the poor quality of the elevation models, in Sweden all such maps were given a notification that they should not be used for detailed planning. Hence, there have been some studies with the specific objective to study how the quality of DEMs affects the accuracy of inundation boundary delineation from 1D hydraulic models, which end products are water levels at each modeled cross section. By comparing these modeled levels with measured levels, several studies have shown that the accuracy of predicting correct levels is surprisingly high, irrespectively of the quality of DEM (e.g. Casas et al. 2006; Yacoub and Sanner 2006; Brandt 2009). Only with poor DEMs (i.e., cell sizes bigger than 10–25 m) together with steep river slopes, or abrupt slope change, the water levels may deviate significantly between modeled and real conditions (Brandt 2009). However, when it comes to the spatial extent, which is important when the inundation extents are transferred to maps, high-resolution DEMs of high quality may also produce inaccurate results.

An early attempt to look at spatial deviations was done by Zhang and Montgomery (1994) on two areas in the USA. They gridded spot elevation data to DEMs of 2, 4, 10, 30, and 90 m resolution. They noticed that better resolution than 10 m lead to improved modeling results. However, the best two DEMs did not produce any significant improvements; most probably due to the catchments being characterized by moderately to steep terrain gradients. Later, Werner (2001) used laser altimetry data for a reach of the river Saar in Germany. The original cell resolution was 2.5 m, which then was aggregated by averaging neighboring cell values to cell sizes of 5, 10, and 25 m. He concluded that a cell resolution of 10 m indicated the break when flood extents started to deviate significantly.

When the modeled areas are big, high-resolution DEMs usually contain enormous amounts of data. Therefore, it is of interest to see how much the original laser data can be filtered, without losing predictability performance. For an area around Leith Creek, North Carolina, Omer et al. (2003) looked at the angle α between two surveyed data points (Fig. 1). If a pre-determined angle is exceeded, the point will be preserved, but if it is not exceeded it will be removed from the dataset. In this way the number of points will be reduced, leading to less computer storage, faster analysis times, but also a DEM of poorer quality. The original dataset had ca 0.0288 points/m^2, equivalent to 5.89 m cell size. By testing different threshold values of α, their recommendation is that α should be less than 4°, which in their case represented about 38 % of the original number of points, i.e., ca 0.0111 points/m^2, equivalent to 9.50 m cell sizes.

Another study was undertaken by Casas et al. (2006). They looked at an area next to the Ter River, Spain, and tested different DEMs ranging from 1 to 4 m in cell size. The DEMs were derived from laser altimetry data, GPS surveyed data, 5 m contour data (scale 1:5000), as well as combinations between them, together with or without bathymetric data.

They concluded that for a 500 m^3/s discharge, the 4 m resolution DEM yielded inundated area differences up to 7.3 %. However, if higher discharges were used (3000 m^3/s), the differences were reduced to 2.6 %. Therefore they argued that coarser resolution will have less consequence in floodplain areas.

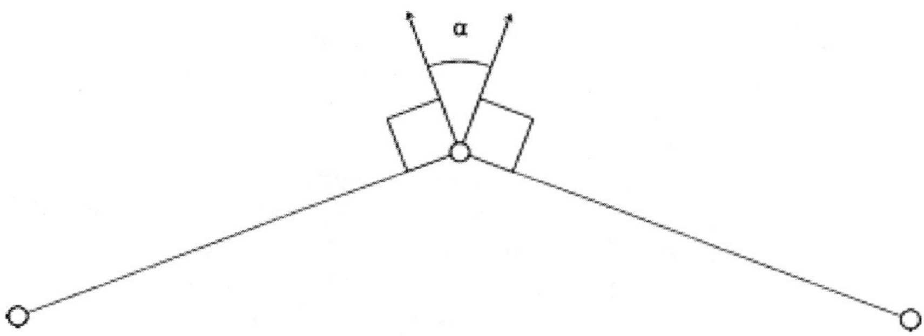

Figure 1. Angle used to determine if point should be filtered away or kept in the DEM (cf. Omer et al. 2003)

Raber et al. (2007) looked at Reedy Fork Creek, North Carolina, and started with laser altimetry data with a mean point distance of 1.35 m, which later were filtered in several steps down to 9.64 m. By comparing statistics over the modeled inundated areas, they concluded that it is enough with 4 m mean point spacing. For better DEMs they did not see any significant differences between the model results.

Cook and Merwade (2009) studied the Brazos River, Texas, and Strouds Creek, North Carolina, for different resolutions (laser altimetry data of 3 m for Brazos River and 6 m for Strouds River, as well as 10 and 30 m USGS data for both rivers) combined with different qualities of cross-section resolutions. Although their research focus was on inundated area differences, they did notice that for the smaller Strouds River (with a width of 9.5 m during normal conditions) the average width of a modeled flood where 25 % wider when poor DEMs were used. Similarly, the larger Brazos River's (with a width of 175 m during normal conditions) average width was 5 % wider. This effect was doubled when laser altimetry data were integrated in the cross-section profiles.

1.3 Aim and objectives

Nowadays flood risk maps are usually based on DEMs with quite high quality. In Sweden, a new national elevation dataset of 2 m resolution is under production, and thanks to the detailed appearance of the maps, many users as well as hydraulic modelers tend to put high confidence in them and consider the results to be very accurate, i.e., with a flood-boundary position accuracy of just one or two raster cells. However, there are a few studies available that have shown that these maps may also suffer severely from DEM-derived uncertainties, but despite the recognition of the problem, it seems that practically no attempts have been made to actually visualize the uncertainties of these maps (cf. Lim et al. 2016). Considering the fact that there still are accuracy and uncertainty issues due to the quality of the DEMs, together with the absence of effective visualization techniques to represent these issues, the general aim of this paper is to provide insights into the importance of DEMs influence on 1D hydraulic modeling. The specific objectives are to produce: (1) a general equation capable of describing the uncertainties related to the DEM resolution and the floodplain characteristics, here represented by the slope perpendicular to the flow direction, and (2) an algorithm capable of illustrating the uncertainties of flood boundary mapping, related to the quality of the DEMs.

2. PRIOR STUDIES OF ESKILSTUNA AND TESTEBO RIVERS

All previously mentioned studies focus on the DEMs global resolution, despite the relatively obvious influence of the local terrain; especially the slope characteristics perpendicular to the flow direction. Where cross sections have steep slopes, the inundation delineation is more certain than for cross sections with gentle slopes. Hence, river side areas with gentle perpendicular slopes call for elevation data of higher resolution to reduce the uncertainty of inundation extent delineation (Brandt 2005). This is also supported by Colby and Dobson (2010) who in their study on rivers in North Carolina concluded that the "extent and internal pattern of flooding in the low-relief coastal plains was found to be especially sensitive to the representation of terrain". Therefore, a first attempt to look into this problem in detail was carried out for two areas of Eskilstuna River, Sweden (reported in Brandt 2009): one with relatively steep side slopes and one with relatively flat side slopes.

These areas were modeled with the 1D hydraulic software HEC-RAS (Hydrologic Engineering Center 2008) for a steady state flow of 198 m^3/s and tested for inundation boundary delineation with several DEMs of varying cell resolution (besides this, that study also looked at the effect of systematic vertical errors of the surrounding terrain as well as the relative importance of errors between roughness and DEM resolution).

The northern investigated area in Eskilstuna River is 2241 m long, consists of two parts with a water power station in between, and has relatively steep side slopes. The original point cloud has a point spacing of 1.36 points/m^2. The southern area, on the other hand, is 1731 m long, has relatively gentle side slopes, and has a point spacing of 1.64 points/m^2. This is equivalent to raster DEMs of 0.86 and 0.78 m cell sizes, respectively. These datasets were then degraded step-by-step through both removal of points and by introducing random errors in order to fully simulate higher flight heights in accordance with the scanner equipment specifications (cf. Klang and Klang 2009, for full details).

When comparing the modeled river widths (based on degraded DEMs) with the width considered to be the truth (based on original point cloud with no introduced errors), the mean disparity for all cross sections was around 0 m (less than ±2 m irrespectively of point density or cell size), except for the poorest DEM of 50 m cell size, where the mean disparity was ca 16 m. The seemingly good results were due to cancelling-out effects of the existence of both negative and positive width differences. Therefore, the mean of the absolute disparity values were also calculated. Then it was clear that the cell resolution negatively affects the results; not only will poorer resolution give poorer spatial accuracy (expressed as width disparities), it is also clear that areas with gentler side slopes have higher width disparities than do areas with steeper side slopes (Fig. 2). However, even if there was a general correlation between slope and width disparity, the width disparity at each cross section varied quite randomly. Some cross sections showed no difference between the different DEMs, while other cross sections for the poorest resolutions differed more than 130 m.

Figure 3a shows the differences between the reference model and two of the degraded models of 3.83 and 50 m cell size, as well as the inundated area when the already existing 50 m DEM (Lantmäteriet's) is used. Therefore, in a later study, to get a more robust analysis, about 400 transects were laid out across the river and checked for disparities between the reference model and the degraded models (cf. Brandt and

Lim2012). The disparity distances were then plotted against the river side slope (Fig. 3b). From this diagram it can be seen that modeled inundation boundaries can be significantly different from the "true" ones, especially at gentle river side slopes.

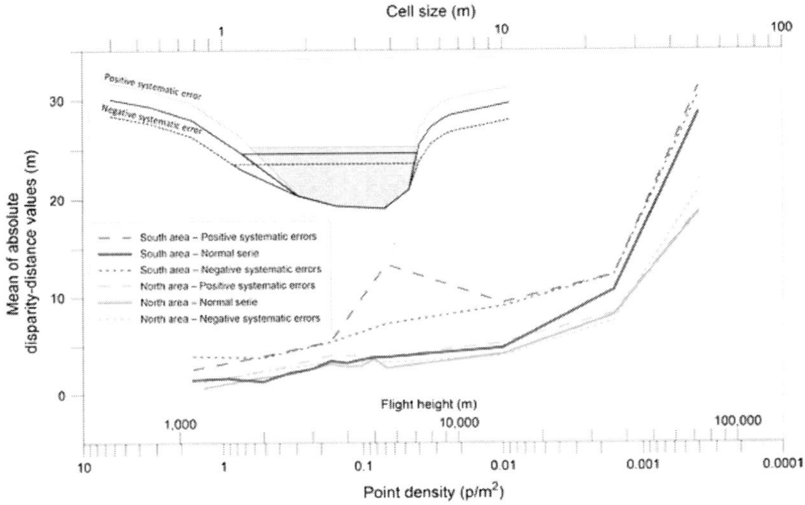

Figure 2. River width disparities between modeled floods using reference DEM and degraded DEMs when the mean of absolute width differences have been used (based on Brandt 2009). *Inset* shows the effect of systematic errors in the DEM of surrounding terrain on water level and inundation width

A follow up on the Eskilstuna study was done on the Testebo River, Sweden (Brandt and Lim 2012; also cf. Lim 2009, 2011). Instead of a reference model of high-resolution laser altimetry data and degenerated DEMs, this area used reference records of inundation extents from the 1977-year flood, together with laser altimetry data and the old 50 m DEM from Lantmäteriet as test models. Furthermore, two separate and independent modeling attempts using the same laser altimetry data were done, one by Brandt and one by Lim, to also see how much the modeler will impact the analysis. As can be seen in Fig. 4a there are places with relatively big disparities between modeled and actual flood extent boundaries. These are mainly located in flat terrain. Also apparent is that the disparities between the modelers are smaller than the disparities between any of the modelers and the reference flood map (Fig. 4b).

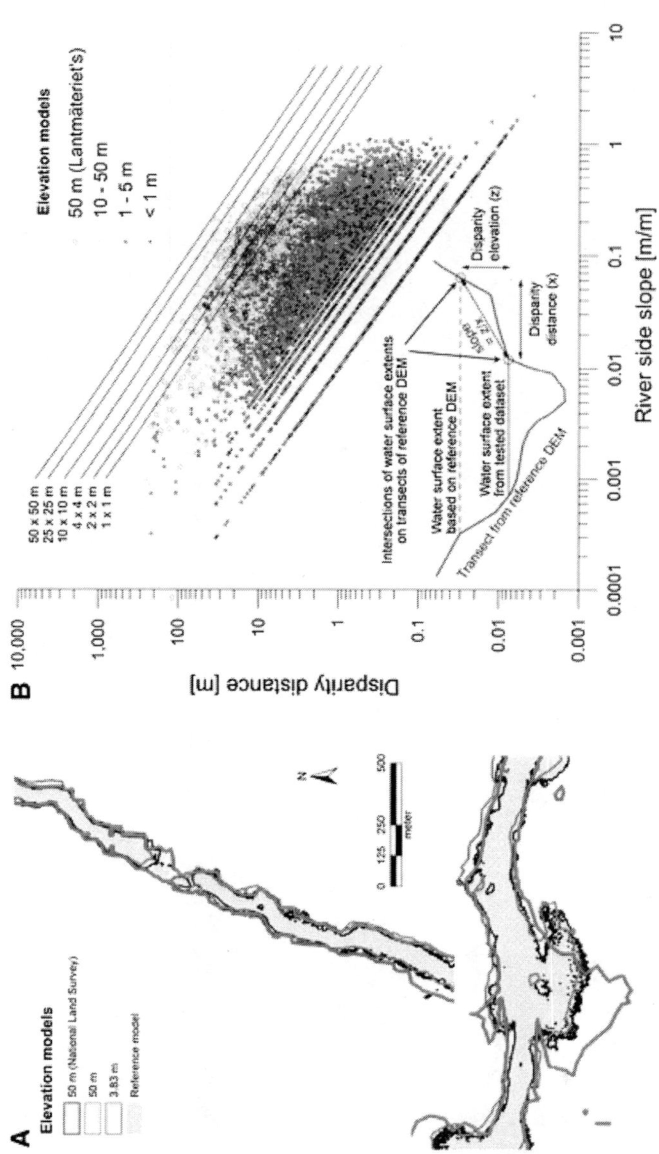

Figure 3. a Disparities between modeled floods using reference DEM and degraded DEMs with 3.83 and 50 m cell size on two areas along the Eskilstuna River, Sweden. The *red line* represents the inundated area based on Lantmäteriet's (denoted National Land Survey) old 50 m DEM (from Brandt 2009). **b** Disparity distances between results based on the reference model and the degenerated DEMs and Lantmäteriet's old 50 m DEM. Also included are visually estimated envelope curves for each DEM resolution range (from Brandt and Lim 2012), as well as an inlay on how disparity distances and the river side slopes were calculated

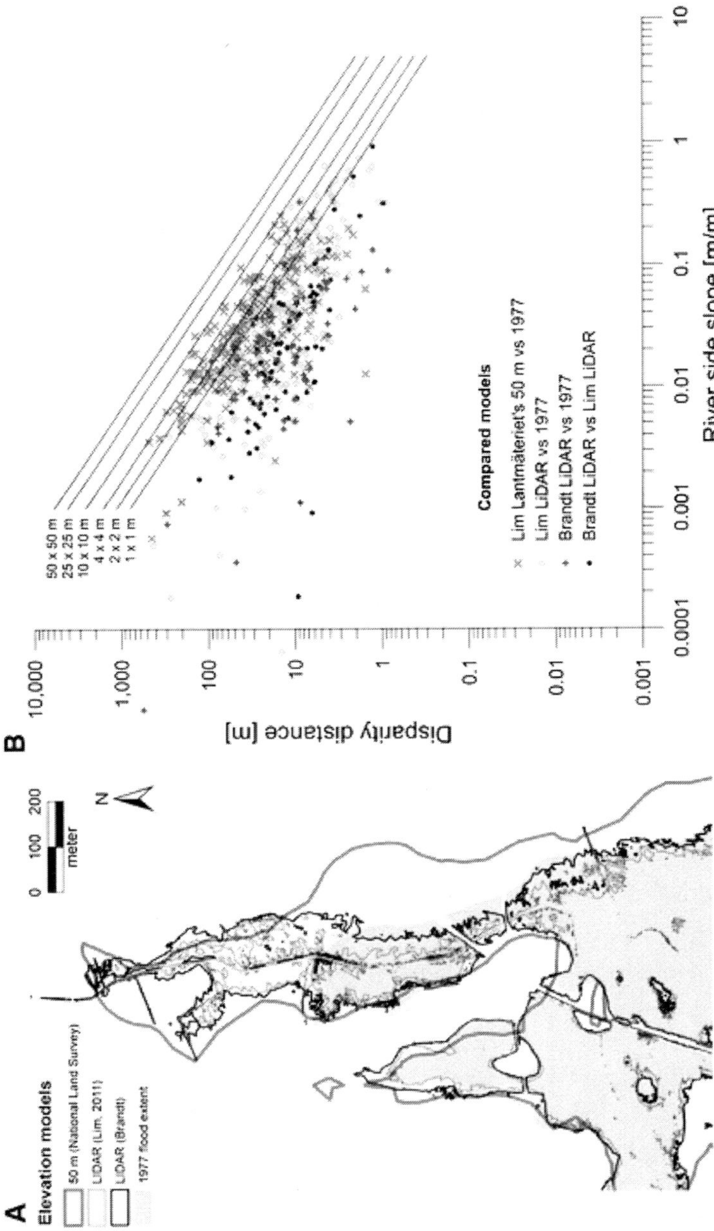

Figure 4. a Modeled floods using laser altimetry data (one by Brandt and one by Lim) and using the 50 m DEM from Lantmäteriet (denoted National Land Survey). The flood in 1977 serves as the reference inundation area. **b**Disparity distances between the different models. From Brandt and Lim (2012)

3. REPRESENTING DEM-DERIVED UNCERTAINTIES OF FLOOD MAPS

This work has focused on two aspects of accuracies and uncertainties related to the elevation models that are used for inundation mapping. One is related to the cell resolution of the DEM and the other is related to the river channel and floodplain slope perpendicular to the flow direction.

3.1 DEMs used for creating an uncertainty equation

To be able to test the influence on resolution, the already existing DEMs from the previous projects were used. In that process, the strategy was to simulate the behavior of the Leica ALS50 LiDAR instrument at different flight heights (cf. Klang and Klang 2009 for full details of this process), and hence these DEMs have the advantage of being more similar to real recordings at different flight heights, than if simple aggregation of cells or re-interpolation of original LiDAR points were to be performed. The original DEMs, with an initial point spacing of 1.64 and 1.36 points/m^2, corresponding to cell sizes of 0.78 and 0.86 m, respectively, had been degraded and downsampled to a number of different DEMs (Table 1) of gradually increased point spacing or cell size. The DEMs were the same as for the two areas shown in Fig. 3a (Brandt2009).

3.2 Estimating the uncertainties due to DEM resolution and floodplain slopes

The previous studies show that it should be possible to create an equation for disparity distances (D_d) according to:

$$D_d = f(S, \delta, P) \tag{1}$$

where S is terrain slope perpendicular to the flow direction, δ is the cell size of the DEM and P is the percentile of interest (confidence) for estimating the uncertainties.

Table 1. Characteristics of the DEMs used in this study (cf. Klang and Klang 2009 and Brandt 2009)

Simulated flight height (m)	No. of points/m^2	Point spacing [Cell size] (m)	Introduced random errors of size 1 σ	
			Planar (m)	Vertical (m)
1122 (Reference for southern area)	1.64	0.78		
1232 (Reference for northern area)	1.36	0.86		
1122 (Southern area)	1.64	0.78	0.17	0.11
1232 (Northern area)	1.36	0.86	0.18	0.12
1500	0.92	1.04	0.20	0.12
2000	0.52	1.39	0.25	0.13
2500	0.33	1.74	0.31	0.14
3000	0.23	2.09	0.36	0.16
3500	0.17	2.43	0.41	0.17
4000	0.13	2.77	0.47	0.19
4500	0.102	3.13	0.52	0.20
5000	0.083	3.47	0.57	0.22
5500	0.068	3.83	0.63	0.24
14,391	0.010	10.0	0.63	0.24
35,979	0.0016	25.0	0.63	0.24
71,960	0.0004	50.0	0.63	0.24

The first step to create a general equation that describes disparity distances involved plotting the disparities against the perpendicular slopes. Using the same transects described in Brandt and Lim (2012), disparity was measured as the absolute value of the distance between the modeled and the "true" flood boundary (based on the reference DEM). The slope was then calculated from the same coordinate pairs of the reference DEM (also cf. Fig. 3b). Next, for each resolution the slopes were divided into seven classes where different percentiles were calculated for each slope class (Table 2), and a regression model was developed linking the percentiles to terrain slope and DEM resolution, i.e., a quantile regression was performed (cf. e.g. Koenker and Hallock 2001). As there were very few observations in the two smallest slope classes, these were excluded from the regression analyses. The total number of observations for each DEM differs; partly because there were a few observations with slopes steeper than 1.78 m/m, i.e., steeper than the slope classes used for the regression analysis (in analogy with the smallest slope classes), and partly because of removal of observations where the slope was exactly 0 m/m (the reasons being: it is not possible to define the logarithm of 0, and these observations would belong to a slope class even smaller than those that were not included in the regression analysis). Then, when the logarithm of the percentile disparities for each slope class were plotted against the logarithm of the slope classes it is clear that they plot as straight lines with high coefficients of determination (R^2) (Fig. 5).

The general equation describing the regression lines, seen in Fig. 5, can be expressed as:

$$D_d = cS^z \quad (2)$$

However, from Fig. 5 it can also be seen that the regression lines vary in both position and degree of slope, i.e., the lines' c coefficients and z exponents vary depending on the resolution and the desired confidence. If the c coefficients are plotted against the resolution it can be seen that c is increasing when resolution gets poorer and confidence percentile gets higher (Fig. 6a).

Although in the figure it can be seen that the matching between the data points and regression line is weaker for the coarse resolution DEMs, relatively strong correlation coefficients are achieved. The lines in Fig. 6a can be represented with the following equation:

$$c = a\delta^b \quad (3)$$

Table 2. Disparity distances for the 1.04, 3.83, and 50 m resolution DEMs

Slope class center (m)	log (Slope class center)	Class boundaries	No. of obs.	Disparity distance (m) for percentile								
				10	20	30	40	50	60	70	80	90
[δ = 1.04 m]												
0.0006	−3.25	(−3.5) to (−3.0)	6	19.53	25.86	28.53	31.48	32.50	33.55	59.86	106.79	107.24
0.0018	−2.75	(−3.0) to (−2.5)	9	6.93	7.67	8.40	8.85	8.89	9.02	12.60	15.97	20.93
0.0056	−2.25	(−2.5) to (−2.0)	39	2.07	2.47	3.55	4.79	5.32	6.17	6.72	9.48	11.65
0.0178	−1.75	(−2.0) to (−1.5)	113	0.78	1.21	1.58	1.99	2.40	2.80	3.59	5.02	6.92
0.0562	−1.25	(−1.5) to (−1.0)	157	0.29	0.42	0.56	0.74	0.99	1.20	1.54	2.15	2.90
0.1778	−0.75	(−1.0) to (−0.5)	261	0.09	0.15	0.22	0.29	0.36	0.46	0.58	0.80	1.14
0.5623	−0.25	(−0.5) to (0)	116	0.05	0.07	0.12	0.15	0.21	0.26	0.31	0.38	0.54
[δ = 3.83 m]												
0.0006	−3.25	(−3.5) to (−3.0)	1	12.67	12.67	12.67	12.67	12.67	12.67	12.67	12.67	12.67
0.0018	−2.75	(−3.0) to (−2.5)	15	5.48	7.73	9.75	17.74	22.77	23.30	24.20	31.65	35.53
0.0056	−2.25	(−2.5) to (−2.0)	50	2.10	2.86	4.33	5.40	8.88	11.93	14.38	19.74	24.99
0.0178	−1.75	(−2.0) to (−1.5)	136	1.09	2.02	3.09	4.17	4.69	5.47	7.05	8.38	11.39
0.0562	−1.25	(−1.5) to (−1.0)	205	0.61	0.97	1.36	1.69	2.20	2.76	3.31	4.58	5.99
0.1778	−0.75	(−1.0) to (−0.5)	234	0.22	0.40	0.64	0.88	1.11	1.35	1.68	2.12	2.77
0.5623	−0.25	(−0.5) to (0)	123	0.11	0.25	0.42	0.55	0.69	0.86	1.02	1.21	1.64
[δ = 50 m]												
0.0006	−3.25	(−3.5) to (−3.0)	1	253.04	253.04	253.04	253.04	253.04	253.04	253.04	253.04	253.04
0.0018	−2.75	(−3.0) to (−2.5)	2	9.58	10.31	11.10	11.94	12.85	13.83	14.88	16.01	17.23
0.0056	−2.25	(−2.5) to (−2.0)	18	14.95	18.42	21.22	22.44	27.93	31.28	33.25	36.74	53.32
0.0178	−1.75	(−2.0) to (−1.5)	96	5.54	7.03	8.56	10.63	15.14	18.59	25.37	33.62	40.17
0.0562	−1.25	(−1.5) to (−1.0)	265	4.00	6.67	9.81	12.58	14.82	18.83	22.51	27.10	32.49
0.1778	−0.75	(−1.0) to (−0.5)	230	2.10	4.80	6.40	7.71	9.62	12.08	14.34	17.36	26.42
0.5623	−0.25	(−0.5) to (0)	30	0.91	2.16	2.57	2.92	3.17	4.90	5.60	6.85	7.89

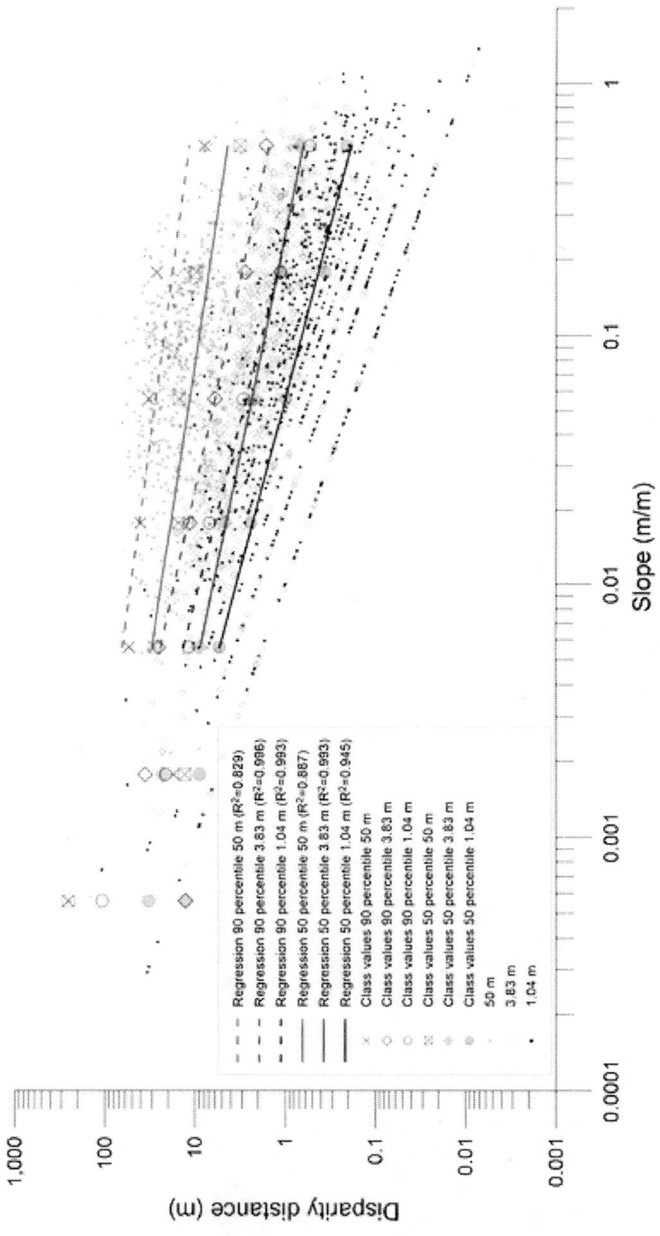

Figure 5. Disparity distances between modeled and true flood boundaries plotted against the perpendicular slopes for 1.4, 3.83, and 50 m DEMs, respectively (*Note* The peculiar linear pattern arises due to the limiting factor of using elevation values with cm precision for slope calculation). Also *plotted* are *percentile values* for each slope class and their corresponding regression lines (for classes 0.0056–0.5623 m/m)

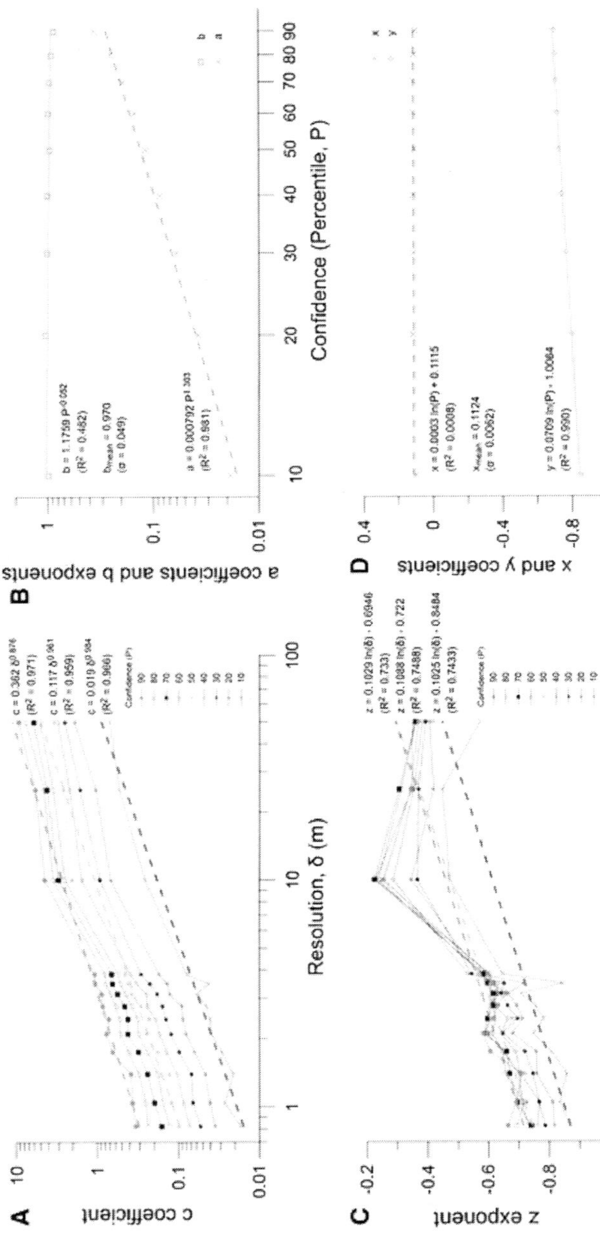

Figure 6. a Variation of the c coefficient depending on resolution (or DEM cell size) and confidence (percentile, P). Regression lines are included for 10, 50, and 90 percentiles, respectively. **b** Variations of a coefficient and b exponent depending on confidence. **c** Variation of the b exponent depending on resolution and confidence. **d**Variation of the x and y coefficients depending on desired confidence

When plotting the a coefficients and b exponents against the confidence percentiles, it appears that a increases exponentially with increasing confidence percentile while b has a relatively weak correlation, where the mean value is 0.970 with a standard deviation of 0.049 (Fig. 6b). Hence the b exponent may be represented by a constant. The c coefficient can therefore be described with the following equation:

$$c = 0.000792P^{1.303}\delta^{0.970} \tag{4}$$

Also the z exponent increases when resolution gets poorer (Fig. 6c), although the correlation coefficients are not as high as for the c coefficient, according to the following general equation:

$$z = x\,ln(\delta) + y \tag{5}$$

However, when plotting the x and y coefficients against the confidence percentiles, it appears that x is constant with a mean value of 0.1124 and a standard deviation of 0.0062, and that y increases with increasing confidence percentile (Fig. 6d). The z exponent can therefore be described with the following equation:

$$z = 0.1124\,ln(\delta) + 0.0709\,ln(P) - 1.0064 \tag{6}$$

The final equation then becomes:

$$D_d = \left[\delta^{0.970}\,0.000792P^{1.303}\right]S^{[0.1124ln(\delta)+0.0709ln(P)-1.0064]} \tag{7}$$

As a check the D_d equation has been plotted against the empirical data (Fig. 7). Also included is a disparity distance example from Tärnsjö (Fig. 7c) (location taken from the study by Vähäkari 2006), which serves as the most extreme disparity known to the author.

4. CONSTRUCTING THE ALGORITHM

4.1 Algorithm for visualizing flood boundary delineation uncertainties

Based on the results from Table 2 and Figs. 5, 6, it is possible to construct an algorithm for direct visualization of the uncertainties related to the quality of the DEM and the perpendicular side slopes. This example will use the modeled inundation polygon, the already existing cross sections in the hydraulic model, and a DEM as input. In short the following procedure is followed (Fig. 8): (1) First the cross sections are prepared for the analysis and different input variables are determined. (2) Then each cross section is checked where (and whether) it is intersected by the modeled flood boundary. From the intersection, an iterative process, going one cross section node at a time, will look both toward the center of the river (determining the inner uncertainty boundary) and in the opposite direction away from the river (determining the outer uncertainty boundary) until the disparity distance from the D_d equation is exceeded. As the areal extent of inundation is derived using a regular grid of raster cells, with individual elevation values, there is a need to translate the disparity distance locations of the cross sections to elevation values. At the coordinates of disparity distance exceedance, elevation values are sampled or calculated based on the $D_d$equation, one lower elevation defining the inner boundary of uncertainty and one higher elevation defining the outer boundary of uncertainty (by using the D_d equation, the algorithm can consider abrupt ends due to e.g. buildings giving a so called "wall effect"). (3) Next the nodes in all cross sections are populated with the uncertainty elevation values and two TINs are created; one with values representing the inner uncertainty boundary and the other the outer uncertainty boundary. These TINs are then rasterized (this point-to-TIN-to-raster conversion will ensure that water levels will always fall in the downstream direction whereas direct raster interpolation from points may produce elevation artifacts). (4) Finally the DEM can be compared with the uncertainty rasters yielding three distinctive areas: not flooded areas (at least according to the percentile used), areas uncertain to be flooded, and flooded areas (at least according to the percentile used).

4.2 Resulting pseudo code

The following variables are used:

Boolean: *flag*

Integers: m_{max} (number of cross sections); *m* (cross section number), *n* (cell or point number in a cross section)

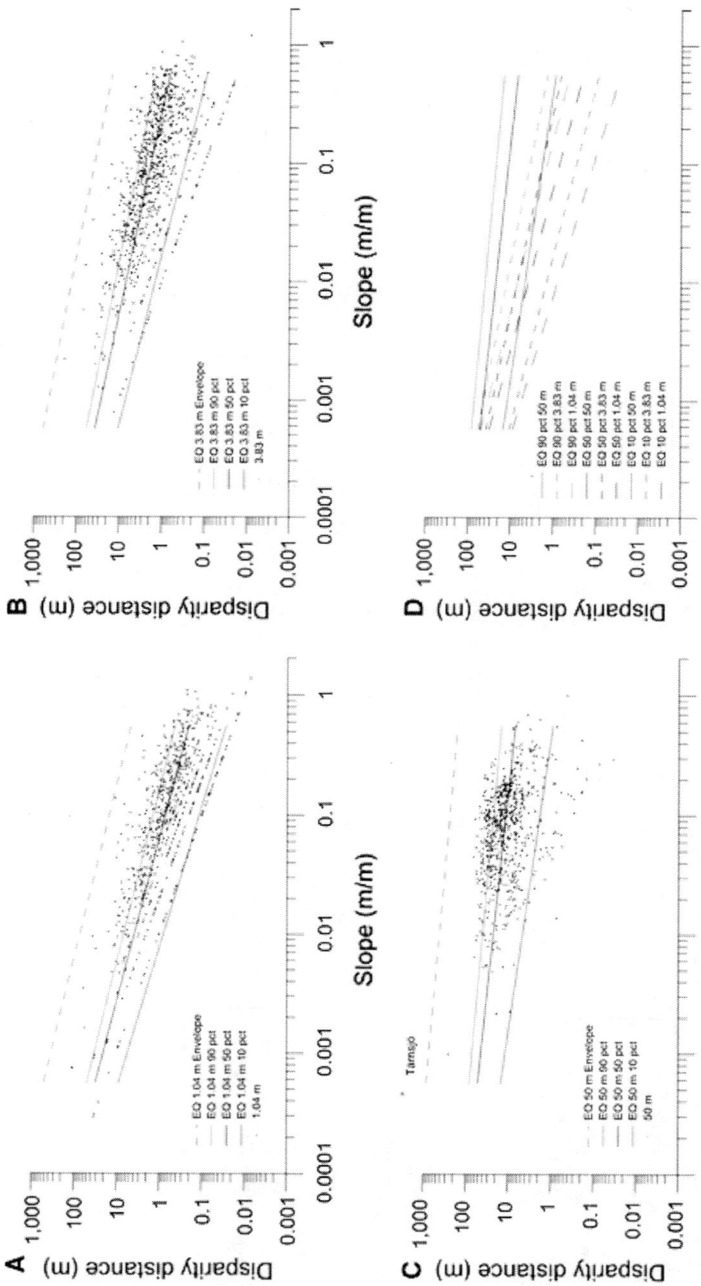

Figure 7. The disparity distance D_d equation for three confidence percentiles compared with the observed disparities of **a** 1.04 m, **b** 3.83 m, and **c** 50 m DEM resolutions, as well as **d** comparison between the D_d equations

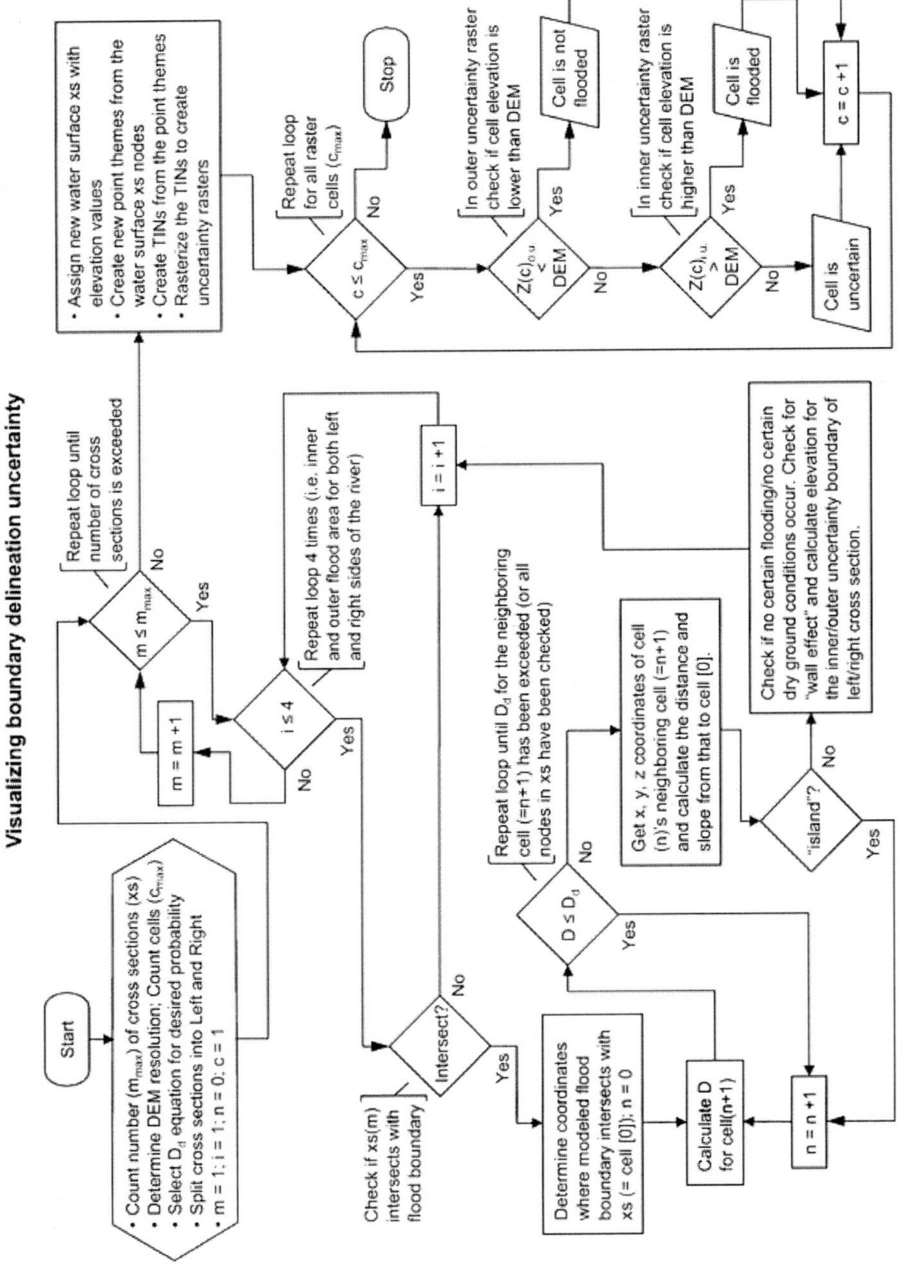

Figure 8. Schema of the uncertainty algorithm

Float: DI_{flag}, DO_{flag}, (previous valid distances of inner and outer uncertain flood area, respectively); *ws*(water surface elevation); *c, z* (coefficient and exponent values taken from the D_d equation); δ (cell resolution)

Float arrays: *x, y, h* (coordinates and elevations of cross section nodes/cells); *DI, DO, SI, SO* (inner and outer uncertainty distances and slopes, respectively); *RXSI, RXSO, LXSI, LXSO* (inner and outer uncertainty elevations for right and left cross sections, respectively

1. Count the number of cross sections (m_{max}). Determine the cell resolution (δ) of the DEM and use the disparity equation (D_d) for the desired probability.

2. Split all cross sections at river channel center. Define cross sections as *LEFT* or *RIGHT*.

3. Let $m = 1$.

4. Repeat until number of cross sections has been exceeded, i.e., $m > m_{max}$.

4.1. If RIGHT cross section (cross section should be seen looking in the downstream direction, where cells are ordered from left (negative cell positions in the river channel) to right (positive cell positions on the ground).

4.1.1. Determine the coordinates, i.e., *x*(0) and *y*(0), and elevation, *h*(0), of the DEM cell, where the modeled flood boundary intersects the cross section, i.e., the raster cell (or point) Cell (0). If no intersection exists, exclude that cross section from the analysis.

4.1.2. Let *n* = 0, *flag* = false, DI(0) = 0.

4.1.3. Repeat until the distance of the inner flood area, *DI*, for Cell (n – 1), is longer than the D_d for the same slope of that between the Cell (n – 1) and Cell (0), or that all cells in the cross section has been treated:

4.1.3.1. Get the coordinates and elevation of Cell (n)'s neighboring cell, i.e. *x*(*n* – 1), *y*(*n* – 1), and*h*(*n* – 1) at Cell (n – 1), and calculate the distance, *DI*(*n* – 1), and slope, *SI*(*n* – 1), from Cell (0) to Cell (n – 1).

4.1.3.2. If *SI*(*n* – 1) > 0 (i.e. the terrain is higher at Cell (n – 1), creating an island) then *n* = *n* – 1, else *flag* = true, DI_{flag} = *DI*(*n*), and compare

$DI(n-1)$ with the D_d for desired probability for the absolute value of slope of $SI(n-1)$. If D_d is exceeded then stop, else $n = n - 1$.

4.1.4. If *flag* = false then $DI_{flag} = \delta/2$ (This accounts for cross sections where no cells in the inner part of the cross section have negative slopes, i.e. no certain flooding seems to occur).

4.1.5. If $ws(m) - \exp[(\ln DI_{flag} - \ln c)/z]*DI_{flag} > h(n-1)$ then inner uncertainty boundary elevation$RXSI(m) = ws(m) - \exp[(\ln DI_{flag} - \ln c)/z]*DI_{flag}$ else $RXSI(m) = h(n-1)$ (This accounts for the wall effect; see Sect. 4.3).

4.1.6. Let $n = 0$, *flag* = false, DO(0) = 0.

4.1.7. Repeat until the distance of the outer flood area, $DO(n+1)$, for Cell (n + 1), is longer than theD_d for the same slope of that between the Cell (n + 1) and Cell (0), or that all cells in the cross section has been treated:

4.1.7.1. Get the coordinates and elevation of Cell (n)'s neighboring cell, i.e. $x(n+1)$, $y(n+1)$, and$h(n+1)$ at Cell (n + 1), and calculate the distance, $DO(n+1)$, and slope, $SO(n+1)$, from Cell (0) to Cell (n + 1).

4.1.7.2. If $SO(n+1) < 0$ (i.e. the terrain is lower at Cell (n − 1), creating a water pond) then$n = n + 1$, else *flag* = true, $DO_{flag} = DO(n)$, and compare $DO(n+1)$ with the D_d for desired probability for the absolute value of slope of $SO(n+1)$. If D_d is exceeded then stop, else$n = n + 1$.

4.1.8. If *flag* = false then $DO_{flag} = \delta/2$ (This accounts for cross sections where no cells in the outer part of the cross section have positive slopes, i.e. no certain dry ground seems to occur).

4.1.9. If $ws(m) + \exp[(\ln DO_{flag} - \ln c)/z]* DO_{flag} < h(n+1)$ then outer uncertainty boundary elevation $RXSO(m) = ws(m) + \exp[(\ln DO_{flag} - \ln c)/z]* DO_{flag}$ else $RXSO(m) = h(n+1)$ (This accounts for the wall effect).

4.2. *Else LEFT cross section (follow 4.1, but beware of sign changes, </> changes, and change of RXSI/RXSO to LXSI/LXSO.*

4.3. Let m = m + 1.

5. Assign cross sections, i.e. both LEFT and RIGHT, with elevation values from XSI and XSO elevation values.

6. Create two point themes where the point locations are taken from the nodes in the cross section lines. One theme with associated inner elevation values and one with outer elevation values.

7. Create two TINs from the point themes. One based on inner elevation values and one based on outer elevation values. Rasterize the TINs to the same extent as the DEM to create uncertainty rasters.

8. Compare the uncertainty rasters with the DEM. If cell elevations in uncertainty raster for outer elevations < DEM then cell is not flooded. If cell elevations in uncertainty raster for inner elevations > DEM then cell is flooded. Other alternatives results in uncertain with respect to flooded/not flooded.

4.3 Calculation of wall effect

Figure 9 illustrates how the water elevations for the cross sections can be calculated (for outer uncertainty boundary elevation in the RIGHT cross section). When the distance to cell C(n + 1) is greater than the D_d for the same slope (i.e., between location of the modeled flood boundary location and the cell C(0)), the uncertainty elevation (from the D_d equation) of cell C(n) is compared with the ground elevation of cell C(n + 1). The elevation can be calculated by adding the water surface elevation to the product of the distance and slope from Eq. 2:

$$h = w_s + e^{\left(\frac{\ln D_d - \ln c}{z}\right)} * D_d \tag{8}$$

If the computed uncertainty elevation for cell C(n) is lower than the ground elevation of cell C(n + 1) the former elevation is used; if not, the ground elevation of cell C(n + 1) is used. By doing that, extremely high uncertainty elevations will not occur, due to for example vertical walls in the cross section. Also, when there is no wall, the resulting uncertainty elevation will be slightly higher (i.e. more conservative, or on the "safe side") compared with using an elevation from a cell farther away.

4.4 Applying the algorithm

To test the D_d equation and the algorithm, two DEMs were chosen, 3.83 m and 50 m, respectively, with a 95 percentile confidence. The results can be seen in Fig. 10, where red areas represent the areal range that is uncertain to be flooded. It is 95 % certain that the flood boundary will be within this area. The remaining 5 % are divided into blue areas that are almost certain to be flooded and areas outside of the red areas that are almost certain not to be flooded.

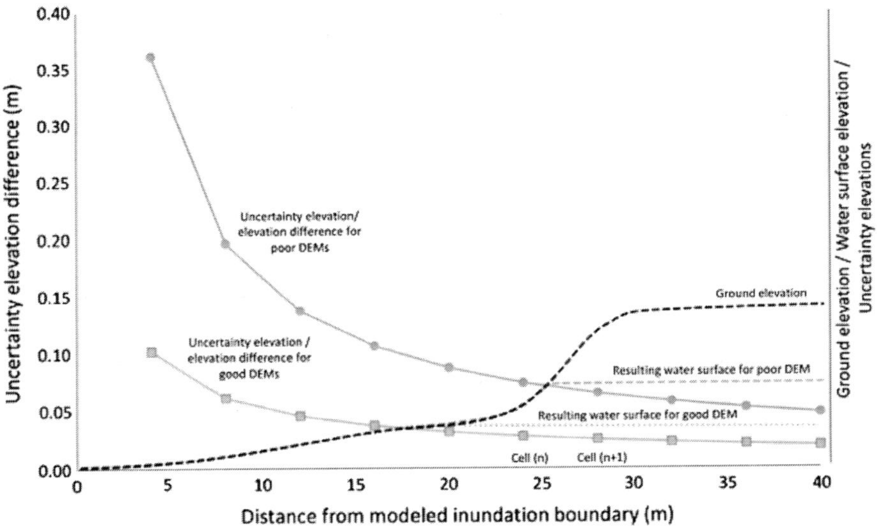

Figure 9. Illustration on how the water elevation is calculated

5. DISCUSSION AND CONCLUSION

Previous research has produced relatively consistent recommendations from 4 to 10 m DEM resolution. Better resolution than that has not produced any significant improvements in inundation boundary prediction. Yet there are numerous real case examples on disparities where the models clearly have failed and where the water boundary may be several hundreds of meters wrong, even when LiDAR data has been used as input (also cf. e.g. Croke et al. 2014 who stressed the importance of geomorphological understanding during flood risk management). Casas et al. (2006) argued that coarser resolution will have less consequence when discharge increases, i.e., floodplain areas will be more accurately mapped than near river areas. This may be true with respect to difference in total surface area of the true and modeled inundation. But, even if the flow rates are bigger, the flood boundary perimeter should be of roughly the same length, provided the river flows through a pronounced river or floodplain valley, and therefore the uncertainty problem will still persist at the modeled inundation boundaries. To overcome unpleasant surprises, extra caution should therefore be taken when either the DEMs are of poor resolution or that the terrain adjacent to the modeled inundation boundary is flat.

It is generally possible to construct an equation for the uncertainty of floodplain boundaries depending on DEM resolution and terrain slope. Except for the relation between z exponent and resolution, high correlation coefficients were yielded for all coefficients and exponents. This is probably due to the topographical character of the areas the modeling is based on; or in other words, the geographical areas of this study are not enough flat for sufficiently long distances. Hence, the main weakness in the data lies in the number of observations for the poorest DEM resolution together with the characteristics of the geographical area. Although there are very flat areas, these are not very big, making it impossible to get big disparities. This will also impact the equation. Most probable, the z exponent is too big, i.e., the slopes of the lines in Fig. 7should be sloping steeper in the negative direction. Therefore, to have a more conservative and "safe" estimate on risks, the z exponent could be set to the value for the highest resolution (~–0.72), irrespectively of DEM resolution. Furthermore, as the same transects, with approximately 10 m spacing, are used for all DEM resolutions, the same DEM cells are used several times for calculation of the transect slopes of the coarse resolution DEMs. This is not the case for fine-resolution DEMs. Another problem may be the usage of retransforming logged values in the regression analysis (cf. Granger and Newbold 1976 or, for a hydrological example, Jansson 1985). No correction has been applied in this work and therefore some of the coefficients and exponents are probably underestimated.

As the disparity distance equation has been developed from elevation and slope values taken from the reference DEM, this may give erroneous estimates when applying the algorithm on DEMs of lower qualities. Therefore, in future research an equation based on the slopes derived from the lower quality DEMs will be created to see if that may affect the results. Also important is not to use the equation for higher confidences than 95 %. For example it is possible to get computed disparity distances that are shorter than those that were actually observed, even when 100 % is used as input. As can be seen in Fig. 7, there are not exactly 10 % of observations above and below the 90 percentile lines for the three DEMs shown. This arises mainly due to the combination of the 14 DEMs. The magnitudes of these errors (which can be seen as the uncertainty of the disparities for a given probability) have not been calculated in this study, as the primary focus have been on modeling and visualizing the uncertainties of the flood boundary lines, not the uncertainties of the uncertainties. However, when more robust data have been gathered for a wider range of geographical settings, it will be of interest to also study this (cf. Roscoe et al. 2012 for such a procedure).

High confidence percentages have to be treated using extreme value statistics. If more data, especially coarse resolution data, had been available, another option would be to use envelope curves. In this study envelope curves were created through increasing the 95 % confidence level by one magnitude. When comparing these envelope curves with the visually determined curves in Brandt and Lim (2012), there are strong indications that the z exponent is too big (cf. Fig. 4b), again calling for a z value closer to −0.72, also for poorer resolutions. Furthermore, the importance of regularly, and not too sparsely, placed cross sections can be seen in Fig. 10. For the 50 m DEM case, the inundated areas do not follow the trunk river in the western (left) part. This can be attributed to cross sections not expanding long enough, making them disqualified in the algorithm which in the end therefore produced too low water levels for this area.

To test the applicability of the equation and resulting algorithm, future research will look at another river where river side slopes are much gentler, as already now it can be seen that the disparities are higher for the Testebo River than the Eskilstuna Rivers, for the same slope. However, the disparities at Testebo River are compared against an actual flood event that may not have been mapped good enough, whereas Eskilstuna River's are compared against a reference DEM. This means that the D_d equation developed for Eskilstuna River provides a measure of uncertainties only related to the DEM, whereas the disparities for Testebo also contain other types of uncertainties. Also, higher precision than cm on elevations has to be used to avoid the linear patterns seen in Fig. 5. This will also reduce the risk of getting 0 m/m slopes that make the use of logarithm functions problematic. More studies on other rivers, following both the Eskilstuna case of totally focusing on the DEM, as well as other cases that are compared with actual flood events, should both verify the approach as well as making adjustment of the equation possible. Finally, as this kind of uncertainty only represents one type of uncertainty, i.e., random errors in the DEM, other types including friction parameter errors, systematic errors of DEMs, rain and reach input of water flow, model structure, operator errors, etc. should also be considered. Therefore, being humble describing the uncertainties of the presented flood risk maps is to be recommended.

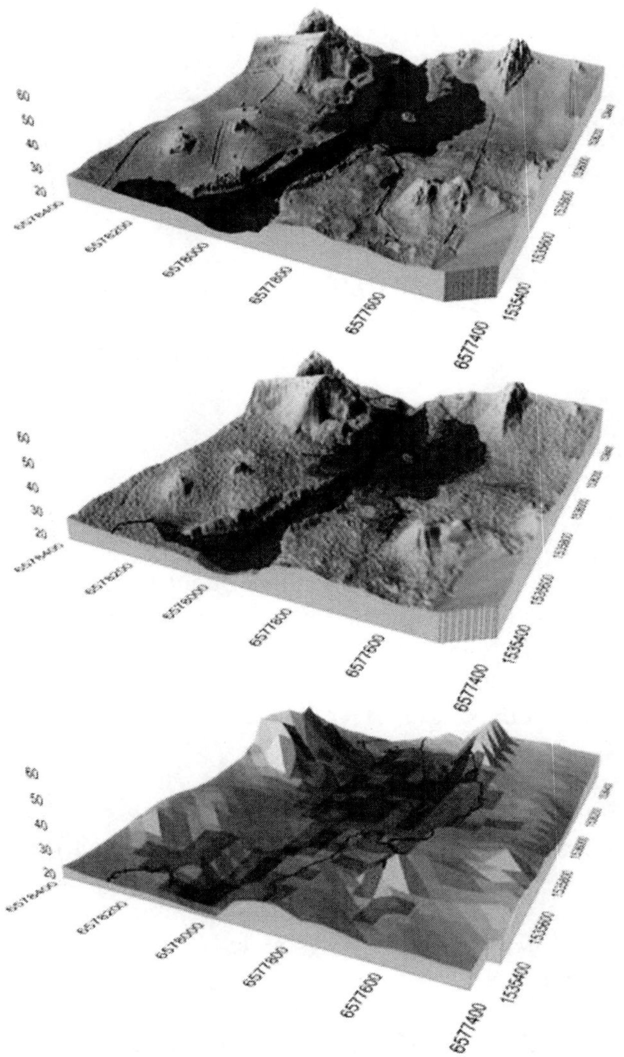

Figure 10. Resulting uncertainty areas for 95 percentile confidence. Upper figure shows the 0.78 m DEM with "true" inundation extents, middle shows the 3.83 m DEM, and the lower figure the 50 m resolution case. The *lines*in the lower two figures represent the "true" inundation extents. *Blue* areas are almost certain to be flooded,*red* areas are uncertain to be flooded (containing 95 % of the possible inundation delineations), and remaining areas are almost certain not to be flooded

ACKNOWLEDGMENTS

Thanks are due to Nancy Joy Lim, both for commenting on the manuscript as well as discussing the entire project, and the anonymous reviewers for their constructive criticism. The study has been financially supported by the EU through Tillväxtverket (Project GLOBES 2, number 170430) and Lantmäteriet (Project: "Kvalitetsbeskrivning av geografisk information vid översvämningskartering").

CONFLICT OF INTEREST

The author declares that he has no conflict of interest.

REFERENCES

1. Apel H, Merz B, Thieken AH (2008) Quantification of uncertainties in flood risk assessments. Int J River Basin Manag 6(2):149–162.

2. Brandt SA (2005) Resolution issues of elevation data during inundation modeling of river floods. In: Jun B-H, Lee S-I, Seo I-W, Choi G-W (eds) Water engineering for the future: choices and challenges. Proceedings of XXXI International Association of Hydraulic Engineering and Research congress, Seoul, September 11–16 2005. Korean Water Association, Seoul, pp 3573–3581

3. Brandt SA (2009) Betydelse av höjdmodellers kvalitet vid endimensionell översvämningsmodellering. FoU-rapport Nr 35, Högskolan i Gävle

4. Brandt SA, Lim NJ (2012) Importance of river bank and floodplain slopes on the accuracy of flood inundation mapping. In: Murillo Muñoz RE (ed) River flow 2012: vol 2. Proceedings of the international conference on fluvial hydraulics, San José, 5–7 Sept 2012. CRC Press/Balkema (Taylor & Francis), Leiden, The Netherlands, pp 1015–1020

5. Casas A, Benito G, Thorndycraft VR, Rico M (2006) The topographic data source of digital terrain models as a key element in the

accuracy of hydraulic flood modelling. Earth Surf Process Landforms 31(4):444–456.

6. Castellarin A, Di Baldassarre G, Bates PD, Brath A (2009) Optimal cross-sectional spacing in Preissmann scheme 1D hydrodynamic models. J Hydraul Eng 135(2):96–105.

7. Colby JD, Dobson JG (2010) Flood modeling in the coastal plains and mountains: analysis of terrain resolution. Nat Hazards Rev 11(1):19–28.

8. Cook A, Merwade V (2009) Effect of topographic data, geometric configuration and modeling approach on flood inundation mapping. J Hydrol 377(1–2):131–142.

9. Croke J, Reinfelds I, Thompson C, Roper E (2014) Macrochannels and their significance for flood-risk minimisation: examples from southeast Queensland and New South Wales, Australia. Stoch Environ Res Risk Assess 28(1):99–112.

10. Granger CWJ, Newbold P (1976) Forecasting transformed series. J R Stat Soc Ser B (Methodological) 38(2):189–203

11. Hydrologic Engineering Center (2008) HEC-RAS: river analysis system. User's manual, version 4.0. US Army Corps of Engineers, Hydrologic Engineering Center, Davis

12. Jansson M (1985) A comparison of detransformed logarithmic regressions and power function regressions. Geogr Ann 67A(1/2):61–70.

13. Klang D, Klang K (2009) Analys av höjdmodeller för översvämningsmodellering. http://www.geoxd.se/GeoXD_Analys%20av%20hojdmodeller.pdf. Accessed 02 April 2015

14. Koenker R, Hallock K (2001) Quantile regression. J Econ Perspect 15(4):143–156.

15. Koivumäki L, Alho P, Lotsari E, Käyhkö J, Saari A, Hyyppä H (2010) Uncertainties in flood risk mapping: a case study on estimating building damages for a river flood in Finland. J Flood Risk Manag 3(2):166–183.

16. Lim NJ (2009) Topographic data and roughness parameterisation effects on 1D flood inundation models. B.Sc thesis, University of Gävle

17. Lim NJ (2011) Performance and uncertainty estimation of 1- and 2-dimensional flood models. M.Sc thesis, University of Gävle

18. Lim NJ, Brandt SA, Seipel S (2016) Visualisation and evaluation of flood uncertainties based on ensemble modelling. Int J Geogr Inf Sci 30(2):240–262.

19. Merwade V, Olivera F, Arabi M, Edleman S (2008) Uncertainty in flood inundation mapping: current issues and future direction. J Hydrol Eng 13(7):608–620.

20. Omer CR, Nelson EJ, Zundel AK (2003) Impact of varied data resolution on hydraulic modeling and floodplain delineation. J Am Water Resour Assoc 39(2):467–475.

21. Pappenberger F, Beven K, Horritt M, Blazkova S (2005) Uncertainty in the calibration of effective roughness parameters in HEC-RAS using inundation and downstream level observations. J Hydrol 302(1–4):46–69.

22. Pappenberger F, Matgen P, Beven KJ, Henry J-B, Pfister L, de Fraipont P (2006) Influence of uncertain boundary conditions and model structure on flood inundation predictions. Adv Water Resour 29(10):1430–1449.

23. Pappenberger F, Beven KJ, Ratto M, Matgen P (2008) Multi-method global sensitivity analysis of flood inundation models. Adv Water Resour 31(1):1–14.

24. Raber GT, Jensen JR, Hodgson ME, Tullis JA, Davis BA, Berglund J (2007) Impact of Lidar nominal postspacing on DEM accuracy and flood zone delineation. Photogramm Eng Remote Sens 73(7):793–804

25. Roscoe KL, Weerts AH, Schroevers M (2012) Estimation of the uncertainty in water level forecasts at ungauged river locations using quantile regression. Int J River Basin Manag 10(4):383–394.

26. Schumann G, Matgen P, Hoffmann L, Hostache R, Pappenberger F, Pfister L (2007) Deriving distributed roughness values from

satellite radar data for flood inundation modelling. J Hydrol 344(1–2):96–111

27. Vähäkari A (2006) Simulering av översvämningar i Nedre Dalälven. M.Sc thesis, Uppsala Universitet

28. Wagener T, Gupta HV (2005) Model identification for hydrological forecasting under uncertainty. Stoch Environ Res Risk Assess 19(6):378–387.

29. Warmink JJ, Straatsma MW, Huthoff F, Booij MJ, Hulscher SJMH (2013) Uncertainty of design water levels due to combined bed form and vegetation roughness in the Dutch River Waal. J Flood Risk Manag 6(4):302–318.

30. Werner MGF (2001) Impact of grid size in GIS based flood extent mapping using a 1D flow model. Phys Chem Earth (B) 26(7–8):517–522.

31. Werner MGF, Hunter NM, Bates PD (2005) Identifiability of distributed floodplain roughness values in flood extent estimation. J Hydrol 314(1–4):139–157.

32. Wilson MD, Atkinson PM (2007) The use of remotely sensed land cover to derive floodplain friction coefficients for flood inundation modeling. Hydrol Proc 21(26):3576–3586.

33. Wu XZ (in press) Probabilistic solution of floodplain inundation equation. Stoch Environ Res Risk Assess.

34. Yacoub T, Sanner H (2006) Vattenståndsprognoser baserade på översiktlig kartering. En fallstudie. SMHI. Hydrologi nr 100

35. Zhang W, Montgomery DR (1994) Digital elevation model grid size, landscape representation, and hydrologic simulations. Water Resour Res 30(4):1019–1028.

8

STOCHASTIC ANALYSIS AND UNCERTAINTY ASSESSMENT OF TSUNAMI WAVE HEIGHT USING A RANDOM SOURCE PARAMETER MODEL THAT TARGETS A TOHOKU-TYPE EARTHQUAKE FAULT

Yo Fukutani [1,2] • Anawat Suppasri [1] Fumihiko Imamura [1]

[1] International Research Institute of Disaster Science, Tohoku University, Aramaki Aoba 468-1, Aoba-ku, Sendai-shi, Miyagi 980-8579, Japan

[2] Risk Modeling Group, Property Risk Engineering Department, Tokio Marine & Nichido Risk Consulting Co., Ltd., 2-1, Marunouchi 1-Chome, Chiyoda-ku, Tokyo 100-0005, Japan

ABSTRACT

We created a fault model with a Tohoku-type earthquake fault zone having a random slip distribution and performed stochastic tsunami hazard analysis using a logic tree. When the stochastic tsunami hazard analysis results and the Tohoku earthquake observation results were compared, the observation results of a GPS wave gauge off the southern Iwate coast indicated a return period equivalent to approximately 1,709 years (0.50 fractile), and the observation results of a GPS wave gauge off the shore of Fukushima Prefecture indicated a return period of 600 years (0.50 fractile). Analysis of the influence of the number of slip distribution patterns on the results of the stochastic tsunami hazard analysis showed that the number of slip distribution patterns considered greatly influenced the results of the hazard analysis for a relatively large wave height. When the 90 % confidence interval and coefficient of variation of tsunami wave height were defined as an index for projecting the uncertainty of tsunami wave height, the 90 % confidence interval was typically high in locations where the wave height of each fractile point was high. At a location offshore of the Boso Peninsula of Chiba Prefecture where the coefficient of variation reached the maximum, it was confirmed that variations in maximum wave height due to differences in slip distribution of the fault zone contributed to the coefficient of variation being large.

KEYWORDS

Stochastic tsunami hazard analysis, Logic tree analysis, Uncertainty assessment, Random source parameter model, Tohoku-type earthquake

1. INTRODUCTION

In earthquake-induced tsunami hazard analysis, it is extremely important to ascertain the effect of earthquake source parameters on the initial displaced water height of tsunamis. Numerous studies on this influence have been conducted (e.g., Hwang and Divoky 1970; Ward 1982; Ng et al. 1991; Pelayo and Wiens1992; Whitmore 1993; Geist and Yoshioka 1996; Geist 1999; Song et al. 2005). Based on observations of the 2004 Indian Ocean tsunami, Song et al. (2005) concluded that slip function is an important factor that governs tsunami intensity and that the fracture rate of the earth's crust determines the spatial pattern of a tsunami. It is now well known that the amount of slip of a rupturing fault strongly affects tsunami wave height, as numerous past studies have indicated.

On the other hand, Freund and Barnett (1976) pointed out that tsunami wave height is underestimated more when tsunami calculations are performed with uniform slip distributions than with heterogeneous slip distributions, but studies on the effect of planar nonuniformity of fault slip on tsunami wave height are limited (e.g., Geist and Dmowska 1999; Geist 2002; McCloskey et al. 2007, 2008; Løvholt et al. 2012). Geist

(2002) calculated tsunami wave height uncertainty for the Pacific coastline of Mexico by fixing the moment magnitude at 8.0 and using a trench fault with 100 cases of random slip. He ascertained that random slip distribution of a trench fault significantly affected the localized coastal wave height and that coastal wave height varied three-fold on average and six-fold at maximum. McCloskey et al. (2007, 2008) examined coastal wave height uncertainty for the Sumatra coast using approximately 100 cases of initial displaced water height values calculated from a trench fault with a random slip distribution that was artificially generated in the same manner. They also confirmed that nonuniformity of slip distribution greatly influenced coastal wave height. Løvholt et al. (2012) quantitatively studied the effect of 500 cases of artificially generated random trench slip distributions on tsunami runup height. They concluded that a hydrostatic pressure model produces artificially high uncertainty and that a nonhydrostatic pressure model produces a decrease in uncertainty.

However, even though these studies describe the effect of planar nonuniformity of slip on a fault on tsunami wave height or tsunami runup height, they do not describe its influence on stochastic tsunami hazard analysis. Thus, we focused on the influence of nonuniformity of slip distribution on a fault on the results of a stochastic tsunami hazard analysis. Various methods of the stochastic tsunami hazard analysis have been proposed by past many studies (e.g., Lin and Tung 1982; Rikitake and Aida 1988; Ward 2002; Geist 2005; Thio 2010; Yong et al. 2013). In this study, we adopted a method of stochastic tsunami hazard analysis using a logic tree (Annaka et al. 2007). Specifically, using a random source parameter model that determined the probability of slip distributions including asperity in a fault zone, we produced multiple giant faults with hypothetical slip distributions for a Tohoku-type fault zone. Then, by treating these multiple giant faults as branches of a logic tree, we treated heterogeneous slip on the fault as epistemic uncertainty. We also examined how the results of the tsunami hazard analysis changed when the number of branches of the logic tree was varied. We used only five patterns of generated random slip distribution, but it appears that they have a large effect on the results of stochastic tsunami hazard analysis when using a logic tree. In this study, we also discuss the regional uncertainty of stochastically assessed tsunami wave height.

2. ASSESSMENT OF SLIP DISTRIBUTION ON A FAULT USING A RANDOM SOURCE MODEL

In this study, we used the correlated random source parameter (CRSP) model (Liu et al. 2006) to generate random slip distributions on a fault plane. The CRSP model is a recently proposed kinematic fault model that smoothly varies slip distributions on fault planes. In this model, a two-dimensional distribution of parameters such as slip, slip angle, rupture propagation rate, and rupture rise time on a fault plane can be estimated based on a probability density function obtained by statistically processing slip distributions estimated from past earthquake records. In this study, we used only the two-dimensional slip distributions that were generated using the CRSP model. The procedure for creating a hypocenter model using this technique is as follows.

First, a completely random two-dimensional Gaussian distribution is generated. Then, the generated random two-dimensional Gaussian distribution is transformed to wavenumber space by a two-dimensional Fourier transform and multiplied by the following filter (Mai and Beroza 2002).

$$F(k_x, k_y) = \left\{1 + \left[(k_x C_L)^2 + (k_y C_W)^2\right]\right\}^{-\nu/2} \tag{1}$$

Here, k_x is the wavenumber of fault strike direction, k_y is the wavenumber of fault dip direction, ν is the fractile dimension, C_L is the correlation length of the fault strike direction, and C_W is the correlation length of the fault dip direction. We set a ν value to 4.0, which was determined by Liu et al. (2006). C_L and C_W are calculated from the following equations, where *Mw* is the moment magnitude of the earthquake fault.

$$\log_{10}(C_L) = -2.5 + M_w/2 \tag{2}$$

$$\log_{10}(C_W) = -1.5 + M_w/3 \tag{3}$$

Next, by Fourier inverse transformation of the two-dimensional distribution that has been multiplied by the filter expressed in Eq. (1), a random Gaussian distribution in which the amplitude in wavenumber space attenuates according to k−ν/2k−ν/2 is obtained. Finally, using the NORTA method (Cairo and Neilson 1997), it is transformed to the following truncated Cauchy distribution in which fault slip distribution follows from the obtained random Gaussian distribution. This probability distribution is based on studies in which past earthquake inversion analyses were statistically processed (Lavallée and Archuleta 2003, 2005; Lavallée et al. 2006).

$$p(D) = C\frac{1}{1 + [(D - D_o/\kappa)]^2}, \quad 0 \leq D \leq D_{max} \tag{4}$$

Here, C is the normalized probability density function, D_o is the Cauchy distribution location parameter, κ is the Cauchy distribution scale parameter, and D_{max} is the maximum slip of fault. κ is a coefficient that is set such that the average value of the slip distribution is equal to the average slip across the fault. The average slip across the fault was taken as D_{ave}, and the following values were used for D_o and D_{max};

$$D_{max} = 2.8 * D_{ave} \tag{5}$$

$$D_o = 0.5 * D_{ave} \tag{6}$$

The coefficient of 2.8 in Eq. (5) uses the ratio of maximum slip and average slip in the shallow part of the 2011 Tohoku earthquake as a reference. In particular, using the results of Yoshida et al. (2011), who studied the rupture process of the Tohoku earthquake using the waveform inversion technique with long-period seismic motion data, Ishii et al. (2013) concluded that the average slip of the shallow part was 16.7 m and that the maximum slip was 46.9 m. Based on these results, we set the ratio of maximum slip to average slip in the truncated Cauchy distribution at 2.8. The coefficient of 0.5 in Eq. (6) is the coefficient of Liu et al. (2006), without modification.

Ishii et al. (2013) focused on a prominent feature of the Tohoku earthquake (i.e., the slip in the shallow part was larger than the slip in the deep part) and proposed a technique by which the average slip of each part is calculated by using the area and modulus of rigidity of each part after the fault is divided into a deep and shallow part. A value of approximately 3.0, derived from analysis of the Tohoku earthquake, has been used as the ratio of average slip of the deep part to that of the shallow part.

In this study, we modeled the hypocenter of the rupture zone of the Tohoku earthquake using the technique of Liu et al. (2006) for generating random slip distributions and the technique of Ishii et al. (2013) for calculating average slip in the shallow and deep parts. A schematic diagram of the modeling is shown in Fig. 1. First, a giant fault was divided into a shallow and deep part, and the average slip of each part was calculated using the technique of Ishii et al. (2013). Because the value of average slip varies with the area of each part according to the scaling law, the depth of the boundary between the shallow and deep parts of the fault model in this study was uniformly set to 15 km so as to align it with the area of the shallow (27,000 km^2) and deep part (73,000 km^2) set by Ishii et al. (2013). If the average slip of each part is determined, it is possible to determine the maximum slip of each part uniquely using Eq. (5). Furthermore, the ratio of average slip and maximum slip of the deep part was assumed to be the same as the ratio of average slip and maximum slip of the shallow part. Specifically, we consider it necessary to concretely determine the ratio of average slip and maximum slip of the shallow part from observation results, but because we could not obtain concrete data, we assumed the ratio of average slip to maximum slip to be the same in both the shallow and deep parts. Next, based on the average slip and maximum slip of each part, the technique of Liu et al. (2006) was applied to both the deep and shallow parts to generate a random slip distribution for each part. Table 1 shows the average slip in the deep and shallow parts and the maximum slip across the entire fault zone (maximum slip of shallow part) calculated in this study and by Yoshida et al. (2011). The results in this study, in which slip was artificially generated using the CRSP model, are slightly higher but express the general trend of numeric values quite well.

For the earthquake fault parameters, we used the longitude, latitude, length, width, upper depth, and strike and dip angle of the fault endpoint of a Tohoku-type earthquake fault published by Fujiwara et al. (2013) (Table 2). Figure 2 shows five patterns of a hypocenter model having different two-dimensional locations of slip distribution generated by the above technique. The figure shows that the large slip region of the shallow part, like that actually produced by the Tohoku earthquake, can be expressed well. The asperity area ratio in the fault for these five cases is shown in Table 3. Somerville et al. (1999) defined asperity as the region having a slip that is at least 1.5 times greater than the average slip across a fault, and, based on analyzing the area of asperities of 15 past earthquakes, demonstrated that asperity accounts for 22 % (average), 10 % (minimum), and 40 % (maximum) of the area of the entire fault. When the fault asperity of the five cases generated in this study is calculated based on the above definition, the area ratio of asperity to total fault area (100,000 km^2) is in the range of 12.5–18.1 %, which agrees relatively well with the results of Somerville et al. (1999); we believe it can appropriately express the heterogeneous slip distribution characteristics of actual fault slip.

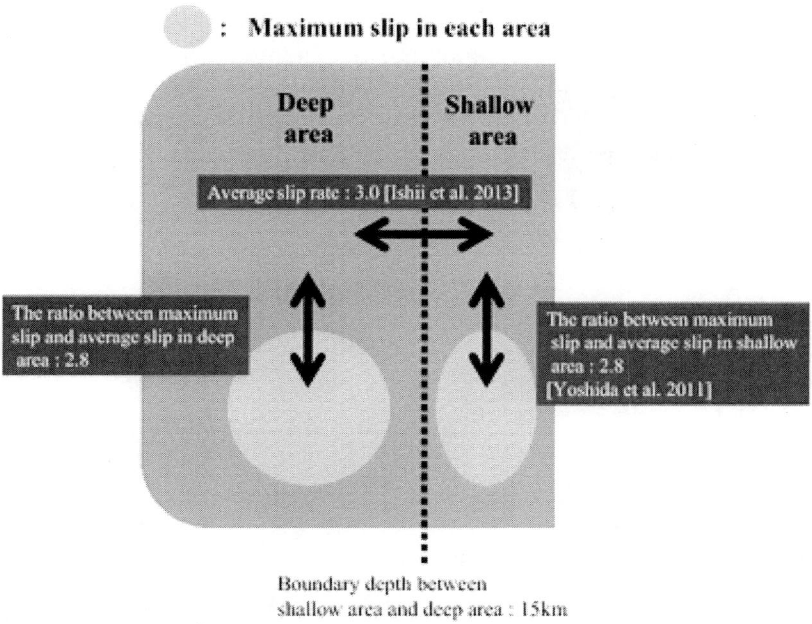

Figure 1. Schematic diagram of modeling a Tohoku-type earthquake fault. When the fault was thought of in terms of a deep part (73,000 km^2) and shallow part (27,000 km^2), the depth of the fault between areas was approximately 15 km. The ratio of average slip of the deep part to that of the shallow part was taken as 3.0 (Ishii et al. 2013). The ratio of maximum slip to average slip in the shallow part was taken as 2.8 (Yoshida et al. 2011). The ratio of maximum slip to average slip in the deep part was assumed to be 2.8, similar to the shallow part

3. STOCHASTIC TSUNAMI HAZARD ANALYSIS

3.1 Numerical tsunami simulation method

Based on the Tohoku-type earthquake fault model generated in Sect. 2, we calculated tsunami wave height at each point as follows. First, the points of a 10-km mesh encompassing a Tohoku-type earthquake fault zone (500 km × 200 km) were generated; then, using the fault parameters in Table 2 and the fault slip artificially generated by the CRSP model as input values, the initial displaced water height was calculated at those points using the formula of Okada (1985). Here, the slip angle was assumed to be a uniform 90°. Using the calculated initial displaced water height as input values, time integration was performed with a time interval of 0.9 s and a grid spacing of 15 s (approximately 450 m) at the longitude/latitude origin by a nonlinear long-wave equation using the TUNAMI program (Tohoku University's Numerical Analysis Model for Investigation of Far-field Tsunamis, IUGG/IOC TIME Project 1997).

Table 1. Comparison of average slip in the deep part and shallow part and maximum slip of the entire fault zone according to this study and Yoshida et al. (2011)

	This study	Yoshida et al. (2011)
Average slip in the deep part (m)	6.0	5.0
Average slip in the shallow part (m)	18.1	16.7
Maximum slip of the entire fault zone (m)	50.7	46.9

3.2 Logic tree construction

Next, to stochastically analyze the tsunami hazard, we constructed a logic tree with five branch levels (range of moment magnitude, position of asperity, earthquake return period, log normal standard deviation expressed as aleatory uncertainty, and truncated uncertainty) based on the technique of Annaka et al. (2007) (Fig. 3).

Table 2. Fault parameters of a Tohoku-type earthquake fault zone

Longitude	Latitude	Length (km)	Width (km)	Upper depth (km)	Strike	Dip	Rake	Slip
144.139°	39.9°	500	200	4.1	193°	13°	90°	[a]

Source Fujiwara et al. (2013)

[a] Slip is applied randomly across the fault using the CRSP model

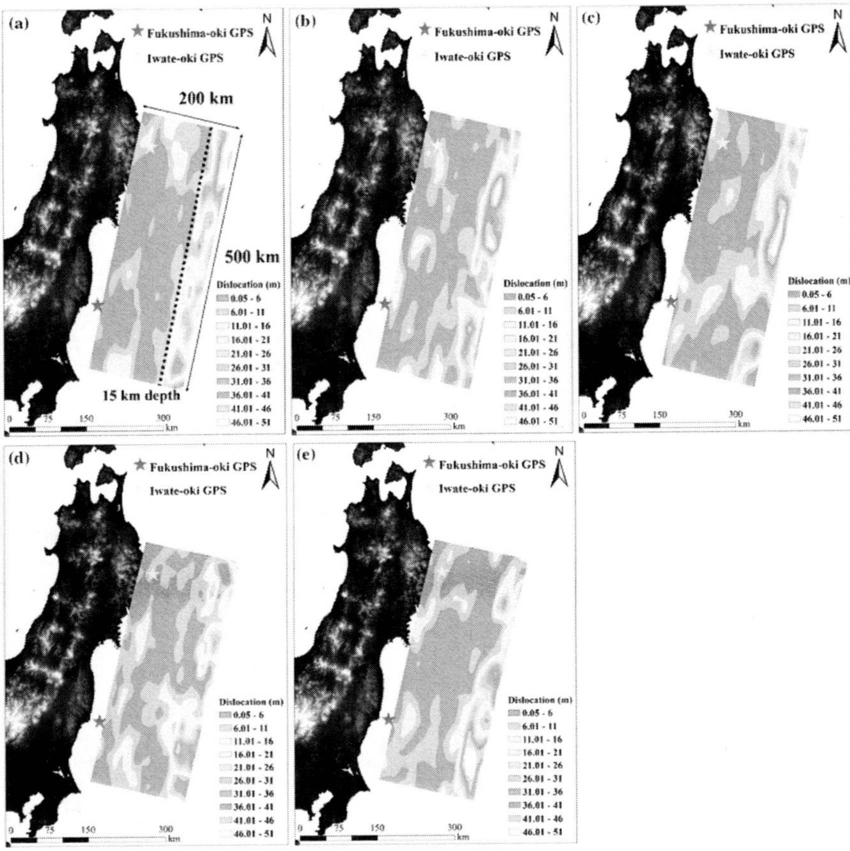

Figure 2. Modeled Tohoku-type earthquake faults having different slip distributions: **a** asperity at the northern edge of the fault, **b** asperity between the northern edge and center of the fault, **c** asperity at the center of the fault, **d**asperity between the southern edge and center of the fault, and **e** asperity at the southern edge of the fault.*Red stars* and *yellow stars* indicate the locations of the Fukushima Prefecture offshore GPS wave gauge and the south Iwate offshore GPS wave gauge, respectively, which were set up by MLIT

Here, we would like to clarify the definitions and handling methods regarding aleatory uncertainty and epistemic uncertainty in this study. Annaka et al. (2007) explain that aleatory uncertainty is due to the random nature of earthquake occurrence and its effects. Its nature can be determined from the variation in the ratios of observed to numerically calculated tsunami heights for historical tsunami sources

(Aida 1978) in their model, which is named “Aida’s kappa (κ)”. They also stated that epistemic uncertainty is due to incomplete knowledge and data about the earthquake process. Uncertainties in various model parameters and various alternatives are treated as epistemic uncertainty using the logic-tree approach.

Table 3. Ratio of asperity (region where the slip is 1.5 times greater than the average slip of the entire fault) generated by the CRSP model to total fault area

Position of asperity in the fault area	Northern edge (%)	Between norther edge and center (%)	Center (%)	Between southern edge and center (%)	Southern edge (%)
Ratio of asperity	18.10	16.70	17.80	12.50	16.60

Aida’s kappa, corresponding to aleatory uncertainty in this study, is determined from a numerical simulation for the optimal fault models of 11 tsunamigenic earthquakes with many historical run-up data (Table 4). However, the uncertainty that can be captured by this calculation only represents the difference between tsunami observation data due to an earthquake that occurred in the past and the tsunami value estimated from the earthquake model using numerical simulation. For the reason stated above, the Japan Society of Civil Engineers (2011) mentioned that parameters, which could vary even though the earthquakes occurred in the same place, should also be considered as “variability due to changing of fault models” (aleatory variability) or “uncertainty of fault models” (epistemic uncertainty). Therefore, in this study, we considered earthquake source models having different slip distributions that have not occurred in the past, as described in the second chapter, the moment magnitude of the earthquakes, and the recurrence interval as epistemic uncertainty. In addition, we treated variability of Aida’s kappa and its truncation as epistemic uncertainty.

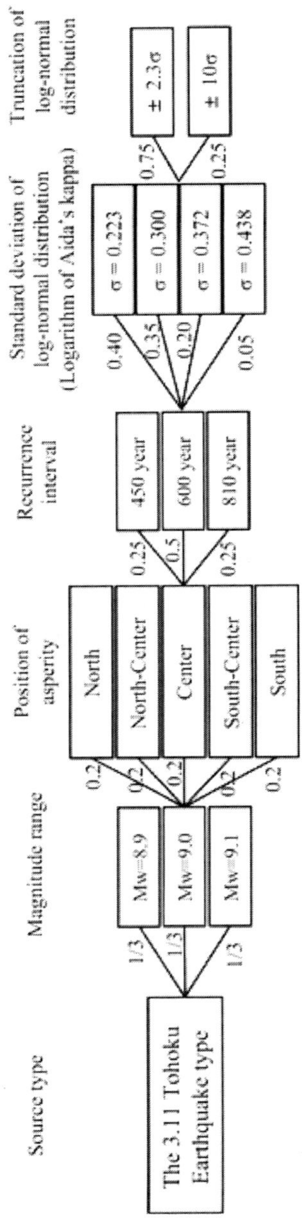

Figure 3. Logic tree for stochastic analysis constructed for the Tohoku-type earthquake fault. Branches are range of moment magnitude, position of asperity, average recurrence interval, log normal standard deviation (logarithm of Aida's kappa), and truncated range of the deviation. The *number* appended to each *branch*indicates the weight of that branch

Table 4. Values of Aida's kappa (κ) and logarithm of kappa (σ) for 11 historical tsunamigenic earthquakes listed by Annaka et al. (2007)

Year	Earthquake name	Number of run-up data	Aida's kappa (κ)	Logarithm of kappa (σ)
1707	Hoei	61	1.35	0.300
1854	Ansei Tokai	89	1.47	0.385
1854	Ansei Nankai	60	1.42	0.351
1896	Meiji Sanriku	257	1.45	0.372
1933	Showa Sanriku	572	1.40	0.336
1944	Tonankai	43	1.58	0.457
1946	Nankaido	159	1.60	0.470
1960	Chile	764	1.37	0.315
1968	Tokachi-oki	273	1.41	0.344
1983	Nihonkai-chubu	209	1.48	0.392
1993	Hokkaido-nansei-oki	216	1.47	0.385

For the moment magnitude (M_w), three cases of M_w 8.9, 9.0, and 9.1 were considered. Variations in moment magnitude are obtained by multiplying the slip across the fault by a coefficient calculated from seismic moment M_0 $(N^*m) = \mu DS$, where μ is the fault rigidity (N/m^2), S is the fault area (m^2), and D is the average displacement (m) across the fault.

For the position of the asperity, we initially generated many two-dimensional distributions of slip on the fault using the CRSP model. From these, we subjectively selected just five patterns of slip distribution such that the position of asperity was dispersed throughout the entire fault. Variations in the tsunami water level accompanying the variations in slip distribution in the strike direction of the fault are greater than the variations in the tsunami water level accompanying the variations in slip distribution in the dip direction (Geist and Dmowska 1999). Therefore, we considered fault patterns in which the variations in slip distribution in the strike direction were expressed more distinctly than were the variations in the slip distribution in the dip direction of the fault. In this study, we dealt with five, three, and one pattern of slip distributions, but we will confirm the large influence on the hazard curve, as will be discussed later.

For the return period of the Tohoku-type earthquake, we used the value of 600 years employed by Fujiwara et al. (2013). In their analysis, a model of the average return period for the Tohoku earthquake was treated as the BPT distribution. In this study, we decided on a deviation of the BPT distribution as 0.3 and set an upper limit and lower limit confidence interval based on its error.

For the log-normal standard deviation (aleatory uncertainty) and truncated uncertainty, we used the values of Annaka et al. (2007), which were determined from a numerical simulation for the optimal fault models of 11 tsunamigenic earthquakes with many historical run-up data (Table 4).

The number appended to each branch of the logic tree in Fig. 3 indicates the weight of that branch. For the branches related to the range of moment magnitude and position of the asperity, the values were equally divided weights. For the weights for each branch of the recurrence interval, we set the basis value as 0.5 and both end values as 0.25 because we set the branch by considering the confidence interval. Finally, regarding the weights for the branches of the standard deviation of the log-normal distribution and truncation of the log-normal distribution, we adapted the weighting values used in the study of Annaka et al. (2007); these values were used when they used a variable fault model in their study.

3.3 Hazard curves

Based on the constructed logic tree and the simulated tsunami wave heights, we created hazard curves (Fig. 4a). Each curve corresponds to a branch of epistemic uncertainty (logic tree) with the aleatory uncertainty for each outcome shown in Fig. 4a. The method by which each hazard curve is generated is as follows. First, we can derive a probability density function of a log-normal distribution, expressed as Eq. (7), when we set a maximum tsunami height value, calculated from the tsunami numerical simulation, using the condition of each branch of the logic tree as a median value (μ) and set a logarithm of Aida's kappa as the log-normal standard deviation (σ):

$$f(x) = \frac{1}{\sqrt{2\pi}\sigma x} \exp\left\{ \frac{-(\log x - \mu)^2}{2\sigma^2} \right\}, \quad 0 \leq x \tag{7}$$

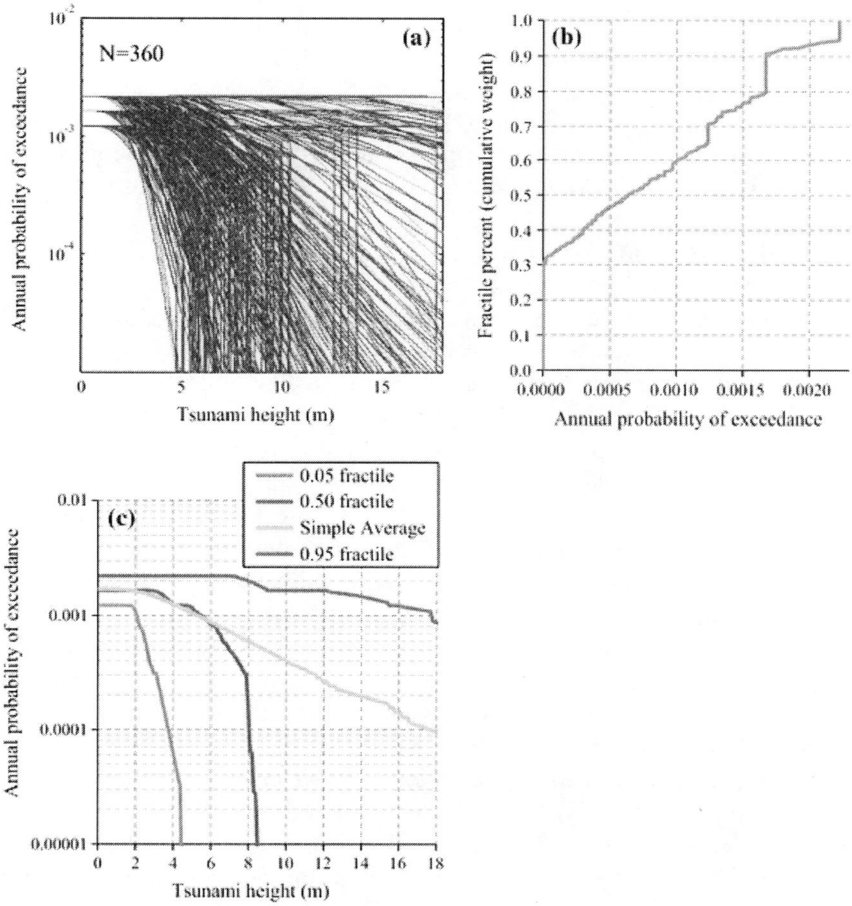

Figure 4. a The 360 *hazard curves* corresponding to all branches of the logic tree, **b** relationship between annual probability of exceedance and cumulative branch weight at a certain wave height, and **c** examples of statistically processed hazard curves (0.05 fractile curve, 0.50 fractile curve, 0.95 fractile curve, and simple average curve)

Second, we can obtain a tsunami hazard curve by converting the probability density function into an exceedance probability distribution on the assumption of an ergodic hypothesis. The ergodic hypothesis is a statistical assumption that spatial variability is equal to temporal variability. Therefore, if we use the log-normal distribution that indicated spatial variability, we can obtain a tsunami hazard curve that shows the relationship between tsunami height and the annual probability of exceedance.

The hazard curves are drawn to correspond to one specific location. Here, for example, the hazard curves at the installation site (39.259°N, 142.097°E) of the south Iwate offshore GPS wave gauge set up by the Ministry of Land, Infrastructure, Transport and Tourism (MLIT) in April 2007 are shown. The same number of hazard curves as the number of branches of the logic tree (3 × 5 × 3 × 4 × 2 = 360) was created.

From these 360 hazard curves, statistical processing was performed in consideration of branch weight. Annual probability of exceedance at a certain tsunami wave height and cumulative branch weight are shown in Fig. 4b. These were calculated in the same manner as the tsunami wave height from 0.0 to 0.1 m, and annual probability of exceedance was determined at the points of 95, 50, and 5 % cumulative weight. From this, the relationship between tsunami wave height and annual probability of exceedance was re-drawn, resulting in the 95, 50, and 5 % fractile curves shown in Fig. 4c. The fractile curves simultaneously include epistemic uncertainty and aleatory uncertainty in earthquake occurrence and tsunami propagation. The analysis of this hazard curve, which includes the various types of uncertainty, enables a stochastic interpretation of a tsunami hazard. Additionally, a curve produced by simply averaging the annual probability of exceedance without considering branch weight was also generated as a simple average curve. We note that the tsunami hazard curves presented in this study are actually the conditional hazard exceedance curves because only 3.11 Tohoku-type earthquake events are included in the logic tree. However, we can say that we conducted stochastic tsunami hazard analysis in terms of considering the variability of the moment magnitude, position of asperity, and recurrence interval of the target earthquake. In addition, there is a possibility that other parameters such as the log-normal standard deviation or weights based on expert judgment strongly influence the shape of the hazard curves (e.g., Annaka et al. 2007; Iwabuchi et al. 2014). However, we would like to note that we focused on how the hazard curves change in response to the different positions of asperity in the logic tree, on the premise that parameters such as the log-normal standard deviation and weights based on expert judgment are fixed.

3.4 Comparison with Tohoku earthquake observation results and a past study

In this section, we compare the results of the stochastic tsunami hazard analysis performed in Sect. 3.1with the observation results of the Tohoku earthquake that occurred in 2011 and a past study performed by Sakai et al. (2006). The measurement data of the south Iwate offshore GPS wave gauge and the Fukushima Prefecture offshore GPS wave gauge set up by MLIT and analyzed by Port and Airport Research Institute (PARI) were used as the Tohoku earthquake observation data (Kawai et al. 2013).

3.4.1 Comparison with the south Iwate offshore GPS wave gauge observation results

Figure 5a shows the waveform observed during the Tohoku earthquake, measured by the south Iwate offshore GPS wave gauge. The position of the south Iwate offshore wave gauge is indicated by the yellow star in Fig. 2a. According to information published by MLIT, the maximum tsunami wave height at the south Iwate offshore GPS wave gauge was 6.7 m. The results of stochastic tsunami hazard analysis at the same location are shown in Fig. 4c. In addition, the relationship between the return period and fractile point at the maximum wave height of 6.7 m observed during the Tohoku earthquake is indicated by the red line in Fig. 6. The results show that the range of the return period is approximately 1,780 years at the 0.50 fractile wave height and 450 years at the 0.95 fractile wave height. Furthermore, the return period at the 0.50 fractile wave height is 1,709 years. In a report by Fujiwara et al. (2013), the return period of the 2011 Tohoku earthquake was estimated at approximately 600 years, but in the results of our analysis, approximately 600 years is equivalent to a fractile from 0.81 to 0.91, which is a relatively high fractile point.

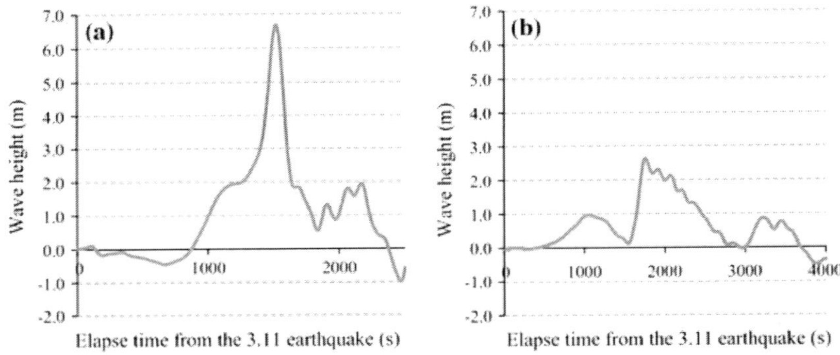

Figure 5. Waveform of the Tohoku earthquake tsunami: **a** measured at the south Iwate offshore GPS wave gauge (39.259°N, 142.097°E)—the maximum wave height recorded was 6.7 m—and **b** measured at the Fukushima Prefecture offshore GPS wave gauge (36.971°N, 141.186°E)—the maximum wave height recorded was 2.6 m (Kawai et al. 2013)

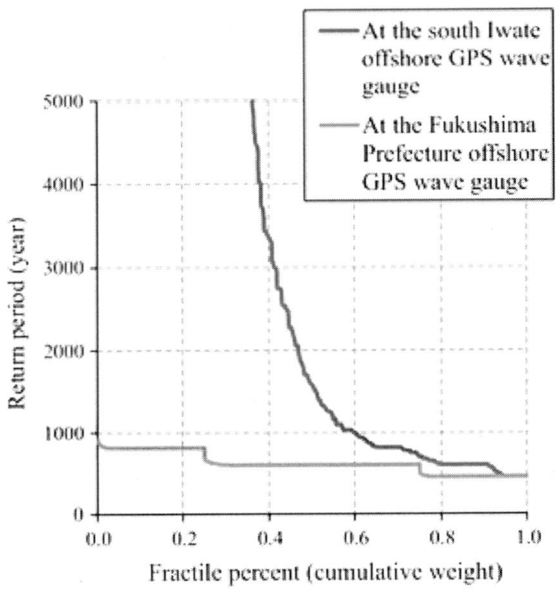

Figure 6. Relationship between the return period and fractile point of a probable wave with a height of 6.7 m at the south Iwate offshore GPS wave gauge (*red line*) and relationship between the return period and fractile point of a probable wave with a height of 2.6 m at the Fukushima Prefecture offshore GPS wave gauge (blue line)

3.4.2 Comparison with the Fukushima Prefecture offshore GPS wave gauge observation results

Figure 5b shows the waveform observed during the Tohoku earthquake, measured by the Fukushima Prefecture offshore GPS wave gauge. The position of the Fukushima Prefecture offshore wave gauge is indicated by the red star in Fig. 2a. According to information published by MLIT, the maximum tsunami wave height at the Fukushima Prefecture offshore GPS wave gauge was 2.6 m. The results of stochastic tsunami hazard analysis at the same location are shown in Fig. 7. In addition, the relationship between the return period and fractile point at the maximum wave height of 2.6 m observed during the Tohoku earthquake is indicated by the blue line in Fig. 6. The results show that the range of the return period is 600 years at the 0.50 fractile wave height and 450 years at the 0.95 fractile wave height. The return period at the 0.50 fractile wave height is 600 years. According to our analysis, the return period of the Tohoku earthquake of approximately 600 years is equivalent to a fractile between 0.35 and 0.75; this is a relatively wide fractile range.

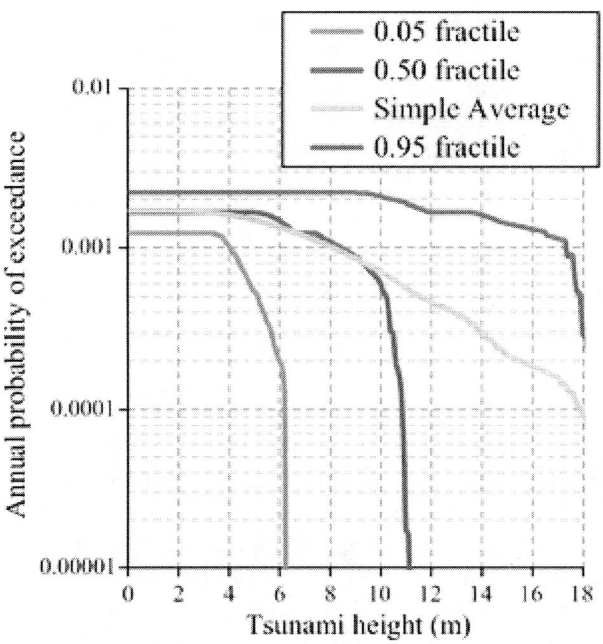

Figure 7. Statistically processed hazard curves at the location of the Fukushima Prefecture offshore GPS wave gauge

3.4.3 Comparison with a past study

Sakai et al. (2006) also conducted probabilistic tsunami hazard analysis using the logic-tree approach, targeting the coast of Fukushima Prefecture. Before comparing our research with their work, it is important to note that one of the fundamental differences between the two studies is that our research considers only the Tohoku earthquake type fault (M_w 8.9–9.1), while their research considered nine earthquake faults (M_w 7.7–8.6) along the Japan trench for the near-field tsunami. In addition, the contents and weights of the branches in the logic tree differ between the two studies. Thus, although there are some differences in the conditions of the studies, we discuss the results of the analyses in the following.

In the long-term averaged hazard curve of Sakai et al. (2006), wave heights with a return period of 1,000 years had a range of about 3.5 m (0.05 fractile wave) to about 5.6 m (0.95 fractile wave). However, in our study, wave heights with a similar return period had a range of about 4.0 m (0.05 fractile wave) to about 17.3 m (0.95 fractile wave). It is indicated that both the calculated stochastic tsunami height and its range in our analysis are larger than in their analysis, which is mainly because the moment magnitude of the assumed earthquake fault in our analysis is larger. Furthermore, the difference of the 0.05 fractile wave between the analyses is about 0.5 m, while the difference of the 0.95 fractile wave is about 13.3 m. We can also understand from the results that the difference in tsunami height between the analyses has a larger value as the fractile point is large.

3.5 Influence of the number of random slip distribution patterns on the results of hazard analysis

In this section, we examine the influence of the number of random slip distribution patterns in a fault zone on the results of stochastic tsunami hazard analysis. In the logic tree, variations in slip distribution in the fault zone are handled by a separate branch as epistemic uncertainty. Thus, we examined the effect of the number of slip distribution patterns in the fault zone on the stochastic tsunami hazard results by comparing the analysis results when using five slip distribution patterns (standard case), three patterns, and one pattern. For the case of three patterns, we considered the cases where the slip distribution was at the northernmost part, in the center, and at the southernmost part of the hypothetical fault. For the one pattern case, we only considered the case where the slip distribution was in the center. Since the number of branches of the logic tree changes if the number of slip distribution patterns is changed, the number of hazard curves also changes. When there are three slip distribution patterns, the number of branches is 216 (3 × 3 × 3 × 4 × 2), and when there is one, the number of branches is 72 (3 × 1 × 3 × 4 × 2).

Figure 8 shows the hazard curves at the site of the south Iwate offshore GPS wave gauge for the cases of three slip distribution patterns and one slip distribution pattern. Figure 9 shows the relationship between the return period and fractile at wave heights of 10 and 2.0 m in addition to the maximum wave height of 6.7 m measured during the Tohoku earthquake. These figures show that although there is almost no influence of the number of slip distribution patterns on the results of the hazard analysis in the case of a relatively small wave height (2.0 m), when the wave height is relatively large (6.7, 10 m), the number of slip distribution patterns greatly influences the results of the hazard analysis. In the case of the 10-m wave height, when there is only one slip distribution pattern, the return period at each fractile point provides the largest estimate (low annual probability of exceedance), whereas in the cases of three and five slip distribution patterns, the return period at each fractile point tends to provide an estimate smaller than that in the case of only one pattern (high annual probability of exceedance). In the case of the 6.7 m wave height, no particular trend is seen, but the variability in the estimated return period becomes relatively large compared to when the wave height is 2.0 m. Based on the approximate wave heights, we speculate that there is a great difference in how the number of slip distribution patterns influences the results of the hazard analysis depending on the hypothesized wave source.

Figure 10 shows the tsunami wave height frequency distribution taking into consideration the uncertainty of a 1,000-year return period at the same location. In the cases of the five and three slip distributions, the standard deviation and coefficient of variation are about the same, but when there is only one slip distribution, the frequency of wave heights 10 m and above becomes extremely small. Accordingly, the standard deviation becomes smaller and the coefficient of variation decreases. This is because, in the cases of the five and three slip distributions, a high tsunami wave height was calculated at the south Iwate offshore GPS wave gauge because a large amount of slip was present in the northern portion of the fault, but in the case of only one slip distribution, a high tsunami wave height was not observed at the south

Iwate offshore GPS wave gauge because a large amount of slip was present only in the center of the fault.

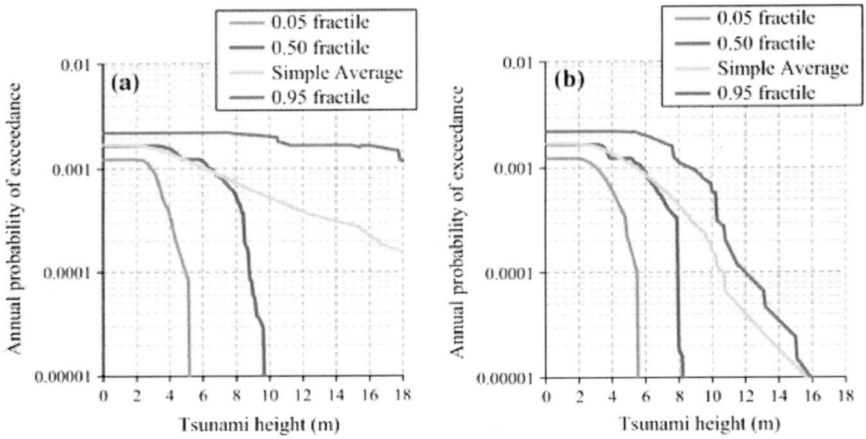

Figure 8. Changes in statistically processed hazard curves in the case of **a** three and **b** one slip distribution patterns of a Tohoku-type earthquake fault

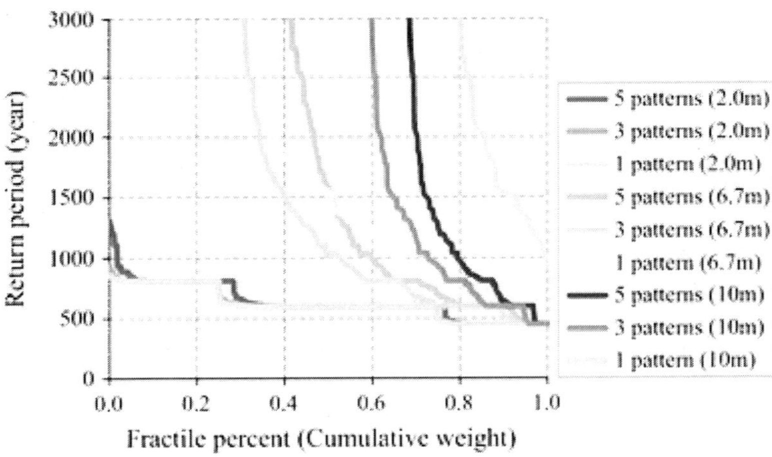

Figure 9. Relationship between the change in the number of slip distribution patterns of a Tohoku-type earthquake fault (five patterns, three patterns, one pattern) and change in the return period seen at typical wave heights (2.0, 6.7, 10 m)

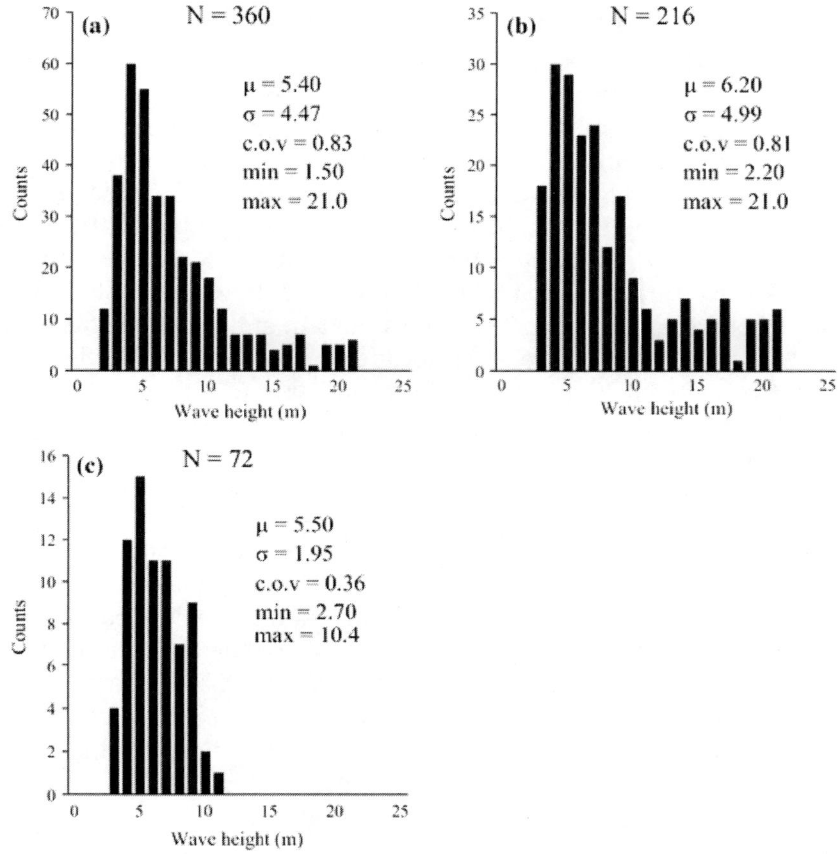

Figure 10. Tsunami wave height frequency distribution when considering the uncertainty of a 1,000-year return period at the location of the south Iwate offshore GPS wave gauge: **a** the number of slip distributions is five, **b** the number of slip distributions is three, and **c** the number of slip distributions is one. Here, *N* is the number of tsunami wave heights for which uncertainty is considered, *μ* is the average, *σ* is the standard deviation,*c.o.v.* is the coefficient of variation, *min* is the minimum value, and *max* is the maximum value

3.6 Quantitative assessment of uncertainty of tsunami wave height

In this section, we discuss the uncertainty of tsunami wave height in each region of the Tohoku earthquake using the results of stochastic tsunami hazard analysis and the logic tree of Fig. 3.

3.6.1 Uncertainty of tsunami wave height in the Tohoku offshore region

Considering the results of stochastic tsunami hazard analysis at the south Iwate offshore GPS wave gauge (Fig. 4), when we consider the tsunami wave height of a certain annual probability of exceedance, we see that various values can be obtained depending on the fractile point. Figure 11a shows the tsunami wave heights at the 5 % fractile point of a 1,000-year return period at a depth of 50 m underwater, offshore from the Tohoku coast, and Fig. 11b shows the tsunami wave heights at the 95 % fractile point at the same locations. For the 5 % fractile wave height, a relatively high wave height from 5.0 m to less than 10 m was calculated primarily offshore from Miyagi Prefecture and Fukushima Prefecture. On the other hand, for the 95 % fractile wave height, a high wave height from 20.0 m to less than 40.0 m was calculated in the coastal regions of the Iwate and Miyagi Prefectures. If we limit the wave height to 5.0 m and above, the offshore area from southern Aomori Prefecture in the north to southern Ibaraki Prefecture in the south was analyzed. In this study, as an index of uncertainty, we defined the wave height equivalent in the range from the 5 % fractile point to the 95 % fractile point (a value obtained by subtracting the 5 % fractile wave height from the 95 % fractile wave height) as the 90 % confidence interval of the tsunami wave height. Figure 11c shows the 90 % confidence interval of tsunami wave height of a 1,000-year return period at the same locations. The figure indicates that, in general, at locations where wave height at each fractile point is high, the 90 % confidence interval is simultaneously high. In particular, a large range in wave height, from 30.0 m to less than 40.0 m, is seen offshore from Iwate Prefecture and at offshore parts in the vicinity of the Oshika Peninsula of Miyagi Prefecture. In the offshore area of Tohoku as a whole, it ranges from a minimum of 0.6 m to a maximum of 36.5 m.

As another index of uncertainty, we defined the coefficient of variation obtained from the standard deviation and average of various tsunami wave heights obtained while focusing on a certain annual probability of exceedance. Figure 11d shows the coefficient of variation of tsunami wave height of a 1,000-year return period. Some very interesting observations can be made from this figure. First, there were multiple locations offshore of the Boso Peninsula of Chiba Prefecture that exhibited high coefficient of variation values, from 0.91 to less than 1.10. When this is considered in combination with Fig. 11a, b, and when we compare the region offshore of the Boso Peninsula of Chiba Prefecture with other regions off the Tohoku coast, the uncertainty of the hypothesized wave height is large, although small wave heights were predicted on average, and there is a high probability of incursion of a wave height divergent from the average wave height. Table 5a shows stochastically processed wave heights at the location at which the maximum coefficient of variation was recorded offshore from the Boso Peninsula of Chiba Prefecture and the maximum wave height of each tsunami calculation. Figure 12a shows the tsunami wave height frequency distribution when considering the uncertainty at the same location. The maximum wave height data calculated directly from the tsunami simulation of Table 5a show a large difference between the maximum wave height when the position of the asperity is placed at the southernmost edge of the fault zone (4.3 m for the case of M_w 9.0) and when the position of asperity is placed at the northernmost edge of the fault zone (1.0 m for the case ofM_w 9.0); we believe this is the cause for the increase in the coefficient of variation. Figure 12a shows a frequency distribution with a relatively wide range at higher wave heights, up to a maximum of 6.8 m with an average of 1.3 m. The second interesting fact is that the point where the coefficient of variation reached a minimum was a point offshore of the Fukushima Prefecture coast. Similarly, Table 5b shows the data at the same location, and Fig. 12b shows the tsunami wave height frequency distribution when considering the uncertainty at the same location. The wave height data for the case of Mw 9.0 calculated directly from the tsunami simulation of Table 5b shows that the wave height was highest when the position of the asperity was placed at the southernmost edge of the fault zone; however, relatively high tsunami wave heights were recorded in all cases. The wave height frequency distribution in Fig. 12b is distributed relatively widely, from a minimum of 4.5 m to a maximum of 18.9 m, but the average wave height was at 10.5 m and the coefficient of variation had the minimum value.

3.6.2 Uncertainty of tsunami wave height at the ria coastline of the Tohoku region

Finally, we look at the results for the ria coastline of Iwate Prefecture. Figure 13a shows 0.05 fractile wave height, Fig. 13b shows 0.95 fractile wave height, Fig. 13c shows 90 % confidence interval wave height, and Fig. 13d shows the coefficient of variation of a 1,000-year return period at a depth of 50 m underwater offshore of the ria coastline of Iwate Prefecture. No noticeable trend is seen for the 0.05 fractile wave height, but the values of the 0.95 fractile wave height, 90 % confidence interval, and coefficient of variation show that the values tend to be higher offshore from the peninsular tips of the ria coastline and lower in the closed-off sections of bays. Table 6a shows the data at the location of the maximum coefficient of variation offshore from the tip of a peninsula of the ria coastline, and Table 6b shows the location of the minimum coefficient of variation in a closed-off section of the bay shown in the yellow boxes in Fig. 13. Figure 14 shows the tsunami wave height frequency distribution when considering the uncertainty at the same locations. These data show that, in the vicinity of the peninsular tip of the ria coastline, the standard deviation is larger and the average is smaller than that in the vicinity of the closed-off section of the bay; however, in the vicinity of the closed-off section of the bay, the standard deviation is smaller and the average is larger, and this determines the magnitude of the coefficient of variation. In the vicinity of the peninsular tip of the ria coastline, there is a trend of tsunami waves with high to low wave heights due to the difference in slip distribution within the fault observed widely, and the uncertainty of the tsunami wave height is large. However, in the closed-off section of the bay, the average tsunami wave height is larger than that near the peninsular tip because of effects of the submarine topography in the vicinity, but the range of the obtained wave height is small, i.e., the uncertainty of the tsunami wave height is small. These results clearly demonstrate the influence of topographical effects specific to ria coastlines.

The above results show that when the results of stochastic tsunami wave height analysis limited to a Tohoku-type earthquake fault are analyzed, regional differences in uncertainty at the 90 % confidence interval and coefficient of variation of tsunami wave height clearly exist. In this study, we analyzed a Tohoku-type earthquake fault, but we believe that if the subject fault is changed, regional variations in the uncertainty of tsunami wave heights will also change.

4. CONCLUSIONS

We draw the following conclusions based on stochastic tsunami wave height hazard analysis of a fault having multiple slip distributions using a random source parameter model with a Tohoku-type earthquake fault zone and logic tree.

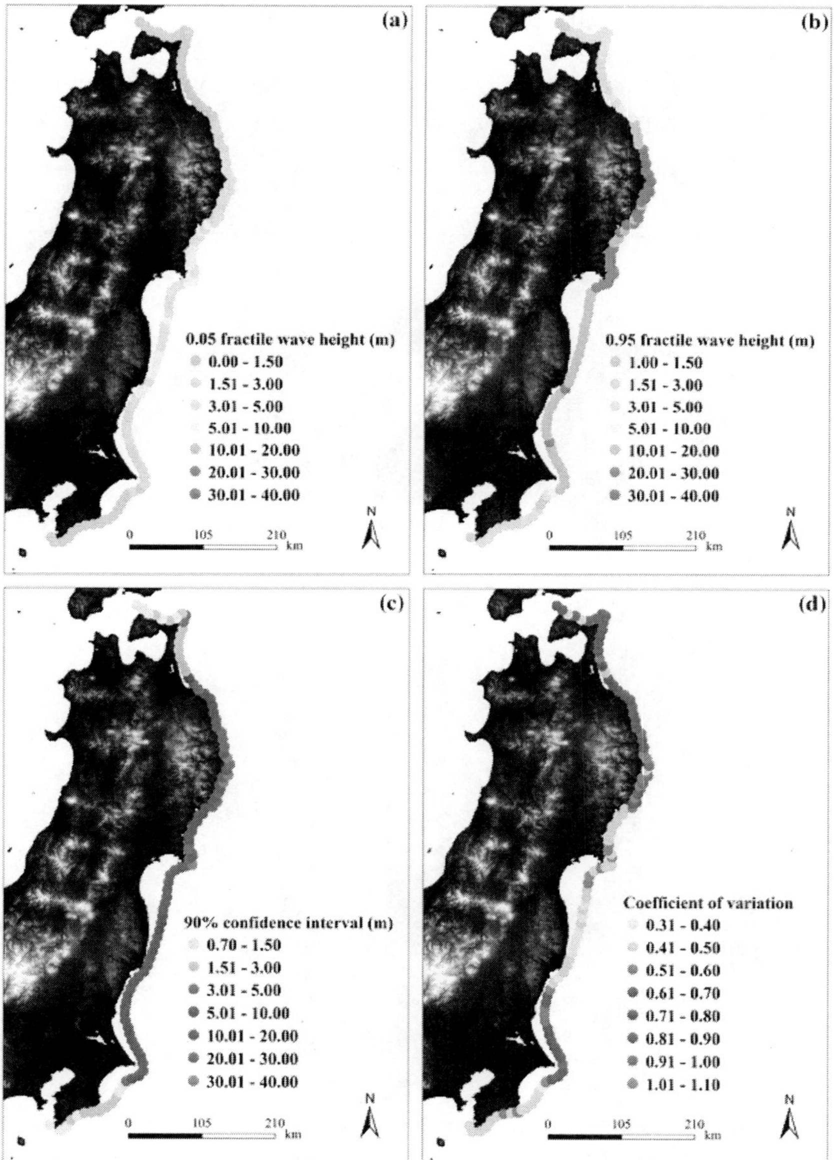

Figure 11. a 0.05 fractile wave height, **b** 0.95 fractile wave height, **c** 90 % confidence interval wave height, and **d**coefficient of variation of a 1,000-year return period at a depth of 50 m underwater offshore from the Tohoku coast

Table 5. Stochastically processed wave heights and maximum wave heights for each case of tsunami simulation at (a) the location of the maximum coefficient of variation and (b) the location of the minimum coefficient of variation

	(a) The location of the maximum coefficient of variation	(b) The location of the minimum coefficient of variation
Latitude	35.505°	37.345°
Longitude	140.75099°	141.1454°
0.05 fractile wave height (m)	0.5	6.6
0.95 fractile wave height (m)	4.9	17.4
90 % confidence interval (m)	4.4	10.8
Average wave height μ (m)	1.3	10.5
Standard deviation σ (m)	1.41	3.45
Coefficient of variation c.o.v.	1.08	0.33
Maximum wave heights calculated from each simulation case (m)		
M_w 8.9		
North	0.7	8.2
North-central	1.8	7.5
Central	0.8	6.8
South-central	0.9	8.8
South	2.7	8.2
M_w 9.0		
North	1.0	12.0
North-central	2.8	11.0
Central	1.2	9.6
South-central	1.4	12.0
South	4.3	12.1
Mw9.1		
North	1.4	15.2
North-central	3.9	13.9
Central	1.6	11.9
South-central	2.1	15.2
South	6.0	16.3

(1) When the observation results of the Tohoku earthquake tsunami were compared with the results of the stochastic tsunami hazard analysis, the return period of the observation results at the location of the south Iwate offshore GPS wave gauge was equivalent to approximately 1,709 years (0.50 fractile wave height), and the return period at the location of the Fukushima Prefecture offshore GPS wave gauge was equivalent to approximately 600 years (0.50 fractile wave height). It was ascertained that the return period statistically determined from stochastic tsunami hazard analysis results differs by a fair amount depending on the region.

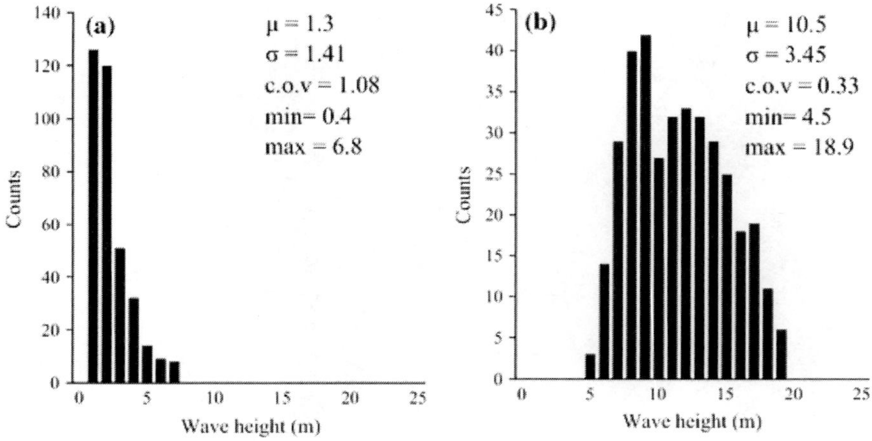

Figure 12. Frequency distribution of tsunami wave height when considering uncertainty at locations where the coefficient of variation reached **a** the maximum of 1.08 and **b** the minimum of 0.33. The total number of calculated tsunami wave heights is 360. Here, μ is the average, σ is the standard deviation, *c.o.v.* is the coefficient of variation, *min* is the minimum value, and *max* is the maximum value

(2) Analysis of the effect of the number of slip distribution patterns on stochastic tsunami hazard analysis results showed that the uncertainty of tsunami wave height increased as the number of slip distribution patterns increased, but for relatively small wave heights, there was almost no effect of the number of slip distribution patterns on the results of the hazard analysis.

(3) When the uncertainty of tsunami wave height of a 1,000-year return period at a depth of 50 m underwater offshore from the Tohoku coast was examined, regional differences in uncertainty were clearly observed. In general, the 90 % confidence interval was high for locations where the wave height of each fractile point was high. Offshore of the Boso Peninsula of Chiba Prefecture, where the coefficient of variation reached the maximum, the difference in maximum wave height due to differences in slip distribution in the fault zone contributed greatly to the large coefficient of variation. At a ria coastline, the coefficient of variation was relatively large at a location offshore from the peninsular tip and relatively smaller at an offshore location in the closed-off section of a bay. We believe this is the effect of the submarine topography in a ria coastline.

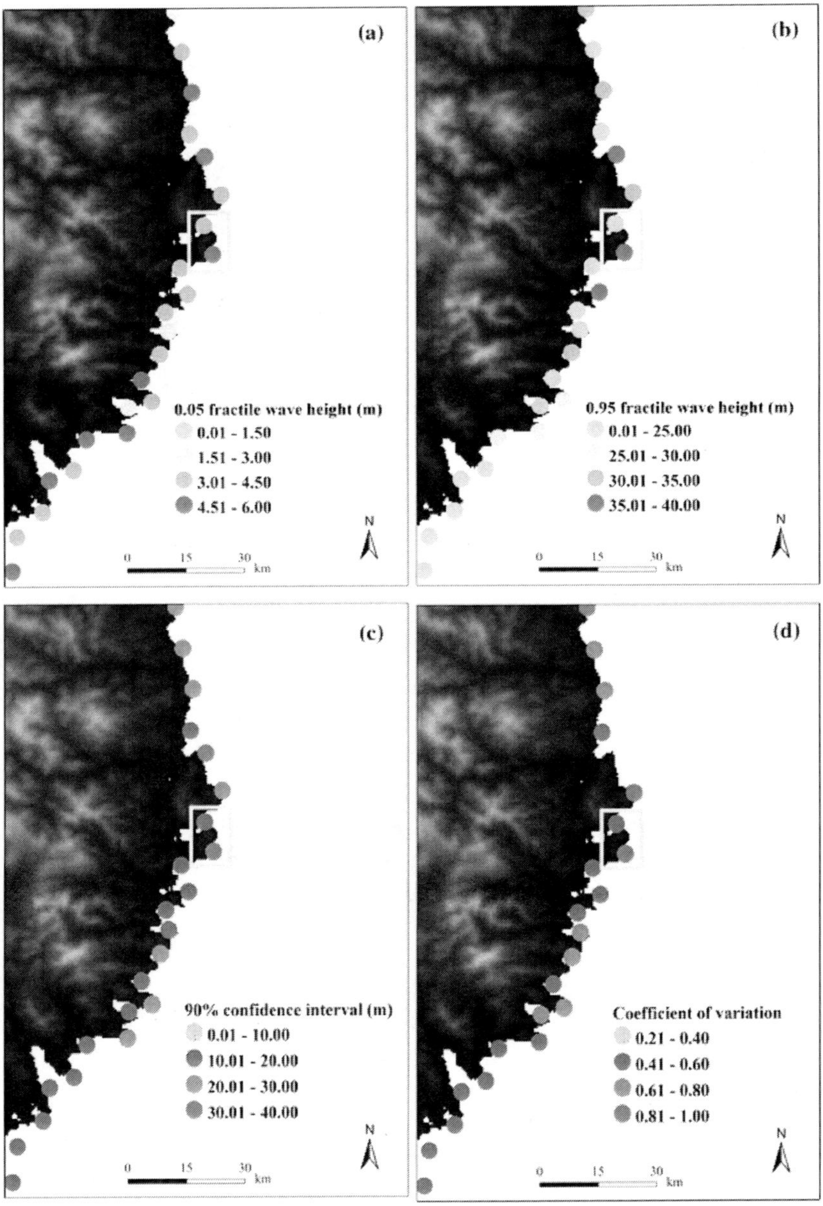

Figure 13. a The 0.05 fractile wave height, **b** 0.95 fractile wave height, **c** 90 % confidence interval wave height, and **d**coefficient of variation of the 1,000-year return period at a depth of 50 m underwater offshore from the ria coastline of Iwate Prefecture. The two locations in the *yellow boxes* indicate a representative location offshore from a peninsular tip of the ria coastline and a representative offshore location in the closed-off section of a bay

Table 6. Stochastically processed wave heights and maximum wave heights for each case of calculation at (a) the location of the maximum coefficient of variation offshore from the peninsular tip of a coastline and (b) the location of the minimum coefficient of variation in the closed-off section of a bay shown in the yellow boxes in Fig. 13

	(a) The location of the maximum coefficient of variation offshore from the peninsular tip of a coastline	(b) The location of the minimum coefficient of variation in the closed-off section of a bay
Latitude	39.6513°	39.4974°
Longitude	142.0329°	142.0303°
0.05 fractile wave height (m)	4.9	4.3
0.95 fractile wave height (m)	39.1	20.3
90 % confidence interval (m)	34.2	16.0
Average wave height μ (m)	9.0	10.7
Standard deviation σ (m)	9.23	4.66
Coefficient of variation c.o.v	1.03	0.44
Maximum wave heights calculated from each simulation case (m)		
Mw8.9		
North	19.7	11.4
North-central	7.1	10.1
Central	6.7	8.8
South-central	5.2	4.7
South	5.4	7.1
Mw9.0		
North	29.6	15.9
North-central	10.4	14.2
Central	9.6	12.4
South-central	7.5	6.7
South	7.7	10.2
Mw9.1		
North	37.6	19.7
North-central	14.3	18.3
Central	13.0	15.8
South-central	9.5	9.0
South	10.3	13.1

As demonstrated in this study, important information for promoting regional disaster prevention activity can be obtained by quantifying and visualizing the uncertainty of tsunami wave heights by stochastic tsunami hazard analysis.

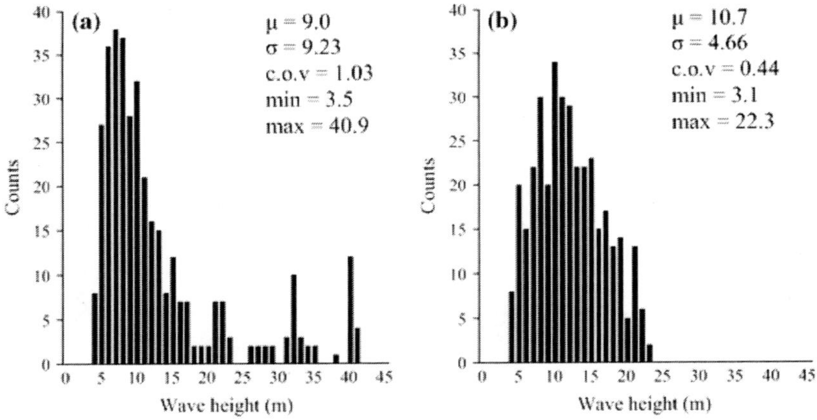

Figure 14. Frequency distribution of tsunami wave height considering uncertainty: **a** at a location offshore from the peninsular tip where the coefficient of variation is large and **b** a location in the closed-off section of a bay where the coefficient of variation is small. These locations are in the *yellow boxes* in Fig. 13. The total number of calculated tsunami wave heights is 360. Here, μ is the average, σ is the standard deviation,*c.o.v.* is the coefficient of variation, *min* is the minimum value, and *max* is the maximum value

ACKNOWLEDGMENTS

We thank the anonymous reviewers who provided us with valuable comments and helped improve the manuscript. This research was partially supported by Specific Project Research (B70) for 2014 from the International Research Institute of Disaster Science (IRIDeS) at Tohoku University. This research was also supported by funding from Tokio Marine & Nichido Fire Insurance Co., Ltd., through the IRIDeS.

REFERENCES

1. Aida I (1978) Reliability of a tsunami source model derived from fault parameters. J Phys Earth 26:57–73

2. Annaka T, Satake K, Sakakiyama T, Yanagisawa K, Shuto N (2007) Logic-tree approach for probabilistic tsunami hazard analysis and

its applications to the Japanese coasts. Pure appl Geophys 164:577–592

3. Cairo MC, Neilson BL (1997) Modeling and generation random vectors with arbitrary marginal distributions and correlation matrix. Technical Report. Department of Industrial Engineering and Management Sciences, Northwestern University, Evanston

4. Freund LB, Barnett DM (1976) A two-dimensional analysis of surface deformation due to dip-slip faulting. Bull Seism Soc Am 66:667–675

5. Fujiwara H, Kawai S, Aoi S, Morikawa N, Senna S, Azuma H, Ooi M, Hao KX, Hasegawa N, Maeda T, Iwaki A, Wakamatsu K, Imoto M, Okumura T, Matsuyama H, Narita A (2013) Some improvements of seismic hazard assessment based on the 2011 Tohoku earthquake. Technical note of the National Research Institute for Earth Science and Disaster Prevention 379 (in Japanese)

6. Geist EL (1999) Local tsunamis and earthquake source parameters. Adv Geophys 39:117–209

7. Geist EL (2002) Complex earthquake rupture and local tsunamis. J Geophys Res 107(B5):2086.

8. Geist EL (2005) Local Tsunami Hazards in the Pacific Northwest from Cascadia Subduction Zone Earthquakes. U.S. Geological Survey Professional Paper 1661-B

9. Geist EL, Dmowska R (1999) Local tsunamis and distributed slip at the source. Pure appl Geophys 154:485–512.

10. Geist EL, Yoshioka S (1996) Source parameters controlling the generation and propagation of potential local tsunamis along the Cascadia margin. Nat Hazards 13:151–177

11. Hwang L, Divoky D (1970) Tsunami generation. J Geophys Res 75:6802–6817

12. Ishii Y, Dan K, Miyakoshi J, Takahashi H, Mori M, Fukuwa N (2013) Source modeling of hypothetical Nankai Trough, Japan, earthquake and strong ground motion prediction using the empirical Green's functions. Shimizu Corporation Research Report 90:17–26 (in Japanese)

13. IUGG/IOC TIME Project (1997), Numerical method of tsunami simulation with the leap-frog scheme, Intergovernmental Oceanographic Commission Manuals and Guides 35

14. Iwabuchi Y, Sugino H, Ebisawa K (2014) Guidelines for Developing Design Basis Tsunami based on Probabilistic Approach. JNES-RE-Report Series, JNES-RE-2013-2041

15. Japan Society of Civil Engineers (2011) A method for probabilistic tsunami hazard analysis, The Tsunami Evaluation Subcommittee, The Nuclear Civil Engineering Committee. Japan Society of Civil Engineers, Tokyo

16. Kawai H, Satoh M, Kawaguchi K, Seki K (2013) Characteristics of the 2011 Tohoku tsunami waveform acquired around Japan by NOWPHAS equipment. Coast Eng J 55(3):1350008.

17. Lavallée D, Archuleta RJ (2003) Stochastic modeling of slip spatial complexities for the 1979 Imperial Valley, California, earthquake. Geophys Res Lett 30(5):1245.

18. Lavallée D, Archuleta RJ (2005) Coupling of the random properties of the source and the ground motion for the 1999 Chi Chi earthquake. Geophys Res Lett 32:L08311.

19. Lavallée D, Liu P, Archuleta RJ (2006) Stochastic model of heterogeneity in earthquake slip spatial distributions. Geophys J Int 165:622–640

20. Lin I, Tung CC (1982) A preliminary investigation of tsunami hazard. Bull Seism Soc Am 72:2323–2337

21. Liu PC, Archuleta RJ, Hartzell S (2006) Prediction of broadband ground-motion time histories: hybrid low/high frequency method with correlated random source parameters. Bull Seism Soc Am 96:2118–2130

22. Løvholt F, Pedersen G, Bazin S, Kühn D, Bredesen RE, Harbitz C (2012) Stochastic analysis of tsunami runup due to heterogeneous coseismic slip and dispersion. J Geophys Res 117:C03047.

23. Mai P, Beroza G (2002) A spatial random-field model to characterize complexity in earthquake slip. J Geophys Res 107(B11):2308.

24. McCloskey J, Antonioli A, Piatanesi A, Sieh K, Steacy S, Nalbant S, Cocco M, Giunchi C, Huang JD, Dunlop P (2007) Near-field propagation of tsunamis from megathrust earthquakes. Geophys Res Lett 34:L14316.

25. McCloskey J, Antonioli A, Piatanesi A, Sieh K, Steacy S, Nalbant S, Cocco M, Giunchi C, Huang JD, Dunlop P (2008) Tsunami threat in the Indian Ocean from a future megathrust earthquake west of Sumatra. Earth Planet Sci Lett 265:61–81.

26. Ng MK, Leblond PH, Murty TS (1991) Simulation of tsunamis from great earthquakes on the Cascadia subduction zone. Science 250:1248–1251

27. Okada Y (1985) Surface deformation due to shear and tensile faults in a half-space. Bull Seism Soc Am 75:1135–1154

28. Pelayo AM, Wiens DA (1992) Tsunami earthquakes: slow thrust-faulting events in the accretionary wedge. J Geophys Res 97(B11):15321–15337

29. Rikitake T, Aida I (1988) Tsunami hazard probability in Japan. Bull Seism Soc Am 78:1268–1278

30. Sakai T, Takeda T, Soraoka H, Yanagisawa K, Annaka T (2006) Development of a probabilistic tsunami hazard analysis in Japan. In: Proceedings of ICONE14, international conference on nuclear engineering, Miami, FL

31. Somerville PG, Irikura K, Graves R, Sawada S, Wald D, Abrahamson N, Iwasaki Y, Kagawa T, Smith N, Kowada A (1999) Characterizing crustal earthquake slip models for the prediction of strong ground motion. Seism Res Lett 70:59–80

32. Song YT, Ji C, Fu LL, Zlotnicki V, Shum CK, Yi Y, Hjorleifsdottir V (2005) The 26 December 2004 tsunami source estimated from satellite radar altimetry and seismic waves. Geophys Res Lett 32:L20601.

33. Thio HK (2010) Probabilistic tsunami hazard analysis in California. Pacific Earthquake Engineering Research (PEER) Report 2010/108. University of California, Berkeley

34. Ward SN (1982) On tsunami nucleation: II. An instantaneous modulated line source. Phys Earth Planet Inter 27:273–285

35. Ward SN (2002) Tsunamis. In: Meyers RA (ed) The encyclopedia of physical science and technology, vol 17. Academic Press, London, pp 175–191

36. Whitmore PM (1993) Expected tsunami amplitudes and currents along the North American coast for Cascadia subduction zone earthquakes. Nat Hazards 8:59–73

37. Yong SC, Yong CK, Dongkyun K (2013) On the spatial pattern of the distribution of the tsunami run-up heights. Stoch Environ Res Risk Assess 27:1333–1346

38. Yoshida K, Miyakoshi K, Irikura K (2011) Source process of the 2011 off the Pacific coast of Tohoku earthquake inferred from waveform inversion with long-period strong-motion records. Earth Planet Space 63:577–582

9

INFLUENCE OF AIR TEMPERATURE ON CHILDREN WATER CONTACTS WITH RESPECT TO SCHISTOSOMIASIS TRANSMISSION RISK IN THE SOUROU VALLEY, BURKINA FASO

Issouf Traore[1,3] • Daniel Karthe2,3 • Ali Sie1 • Victor Coulibaly1 • Martin Kappas3

[1]Centre de Recherche en Sante´ de Nouna, BP 02, Nouna, Burkina Faso

[2] Helmholtz Centre for Environmental Research, Magdeburg, Germany

[3] Department of Geography, Georg-August Universita¨t, Go¨ttingen, Germany

ABSTRACT

Climate and water contacts are among the most important risk factors of human schistosomiasis transmission. This study aimed to determine—through direct field observations—the influence of air temperature on the frequencies of children´s water contacts with respect to human schistosomiasis transmission risk in the Sourou Valley, located in the North-western part of Burkina Faso. The objectives of the study were: (i) to study the air temperature at which children have the sensation of heat and start looking for water as a means for natural cooling; and (ii) to study whether high frequencies of swimming or bathing may lead to a higher risks of schistosomiasis transmission. To do so, swimming or bathing were observed during two different periods: a cold one from 3rd January to 2nd February 2011 and a hot one from 3rd April to 3rd May 2011 in Toma-Île, a village totally surrounded by water. The results showed that daily mean air temperatures never exceeded 30 °C during the coldest period, while they were above this value during the hottest period. In total, swimming activities were observed 11/31 days during the coldest period and 31/31 during the hottest period. Bathing of children below the age of 5 years occurred on 55/62 days. 25 °C was the day mean air temperature from which children felt hot and began playing in water. The hourly occurrence of swimming and bathing showed peaks at 2 PM. This coincides with the diurnal maximum of infective cercariae present in the same water. Thus, in order to freshen up during hot periods of the day, children used to look for freshwater and thus expose themselves to a high schistosomiasis transmission risk.

KEYWORDS

Air temperature, Water contacts, Human schistosomiasis, Children, Sourou valley, Burkina Faso

1. INTRODUCTION

Human schistosomiasis (also known as Bilharzia) is a disease caused by infection with blood flukes of the*Schistosoma* genus. Most cases of the disease result from infection with *Schistosoma japonicum* and *S. mansoni,* both of which cause intestinal schistosomiasis and *S. haematobium,* the causative agent of urinary schistosomiasis. The larvae of the organism (cercariae) are released into water by infected snail as intermediate hosts. Human contact with water inhabited by snails is the source of the persistence of the transmission of schistosomiasis (WHO 2013). As a water-based disease, schistosomiasis comes second only to malaria in terms of public health impact. Globally, there are approximately 779 million persons at risk and 207 million persons infested (Van der Werf et al. 2003; Gryseels et al. 2006; Steinmann et al. 2006; Utzinger et al. 2009; WHO 2013). Higher prevalence and intensity of infections are concentrated in childhood (Butterworth and Hagan 1987; Fulford et al. 1998; Kabatereine et al. 1999). In the case of urinary schistosomiasis, the eggs cause damage to the urinary tract and blood appears in urine (haematuria). Urination becomes painful and there is progressive damage to the bladder, ureters and kidneys. Bladder cancer is a common complication (van der Werf et al. 2003; Gryseels et al. 2006; Steinmann et al. 2006; Utzinger et al. 2009; WHO 2013). In the case of intestinal schistosomiasis, the eggs bring about progressive enlargement of the liver and spleen as well as damage to the intestine and hypertension of the abdominal blood vessels. Symptoms such as bloody diarrhoea, abdominal pains and fatigue are commonly reported (Gryseels and Nkulikyinka 1990). A very significant proportion of death due to schistosomiasis is concentrated in Sub-Saharan Africa (SSA). There are approximately 150,000 cases of death from urinary schistosomiasis and 130,000 from *S. mansoni* infections in SSA, per annum (van der Werf et al. 2003; King and Dangerfield-Cha 2008; Hotez and Kamath 2009).

Climate, environment and human behaviour remain the most important risk factors of human schistosomiasis transmission (Weil and Kvale 1985; Mayer 1983; Ernould 2000; Scott et al. 2003; Gazzinelli et al. 2006; Mangal et al. 2008; Zhou et al. 2008; Koukounari et al. 2011; Tay et al. 2011). Much is known about the influence of temperatures on parasites and intermediate hosts through numbers of studies by biologists. Water temperatures between 10 and 30 °C are a prerequisite for egg hatching. The ability of the hatching miracidia (the first development stage of the parasite) to penetrate snails is also influenced by water temperature with an optimum occurring between 26 and 28 °C. Below 10 °C, miracidia are very inactive (DeWitt 1955; Chu et al. 1966). The parasites within the snail intermediate host are profoundly influenced by habitat temperatures, particularly the duration of the pre-patent period in snails (Stirewalt 1954; DeWitt 1955). Similar results were obtained with the Egyptian *S. mansoni* in *Planorbis pfeifleri*, where the pre-patent period lasted 15 days at 32–33 °C; 19–22 days at 26–28 °C; and 33–37 days at 20–22 °C. Comparable periods for*S. haematobium* in *Physopsis globosa* were 22–23 days at 32–33 °C; 36 days at 26–28 °C and 66–68 days at 20–22 °C. The infectivity rate or the cercariae capacity to penetrate the skin depends on temperature with an optimum ranging from 26 to 28 °C (Stirewalt 1954).

Human behaviour plays a key role in the schistosomiasis transmission process. Through various water contacts, the human being ensures the success of the circulation of parasites (Weil and Kvale 1985; Chandiwana 1987; Takougang et al. 1993; Amazigo et al. 1997; Kloos et al. 1998; Watts et al. 1998; Ernould2000; Scott et al. 2003; Grisorio et al. 2005; Gazzinelli et al. 2006). Various studies addressed the role of different water contact activities and relevant age groups but their results vary with regard to the incidence or prevalence of schistosomiasis infections. In Sub-Saharan Africa, the most important water contact activity in childhood is swimming or bathing, the frequencies of which depend on the variability of air temperature. Understanding the short and mid-term influence of air temperature on children exposure to contaminated water may help develop innovative strategies towards effective control of schistosomiasis. To date, such information is not available.

This study tested the short-term (hourly) and the mid-term (diurnal scale) influence of air temperature on the frequencies of children's water contacts with respect to human schistosomiasis transmission risk. The objectives of the study were the following: (i) to study the air temperature at which children have the sensation of heat and start looking for water as a means for natural cooling; and (ii) to study whether daily and hourly occurrence of swimming or bathing leads to higher risks of human transmission of schistosomiasis.

2. MATERIALS AND METHODS

2.1 Study site

This direct field observation was conducted in Toma-Île (N13°08.807′, W3°27.056′), a village totally surrounded by water throughout the year, in the Sourou Valley, in the North-western part of Burkina Faso. The Sourou River is surrounded by a large scale irrigation scheme that is typical in size and character for similar developments in Sub-Saharan Africa. Since the construction of a dam in 1976, water flow in the river is regulated and the river is de facto turned into a reservoir storing up to 200 million cubic meters of water (Karthe et al. 2011). The Sourou River marks the boundary between the provinces of Kossi and Sourou (the umbrella province of Toma-Île). The region is characterized by a semi-arid tropical climate, difficult access even to basic health infrastructure and high prevalence of water associated diseases, the most relevant which are malaria and schistosomiasis (Karthe and Traoré 2009; Karthe 2010; Traoré et al. 2011; Karthe et al. 2012; Traoré et al. 2012; Traoré 2013). A canoe is always needed to access Toma-Île. The Sourou River, a perennial one, is the main water source for the community. Safe water supply and wastewater treatment facilities were non-existent during the study period (2011). The absence of trees on the island suggests that the most common way to freshen up during the intense midday heat is bathing or swimming in the river. Previous epidemiologic and biologist studies showed that human schistosomiasis is widely spread in Burkina Faso (Poda et al. 2004; Koukounari et al. 2007; Clements et al. 2008; Koukounari et al. 2011; Zongo et al. 2012). Snails intermediate hosts collected and identified in the Sourou Valley were: *Biomphalaria pfeifferi* for the intestinal schistosomiasis (*S. mansoni*) and *Bulinus truncatus* and *B. senegalensis* for the urinary schistosomiasis (*S. haematobium*) (Dianou et al. 2004; Poda et al. 2004; Zongo et al. 2012). The levels of prevalence in Toma-Île village were 51.9 % for the urinary schistosomiasis and 50.6 % for the intestinal schistosomiasis among school aged children. Two years after in 2002 prevalence were 43.4 and 90.8 %, respectively (Poda et al. 2004). Zongo et al. (2012) indicated a global prevalence of 3 % for *S. haematobium* among school aged children in the Sourou Valley and within a particular context of mass drug administration with praziquantel. These authors have also observed that in general in Burkina Faso, prevalence of *S. haematobium* is decreasing while that for *S. mansoni* is increasing. Schistosomiasis remains a major public health problem in endemic settings to both *S. haematobium* and *S. mansoni* like Toma-Île in the Sourou Valley.

2.2 Observation periods

For comparison purpose, both hot and cold periods of 1 month (31 days) were selected. In order to identify these two periods, we used a 31-year monthly mean air temperature dataset obtained from the Di-Sourou weather station (located 4 km from Toma-Île). We determined that for an average year (Fig. 1), the coldest period runs from 3rd January to 2nd February (31 days) and the hottest period from 3rd April to 3rd May 2011 (31 days). During these two observation timeframes, we recorded all activities affecting a partial or total immersion into water. Daily observations were carried out from 6 AM to 6 PM (12 h/day).

The study schedule, the data collection method and two fieldworkers supporting the observation were introduced to the community through an informative meeting held in December 2010. The community members in general specifically women were made aware of the study by its supervisor and a delegated administrative representative.

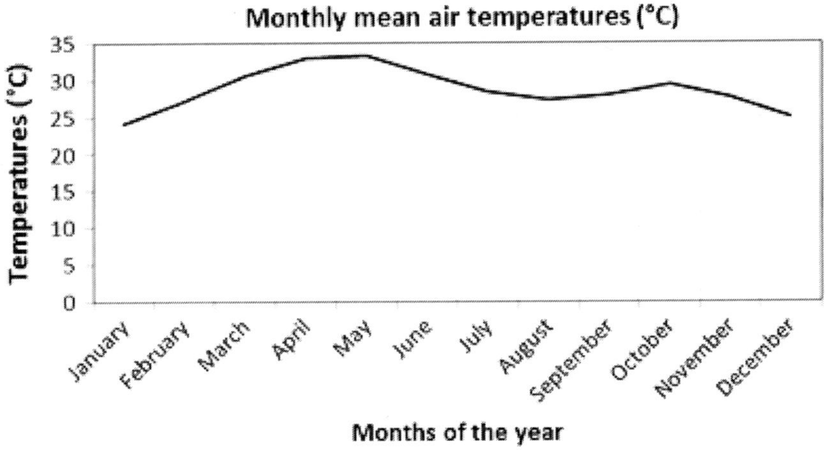

Figure 1. A 31-year (1980–2010) curve of mean air temperatures (data from Di-Sourou weather station)

2.3 Activities tracked and determination of children's ages

Two water contact activities were of concern: recreational swimming and bathing. Children were the target population. In the context of the paper, swimming refers to a recreational water activity carried out by children aged between 5 and 14 years, typically consisting in a total immersion of the body (Chandiwana1987; Takougang et al. 1993). Conversely, bathing refers to children below the age of 5 years being brought to and washed by their mothers or other persons into the river. Children were not followed up individually. Fieldworkers were asked to report daily entries into contact with water. A child or a woman can have several such contacts per day. Only the hourly water contact frequencies at the observed sites were recorded in the day.

To determine the age of children, fieldworkers asked directly to children and requested confirmation from their parents. To identify children less than 5 years, any children seen with their parents in the river were asked about their ages and confirmations were given by the mothers, if any. Since the fieldworkers were from the local community it was easy to estimate the ages of children.

2.4 Identification of the mean air temperature threshold

This threshold separates air temperature conditions where a water-contact is likely to be observed versus those where it is likely to be absent. We used a simplest method of breakpoint-based threshold which consists of visually estimating the location of the threshold from plotted data (Ficetola and Denoël 2009). In the presence study, the identification of the breakpoint-based threshold was easy because no swimming occurred over the first 20 days of the direct field observations. Swimming was observed over the rest of 42 days of direct observation. Therefore, the threshold was located at the 21st day of observation. Using a graph, we plotted the 62 days of observation on the X axis and the corresponding mean air temperatures on the Y axis. Then we projected the 21st day on the Y axis to determine the mean air temperature as the threshold.

3. RESULTS

3.1 Overall outcomes of recording

3.1.1 Temperatures variability during the direct observation period

Figure 2a shows minimal air temperatures daily profile for both coldest and hottest periods. The extremes varied between 11.5 and 19.8 °C (range 8.3 °C), and 20.0 and 30.5 °C (range 10.5 °C), respectively. Figure 2c compares daily maximal air temperatures of the coldest and hottest periods. Extreme values varied between 29.0 and 40.3 °C (range 11.3 °C), and 35.5 and 45.0 °C (range 9.5 °C), respectively. Means of minima and maxima per period are plotted in Fig. 2b. Daily mean air temperatures never exceeded 30.0 °C during the coldest period, as opposed to the hottest period where they exceeded this value. Mean temperatures on the coldest day of the hot period exceeded those of the hottest day of the cold period. The same applies to minimum temperatures and, with minor exceptions, to the daily maxima (Fig. 2).

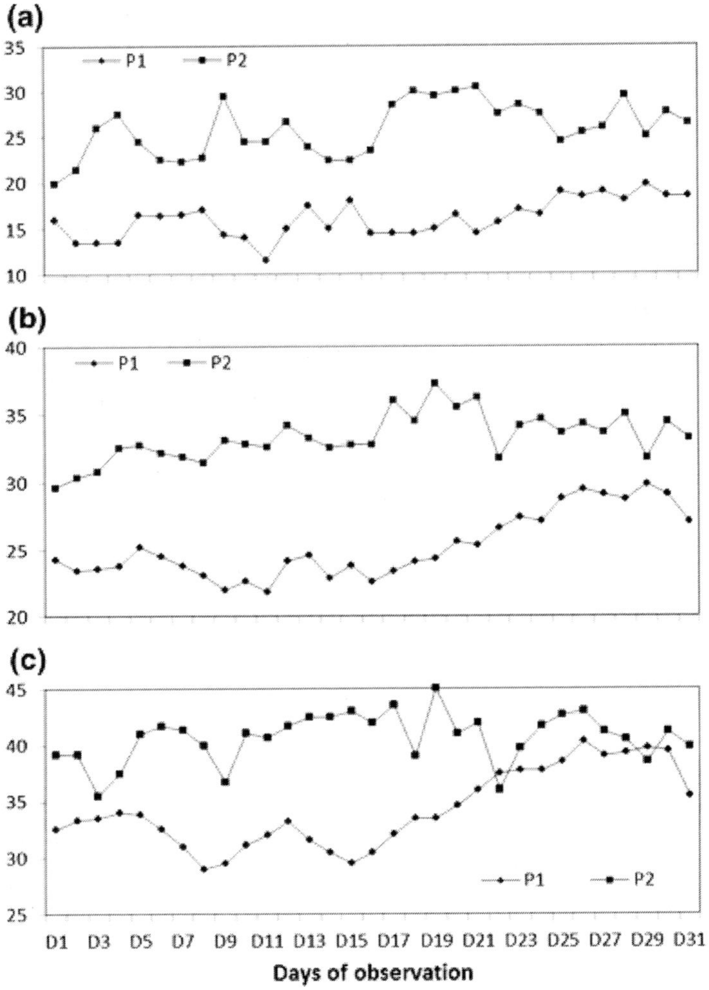

Figure 2. Profiles of air temperatures during the periods of observations (Data from Di-Sourou weather station). **a**Minima, **b** Mean, **c** Maxima; *P1* coldest period (*D1* 3rd January, *D31* 2nd February 2011); *P2* hottest period (*D1* 3rd April, *D31* 3rd May 2011)

3.1.2 Water contact frequencies during the period of direct observation

Figure 3a shows that swimming was observed during 11/31 days and extreme numbers of water contacts varied between 0 and 42 (average of 9.3 contacts/day, 95 % CI 7.5–11.1) during the coldest period. By contrast, during the hottest period swimming occurred during 31/31 days, with extreme numbers ranging from 18 to 192 (average of 77.6 contacts/day, 95 % CI 73.1–82.0). Over the entire observation period, swimming activities were observed during 42/62 days. This means that there were 20 days without recreational swimming.

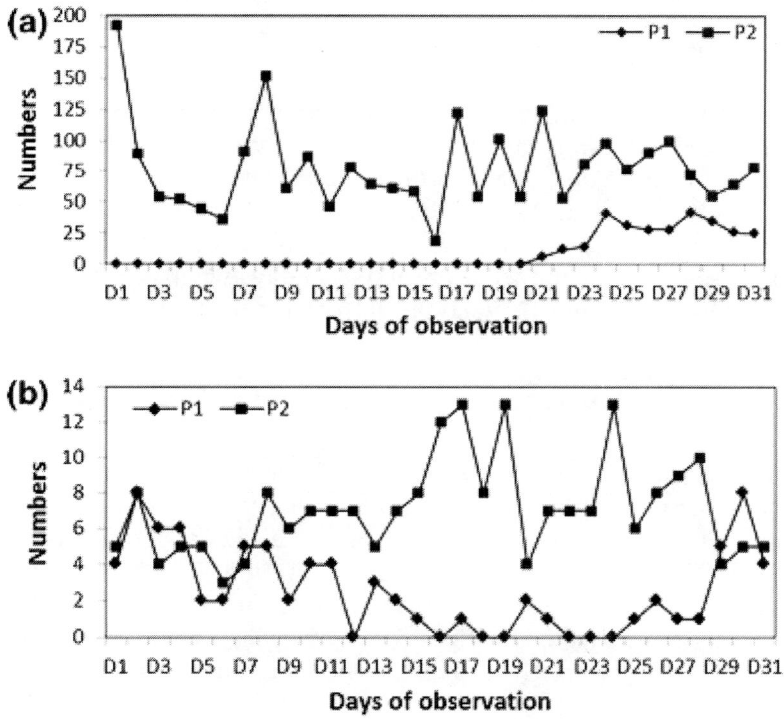

Figure 3. Water-contact frequencies. **a** Frequencies of swimming, **b** Frequencies of bathing of children less than 5 years. *P1* Coldest period (*D1* 3rd January, *D31* 2nd February 2011); *P2* hottest period (*D1* 3rd April, *D31*3rd May 2011)

Figure 3b shows the daily water contacts through bathing of children less than 5 years directly in the river by their mothers. During the coldest period this activity was observed during 24/31 days and extreme numbers of contacts varied between 0 and 8 (average of 3.0 contacts/day, 95 % CI 2.7–3.3). Similarly to swimming, bathing was recorded all 31/31 days with extreme numbers ranging from 3 to 13 contacts (average of 7.0 contacts/day, 95 % CI 6.6–7.4). Over of the entire observation period, bathing occurred during 55/62 days. Therefore, there were 7 days without any contact with the river.

3.2 Influence of air temperatures on water-contact frequencies

3.2.1 Mean term influence (day scale)

Figure 4 shows results of the influence of air temperature on children recreational swimming in the river. Graphs a1, b1 and c1 show the patterns of the coldest period; graphs a2, b2, and c2 show the patterns of the hottest period. Results show that swimming never occurred on the days where the minimal air temperature was less than 15 °C. The first cases of swimming were observed on days where the maximal air temperature reached 35 °C, typically coinciding with mean daily air temperatures of more than 25 °C. The values of the coefficients of determination (R^2) for the coldest period were higher than those for the hottest period.

Figure 5 displays the results of the influence of air temperature on frequencies of bathing children less than 5 years directly in the river. Graphs a1, b1 and c1 show the patterns of the coldest period; graphs a2, b2, and c2 show the patterns of the hottest period. Women used to bring and wash their children less than 5 years even when the day minimal of air temperature was below 15 °C. Here, the values of the coefficients of determination (R^2) for the coldest period were lower than those for the hottest period.

3.2.2 Hot days versus cold days

Based on results in Fig. 4, we used the daily mean air temperature of 25 °C as cut-off point to demarcate cold days (<25 °C) from hot days (≥25 °C) over a 31-year timeframe (1980–2010) as indicated in Fig. 6. The results indicate that 100 % days were hot in 3 months (April, May and October); 99 % days were hot in 2 months (March and June). Contrastingly, more than 50 % days were cold over 2 months (January and December).

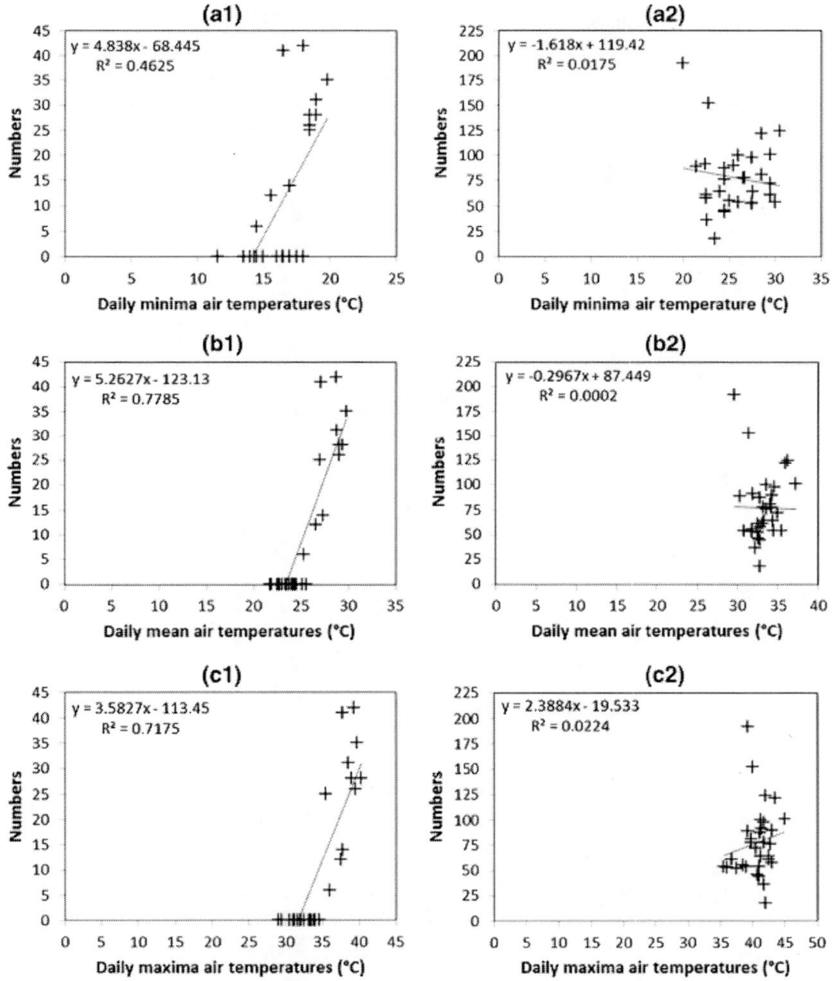

Figure 4. Plot of daily air temperatures and frequencies of recreational swimming. Graphs *a1, b1, c1* show the situations of the coldest period; and graphs *a2, b2, c2* indicate the situations of the hottest period

3.2.3 Short-term influence (hour scale)

Figure 7a shows the hourly occurrence of swimming. During the coldest period water contacts were recorded in 8/12 h (10 AM to 5 PM). The total number of contacts was 288 and the highest peak (28 %) occurred at 4 PM. During the hottest period swimming occurred in 10/12 h (8 AM to 5 PM). The total number of contacts was 2405 with the highest peak (21 %) just at 2 PM.

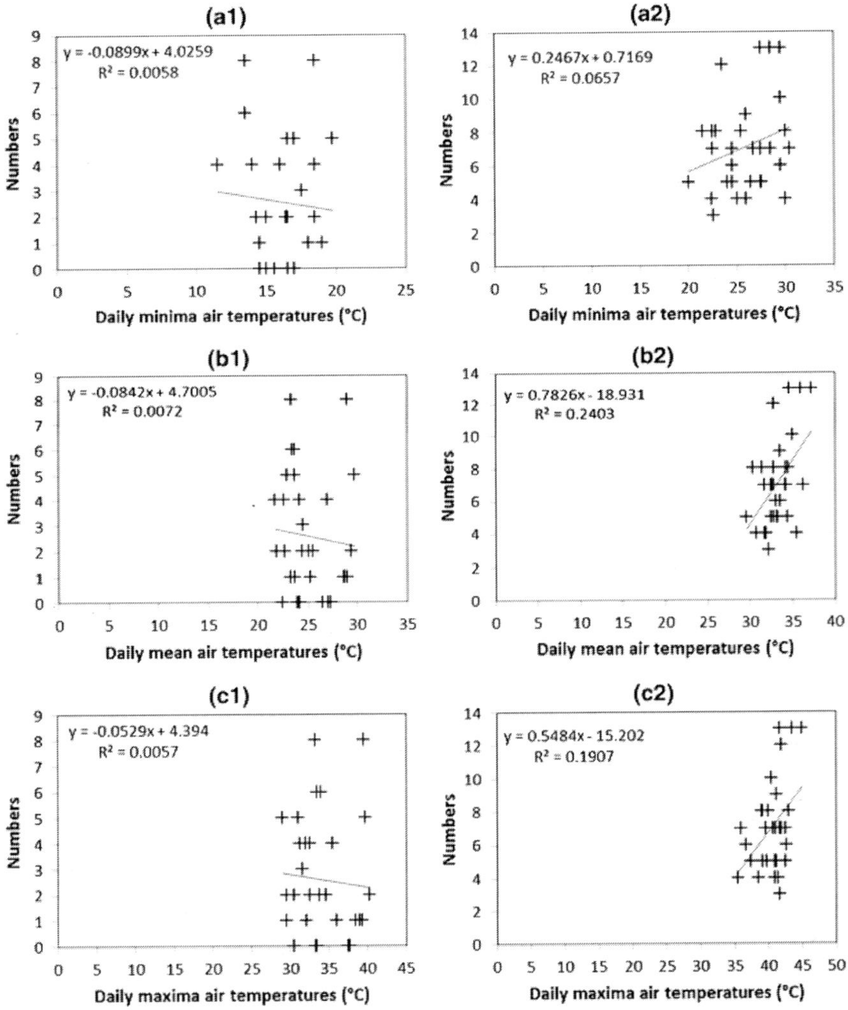

Figure 5. Plot of daily air temperatures and frequencies of washing children below 5 years in the river. Graphs *a1, b1,c1* show situations during the coldest period; graphs *a2, b2, c2* indicate situations during the hottest period

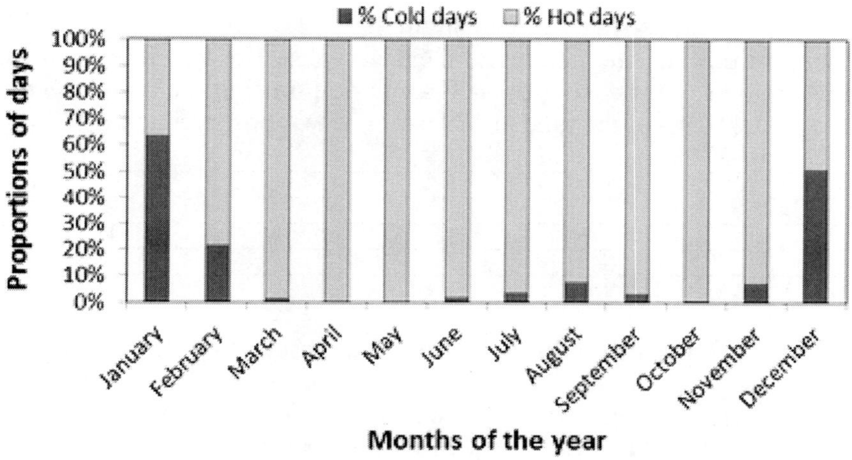

Figure 6. Distribution (%) of cold days (<25 °C) and hot days (≥25 °C) per month using daily data from 1980 to 2010 (*Source of data* Di-Sourou weather station)

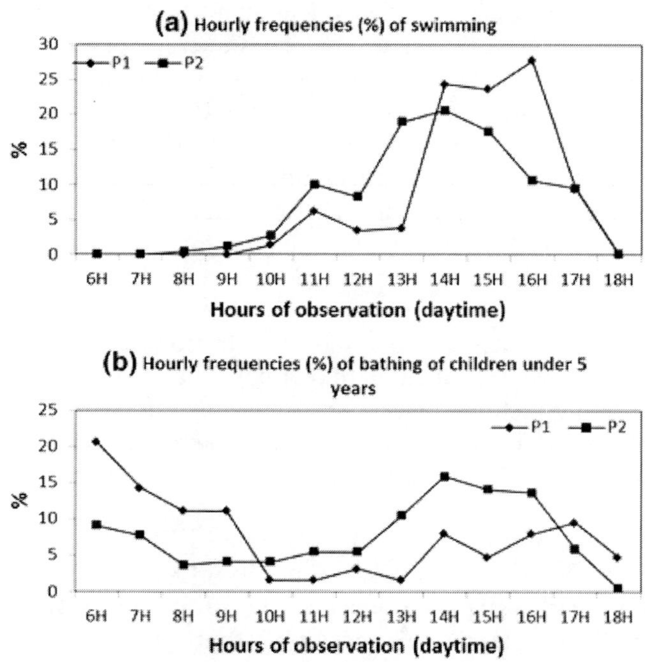

Figure 7. Profiles of the hourly occurrence of water contacts. *P1* Coldest period (3rd January to 2nd February 2011);*P2* hottest period (3rd April to 3rd May 2011)

Figure 7b shows that the maximum (21 %) of the total of 80 water contacts through the bathing of children below the age of 5 years was observed at 6 AM during the coldest period. During the hottest period the peak (16 %) of the total of 220 contacts was observed at 2 PM.

4. DISCUSSION

This study determined through direct field observations the influence of air temperature on children water contact frequencies with respect to human schistosomiasis transmission risk in Toma-Île village which is located in the Sourou Valley, North-western Burkina Faso.

4.1 Study approach and data collection

Two approaches were used to measure individual exposure to the risk of transmission of schistosomiasis. The quantitative approach based on direct field observation and the qualitative approach based on individual report of water contacts (Ernould 2000). The first method was used in this study, consisting in tracking contact activities namely swimming and bathing. The link to the risk of schistosomiasis transmission in the context of this study was limited to the contact with the Sourou River standing as an excellent habitat both for *Bulinus truncatus*, the snail intermediate hosts for *S. haematobium*, and *Biomphalaria pfeifferi*, snail intermediate host for *S. mansoni* (Poda et al. 2004; Dianou et al. 2004; Poda et al. 2006; Zongo et al. 2012). Since the schistosomiasis transmission route is the contact with water, any contact with the river is deemed to be risky. The magnitude of the risk is driven by many factors such as the time the contact occurs (morning, noon, afternoon), the duration and the degree of immersion of the body (Takougang et al. 1993; Ernould 2000).

Toma-Île, the site of this study, is a lakeside village totally surrounded by water inhabited by a small community of 40 compounds and 460 inhabitants with slight difference in the socioeconomic status (Traoré2013). Children (<15 years old) represent about 51 % of the population (INSD 2008). In this study, children were not followed up individually, and since our aim was not to quantify water contacts per child, no additional census was needed. The purpose of the work is to link child behaviour to temperature variability. When coming from outside the village, a canoe is needed to reach the water contact points we observed. Therefore, all children observed in contact with the river were dwelling in the village. This means that the children were observed in their natural environment. Willing to map the urinary schistosomiasis prevalence among schoolchildren in 2011 in the study area, Traoré et al. (2012) realized that no school-goers were coming from Toma-Île. Therefore, school holidays did not have any influence on the results of this study. In the context of Toma-Île, swimming or bathing are diurnal activities. Therefore observations were carried out in the daytime, from the sunrise to the sunset. During this time, we recorded all water contacts. The selection of two observation periods of 31 days each enabled to compare swimming or bathing frequencies within and among the observation periods in order to better understand the influence of air temperature on water contact frequencies. Children with different socioeconomic status and residential locations may have different water-contact behaviours (Gazzinelli et al. 2006). In Senegal, Scott et al. (2003) found that that age, sex and place of residence do determine exposure without suggesting that exposure has an influence on the relationship across/between these factors and the intensity of the infection. The socioeconomic status plays a minor role in the case of Toma-Île where the entire community have contacts with the river because of the lack of hydraulic facilities.

4.2 Influence of air temperature on children behaviour

The study was designed to catch the role of the heat in triggering swimming or bathing activities. Toma-Île, being a lakeside setting without any safe water supply facilities and where the Sourou River remains the unique water source point, this provides an opportunity to observe such behaviour. The sample scheme targeted children and consisted in recording all entries into the river for swimming or bathing on a daily basis. Even though the perception of heat is different among individuals, extreme heat typically leads to avoidance mechanisms. For children in the Sourou Valley, swimming comes out as a reaction to hot days. In the context of our study, hot days leading to swimming were characterized by daily mean temperatures of 25 °C and more. The first 20 days of observation could be categorized as cold days. Consequently, children did not go swimming in the river. Hot days totalled the remaining 42/62 days of observation and children generally reacted by swimming, not continuously though. The number of daily water contacts varied unevenly during these 42 hot days. This fact explains the lower values of R^2 obtained during the hottest period (Fig. 4). Chandiwana (1987) observed seasonal variation of water contacts with intense contacts during the hot dry season in Zimbabwe. The author considered swimming activities as 'pressures' since with every individual swimming increased the chance of introducing parasites into the river. On an annual scale, considerable pressures on the river due to swimming were recorded in 5 months (March, April, May, June and October) since children went swimming every day. Lower pressures on the river were observed during 3 months (January, February and December) because children did not go swimming every day. However, whatever the period and temperature, children never went swimming during the entire daytime. Peaks of recreational swimming were observed by Watts et al. (1998) in summer. The hourly occurrence of swimming showed a shift of 2 h between the coldest period and the hottest one. Children started to play in the river from 10 AM during the cold period while, as compared to an earlier starting time during the hot period, around 8 AM. The same lag could be observed for peaks of swimming times: 4 PM versus 2 PM, respectively (Fig. 7a). Higher air temperatures led to higher swimming frequencies, earlier starts and a later end of the swimming period. The total absence of clean water facilities explains why women wash children less than 5 years old directly in the river. Peaks of bathing observed at 6 AM can be explained by the fact that during the night, children below the age of 5 years often defecate. Therefore, daily bathing is carried out in the morning, irrespective of temperatures. This explains why values of R^2 for temperature and bathing were very low (Fig. 5). The influence of air temperature appeared more clearly at the hourly level. The curve representing the frequencies of bathing during the hottest period is above that of the coldest period from 10 AM to 4 PM (Fig. 7b). Similar results were found by Takougang et al. (1993) in the extreme North of Cameroon and by Zongo et al. (2012) over 10 sites in Burkina Faso.

At the Di-Sourou weather station hourly data are not measured. Temperatures are recorded three times a day: 6 AM, 12 and 17 PM. Therefore, local data about daily minimum and maximum air temperatures were missing. Since all curves representing hourly frequencies of swimming and bathing showed a peak at 2 PM (Fig. 7), we assumed this time to be the hottest in the daytime.

4.3 Schistosomiasis transmission risk

Our field observation was not coupled with parasitological examinations of urines and faeces to check the relations between water contact and infection. However, such relationship derived from empirical studies. Tay et al. (2011) found a significant association between specific water contacts including bathing and playing in streams/ponds in Ghana. Kloos et al. (1990) indicated overall significant correlations between water contacts and schistosomiasis infection. But swimming was the only activity positively correlated with egg-counts, all other activities being negatively correlated. Swimming leads to exposing the whole body, which increases the risk of infection (Takougang et al 1993). This water contact occurs at the time when cercariae are very active with increase of infectivity (Stirewalt 1954; N'Goran et al. 1997). The duration of the exposure to contaminated water can increase the risk of infection since several cercariae can penetrate the swimmer's body. The intensity of penetration will also depend on the density of the cercariae at the contact point (Ernould 2000).

Swimming or bathing is a key cause of schistosomiasis transmission in the water points inhabited by snail species which are the intermediate hosts of the parasite. *S. mansoni* and *S. haematobium* cercariae are usually released from their intermediate hosts during the daytime; they showed mean shedding time between 1 PM and 4 PM (Nojima and Sato 1982; Théron 1984; N'Goran et al. 1997; Wolmarans et al. 2002). The daily peak of human pressure of water exposure was observed at the hottest hours of the daytime (1 PM and 3 PM). Although many parameters govern cercarial shedding, rhythms of cercarial emergence have been found as adaptive behaviours and shaped under the selective pressures exerted by the behaviour of the human host. The early or late cercarial emergence pattern is influenced by the epidemiological predominance of the human host in the transmission zone (Théron 1984). Numerous studies found that the circadian cercarial shedding shows low intensities to null shedding before 9 AM and after 5 PM, and high intensity around noon (Théron 1984; N'Goran et al. 1997; Wolmarans et al. 2002). According to our observations (Fig. 7), there is a coincidence of higher density of children with a higher density of infecting cercariae in the same water between 1 PM and 4 PM. Subsequently, there is increasing likelihood to encounter the cercariae. Contrastingly, bathing occurring at 6 AM is associated with lesser likelihood; hence a lower risk. Since, the exposure risk is function of the presence of cercariae in the water, low cercarial shedding periods can be qualify as periods of low risk.

Water contact per se is not a direct measure of exposure. Many aspects of contact (such as the frequency or total duration of contact, the proportion of the body being exposed and the moment of the occurrence) contribute to the likelihood of encountering infective cercariae. However, air temperature appears to be a good proxy indicator in terms of temporal variation of schistosomiasis transmission risks. Nevertheless, the key drivers may include several other socioeconomic factors (Chandiwana 1987; Takougang et al. 1993; Amazigo et al. 1997; Kloos et al. 1998; Watts et al. 1998; Ernould 2000; Scott et al. 2003; Grisorio et al.2005).

Swimming or bathing may also lead to the contamination of waterbodies. The circadian rhythm of*schistosome* eggs excretion in the urine was reported from various geographical areas in Egypt, in East and West Africa and even out of African (McMahon 1976; Doehring et al. 1985). Traoré (2013) reported urination of children during swimming in Toma-Île. Infected children may deposit eggs contained in urine directly in the water. Fragments of faeces on the bodies of young children may also contain eggs. Bathing directly in the river may bring about the contamination of water.

4.4 Potential impact of climate variability and climate change on schistosomiasis transmission in the Sourou Valley

There is evidence of the impact of climate variability and climate change on the three actors (parasite—snail intermediate host—human-being) of the schistosomiasis development cycle in literature (Mangal et al. 2008; Zhou et al. 2008; Tay et al. 2011). Our findings show that increase in air temperature leads to massive swimming or bathing. An increase in the mean air temperature of 0.8 °C by 2025 and of 1.7 °C by 2050 is forecasted at the country scale by the Program of national action of adaptation to the variability and to climate change (PANA 2007). Such a rise in temperature will affect the monthly distribution of the proportions of cold and hot days (Fig. 6). Breaks of water contacts observed during the coldest period will became shorter since the number of hot days will dramatically increase. Exposure to schistosomiasis through swimming or bathing will occur more frequently and continually.

4.5 Health policy implications to reduce schistosomiasis risk based on our findings

Based on our findings, the safe water and adequate sanitation supply constitutes a health policy implication on how to reduce schistosomiasis. A major strategy for the resilience to climate change and the control of schistosomiasis will be the supply of safe water for children recreational activities. While the politically agenda prioritising drinking and cooking water (DGRE 2006), safe water for ludic activities should be included in national programs to enhance the water supply and sanitation services. Several studies reported that an excellent water and sanitation supply can break the transmission cycle (White et al. 1972; Esrey et al.1990). But to be effective in controlling human schistosomiasis, water supply and sanitation must be convenient enough to successfully discourage water activities in contaminated water (Cairncross and Valdmanis 2006). Water and sanitation supply should be accompanied by massive drug administration campaigns with praziquantel at least ones per year particularly in lakeside settlements like Toma-Île.

5. CONCLUSION

This study determined for the first time the threshold of air temperature at which children have the sensation of heat and react by swimming in contaminated water. In their desire to fight against extreme heat, children look for freshwater as a means for natural cooling and, by so doing, expose themselves to the risk of the transmission of schistosomiasis. Despite the relative short period (62 days) of direct field observation, the results showed a strong influence of air temperature on children's recreational swimming temporal patterns. We suggest that future research should undertake a long-term observation (at least 1 year) in order to confirm or refute the threshold of daily mean air temperature of 25 °C we found to be the starting point of children's exposure to contaminated water through swimming and increases in bathing frequency. Our findings indicate that information on the temporal variation in water contact and schistosomiasis transmission can be useful in strategies to reduce transmission by regulating swimming or bathing during high risk periods, possibly by limiting access to the river at such times. The figures provided in the paper may be used to inform models of transmission and optimise control programmes, especially while delivering mass drug administration programmes just after hot seasons.

In the way forwards, an integration of environmental data and information on human behaviour can be valuable in designing and improving models of vector-borne disease transmission. Such models are increasingly used to predict disease outbreaks of high or emerging public health relevance (Aldstadt 2007; Hollingsworth et al. 2014) such as malaria (Hoshen and Morse 2004) and dengue fever (Liao et al. 2014; Yu et al. 2011).

ACKNOWLEDGMENTS

We thank the children, parents and the communities of Toma-Île for their support and willingness to participate in this study. We also thank fieldworkers who undertook the survey.

CONFLICT OF INTEREST

The authors declare that they have no conflict of interest.

REFERENCES

1. Aldstadt J (2007) An incremental Knox test for the determination of the serial interval between successive cases of an infectious disease. Stoch Environ Res Risk A 21(5):487–500

2. Amazigo UO, Anago-Amanze CI, Okeibunor JC (1997) Urinary schistosomiasis among school children in Nigeria: consequences of indigenous beliefs and water contact activities. J Biosoc Sci 29:9–18

3. Butterworth AE, Hagan P (1987) Immunity in human schistosomiasis. Parasitol Today 3(1):11–16

4. Cairncross S, Valdmanis V (2006) Water supply, sanitation, and hygiene promotion. In: Janison DT, Breman JG, Measham AR (eds) Disease control priorities in developing countries, 2nd edn. World Bank, Washington, DC, pp 771–792

5. Chu KY, Massoud J, Sabbaghian H (1966) Host-parasite relationship of *Bulinus truncates* and *Schistosoma haematobium* in Iran. 3. Effect of water temperature on the ability of miracidia to infect snails. Bull World Health Organ 34:131–133

6. Clements ACA, Garba A, Sacko M, Toure S, Dembele R, Landoure A, Bosque-Oliva E, Gabrielli AF, Fenwick A (2008) Mapping the probability of schistosomiasis and associated uncertainty, West Africa. Emerg Infect Dis 14(10):1629–1632

7. DeWitt WB (1955) Influence of temperature on penetration of snail hosts by *Schistosoma mansoni* miracidia. Exp Parasitol 4(3):271–276

8. DGRE (2006) Programme National d'Approvisionnement en Eau Potable et d'Assainissement à l'horizon 2015—PN-AEPA 2015. Direction Générale des Ressources en Eau (DGRE), Ouagadougou

9. Dianou D, Poda JN, Savadogo LG, Sorgho H, Wango SP, Sondo B (2004) Parasitoses intestinales dans la zone du complex hydroagricole du Sourou au Burkina Faso. VertigO 5(2):8p

10. Doehring E, Vester U, Ehrich HHJ, Feldmeier H (1985) Circadian variation of ova excretion, proteinuria, hematuria, and leukocyturia in urinary schistosomiasis. Kidney Int 27:667–671

11. Ernould JC (2000) Importance du comportement humain dans la transmission des Schistosomoses. In: Chippaux JP (ed) La lutte contre les schistosomoses en Afrique de l'Ouest. Colloques et Séminaires, éditions IRD, pp 43–46

12. Esrey SA, Potash JB, Roberts L, Shiff C (1990) Health benefits from improvements in water supply and sanitation: survey and analysis of the literature on selected diseases. WASH technical Report no. 66. United States Agency for International Development, Washington, DC

13. Ficetola GF, Denoël M (2009) Ecological thresholds: an assessment of methods to identify abrupt changes in species–habitat relationships. Ecography 32(6):1075–1084.

14. Fulford AJC, Webster M, Ouma JH, Kimani G, Dunne DW (1998) Puberty and age-related changes in susceptibility to schistosome infection. Parasitol Today 14(1):23–26

15. Gazzinelli A, Velasquez-Melendez G, Crawford SB, LoVerde PT, Correa-Oliveira R, Kloos H (2006) Socioeconomic determinants of schistosomiasis in a poor rural area in Brazil. Running short title: Socioeconomic determinants of schistosomiasis in Brazil. Acta Trop 99(2–3):260–271

16. Grisorio E, Bremond P, Mietton M (2005) Les différentes dimensions d´un risque à l´interface environnement-santé. Approche écogéographique d´une maladie hydro-dépendante: La schistosomose urogénitale à Madagascar. Environ Risques Santé 4(5):335–340

17. Gryseels B, Nkulikyinka L (1990) The Morbidity of *Schistosomiasis mansoni* in the Highland focus of Lake Cohoha, Burundi. Trans R Soc Trop Med Hyg 84:542–547

18. Gryseels B, Polman K, Clerinx J, Kestens L (2006) Human Schistosomiasis. The Lancet 368(9541):1106–1118

19. Hollingsworth TD, Pulliam JRC, Funk S, Truscott JE, Ishami V, Lloyd AL (2014) Seven challenges for modelling indirect transmission: vector-borne diseases, macroparasites and neglected tropical diseases. Epidemics.

20. Hoshen MB, Morse AP (2004) A weather-driven model of malaria transmission. Malar J 3:32.

21. Hotez PJ, Kamath A (2009) Neglected tropical diseases in Sub-Saharan Africa: review of their prevalence, distribution, and disease burden. PLoS Negl Trop Dis 3(8):e412

22. INSD (2008) Recensement général de la population et de l'habitation de 2006. Résultats définitifs. Institut National de la Statistique et de la Démographie (INSD), Ouagadougou

23. Kabatereine NB, Vennervald BJ, Ouma JH, Kemijumbi J, Butterworth AE, Dunne DW, Fulford AJC (1999) Adult resistance to schistosomiasis mansoni: age dependence of reinfection remains constant in communities with diverse exposure patterns. Parasitology 118:101–105

24. Karthe D (2010) Geographic determinants of malaria transmission. A case study from Kossi Province, Burkina Faso. Ph.D. dissertation, University of Göttingen, Germany. http://webdoc.sub.gwdg.de/diss/2010/karthe/karthe.pdf

25. Karthe D, Traoré I (2009) Geographic pattern of malaria transmission: a case study from Kossi Province, Burkina Faso. Geoöko 30(1–2):43–63

26. Karthe D, Traoré I, Reeh T, Ouédraogo F, Pörtge KH (2011) Das Bewässerungsprojekt im Sourou-Tal/Burkina Faso—eine nachhaltige Maßnahme zur Ernährungssicherung? Zentralblatt für Geologie und Paläontologie 1(1–2):15–37

27. Karthe D, Traoré I, Sié A, Kappas M (2012) Malaria in Kossi Province, Burkina Faso: an investigation of spatio-temporal incidence pattern. In: Kappas M, Gross U, Kelleher D (eds) Global health: a challenge for interdisciplinary research. Universitätsverlag Göttingen, Göttingen, pp 305–320

28. King CH, Dangerfield-Cha M (2008) The unacknowledged impact of chronic schistosomiasis. Chronic Illn 4:65–79

29. Kloos H, Higashi IG, Schinski VD, Mansour NS, Murrell KD, Miller FD (1990) Water contact and *Schistosoma haematobium* infection: a case study from an upper Egyptian village. Int J Epidemiol 19(3):749–758

30. Kloos H, Gazzinelli A, Van Zuyle P (1998) Microgeographical patterns of schistosomiasis and water contact behavior; examples from Africa and Brazil. Mem Inst Oswaldo Cruz, Rio de Janeiro 93(I):37–50

31. Koukounari A, Gabrielli AF, Toure S, Bosque-Oliva E, Zhang Y, Sellin B, Donnelly CA, Fenwick A, Webster JP (2007) Schistosoma haematobium infection and morbidity before and after large-scale administration of Praziquantel in Burkina Faso. J Infect Dis 196:659–669

32. Koukounari A, Toure S, Donnelly CA, Ouedraogo A, Yoda B, Ky C, Kaboré M, Bosqué-Oliva E, Basáñez MG, Fenwick A, Webster JP (2011) Integrated monitoring and evaluation and environmental risk factors for urogenital schistosomiasis and active trachoma in Burkina Faso before preventative chemotherapy using sentinel sites. BMC Infect Dis 11(191):12

33. Liao C-M, Huang T-L, Lin Y-J, You S-H, Cheng Y-H, Hsieh NH, Chen W-Y (2014) Regional response of dengue fever epidemics to interannual variation and related climate variability. Stoch Environ Res Risk A 29(3):947–958

34. Mangal TD, Paterson S, Fenton A (2008) Predicting the impact of long-term temperature changes on the epidemiology and control of schistosomiasis: a mechanistic model. PLoS One 3(1):e1438.

35. Mayer JD (1983) The role of spatial analysis and geographic data in the detection of disease causation. Soc Sci Med 17(16):1213–1221

36. McMahon JE (1976) Circadian rhythm in *Schistosoma haematobium* egg excretion. Int J Parasitol 6:373–377

37. N'Goran E, Bremond P, Sellin E, Sellin B, Theron A (1997) Intraspecific diversity of *Schistosoma haematobium*in West Africa: chronobiology of cercarial emergence. Acta Trop 66:35–44

38. Nojima H, Sato A (1982) *Schistosoma mansoni* and *Schistosoma haematobium*: emergence of schistosome cercariae from snails with darkness and illumination. Exp Parasitol 53(2):189–198

39. PANA (2007) Programme d'action national d'adaptation à la variabilité et aux changements climatiques (PANA du Burkina Faso). Secrétariat Permanent du Conseil National pour l'Environnement et le Développement Durable/Ministère de l'Environnement et du Cadre de Vie/ Burkina Faso

40. Poda JN, Wango SP, Sorgho R, Dianou D (2004) Evolution récente des schistosomoses dans le complexe hydroagricole du Sourou au Burkina Faso. Bull Soc Pathol Exot 97(1):15–18

41. Poda JN, Mwanga J, Dianou D, Garba A, Ouedraogo FC, Zongo D, Sondo KB (2006) Les parasitoses qui minent les nouveaux pôles de développement au Burkina Faso : cas des schistosomoses et des

géohelminthes dans le complexe hydroagricole du Sourou. VertigO 7(2):1–7

42. Scott JT, Diakhate M, Vereecken K, Fall A, Diop M, Ly A, De Clercq D, de Vlas SJ, Berkvens D, Kestens L, Gryseels B (2003) Human water contacts patterns in Schistosoma mansoni epidemic foci in northern Senegal change according to age, sex and place of residence, but are not related to intensity of infection. Trop Med Int Health 8(2):100–108

43. Sk Chandiwana (1987) Community water-contact patterns and the transmission of *schistosoma haematobium*in the highveld region of Zimbabwe. Soc Sci Med 25(5):495–505

44. Steinmann P, Keiser J, Bos R, Tanner M, Utzinger J (2006) Schistosomiasis and water resources development: systematic review, meta-analysis, and estimates of people at risk. The Lancet 6(7):411–425

45. Stirewalt MA (1954) Effect of snail maintenance temperatures on development of *Schistosoma mansoni*. Exp Parasitol 3(6):504–516

46. Takougang I, Louis JP, Migliani R, Noumi E, Mohome N, Ekobo AS (1993) Quelques aspects comportementaux de l´exposition à la bilharziose dans les aménagements hydroagricoles en zone sahélienne (extrême nord du Cameroun). Cahier Santé 3:457–463

47. Tay SCK, Amekudzi LK, Tagoe G (2011) Comparative study of the impact of climate variability on prevalence of urinary schistosomiasis: cases at Sunyani Regional Hospital and among School Children in Atronie, Sunyani. J Environ Sci Eng 5:1474–1482

48. Théron A (1984) Early and late shedding patterns of *Schistosomiasis mansoni* Cercariae: ecological significance in transmission to human and murine hosts. J Parasitol 70(5):652–655

49. Traoré I (2013) Geographic determinants of human schistosomiasis transmission in the Sourou Valley, Burkina Faso. Ph.D. dissertation, University of Göttingen, Germany. http://hdl.handle.net/11858/00-1735-0000-0001-BB25-F

50. Traoré I, Ouédraogo FC, Yé M, Sié A, Karthe D (2011) Distribution spatiale de la mortalité palustre dans le district sanitaire de Nouna

au nord-ouest du Burkina Faso. J African des Sci de l'Environnement 1:27–49

51. Traoré I, Sié A, Coulibaly B, Yé M, Karthe D, Kappas M (2012) Rapid screening and mapping of urinary schistosomiasis prevalence among primary school children in Sourou Valley, Burkina Faso. In: Kappas M, Gross U, Kelleher D (eds) Global health: a challenge for interdisciplinary research. Universitätsverlag Göttingen, Göttingen, pp 255–281

52. Utzinger J, Raso G, Brooker S, De Savigny D, Tanner M, Ørnbjerg N, Singer BH, N'Goran EK (2009) Schistosomiasis and neglected tropical diseases: towards integrated and sustainable control and a word of caution. Parasitology 136:1859–1874

53. Van der Werf MJ, de Vlas SJ, Brooker S, Looman CWN, Nagelkerke NJD, Habbema JDF, Engels D (2003) Quantification of clinical morbidity associated with schistosome infection in sub-Saharan Africa. Acta Trop 86:125–139

54. Watts S, Khallaayoune K, Bensefia R, Laamrani H, Gryseels B (1998) The study of human behavior and Schistosomiasis transmission in an irrigated Area in morocco. Soc Sci Med 46(6):755–765

55. Weil C, Kvale KM (1985) Current research on geographical aspects of schistosomiasis. Geogr Rev 75(2):186–216

56. White GF, Bradley DJ, White AU (1972) Drawers of water. Domestic water use in East Africa. University of Chicago Press; reproduced by Bulletin of the WHO, 2002, 80(1): 63–73

57. WHO (2013) Sustaining the drive to overcome the global impact of neglected tropical diseases: second WHO report on neglected tropical diseases. WHO/HTM/NTD/2013.1

58. Wolmarans CT, de Kock KN, Strauss HD, Bornman M (2002) Daily emergence of *Schistosoma mansoni* and *S. haematobium* cercariae from naturally infected snails under field conditions. J Helminthol 76:273–277

59. Yu H-L, Yang S-J, Yen H-J, Christakos G (2011) A spatio-temporal climate-based model of early dengue fever warning in southern Taiwan. Stoch Environ Res Risk A 25(4):485–494

60. Zhou XN, Yang GJ, Yang K, Wang X-H, Hong Q-B, Sun L-P, Malone JB, Kristensen TK, Bergquist NR, Utzinger J (2008) Potential impact of climate change on schistosomiasis transmission in China. Am J Trop Med Hyg 78(2):188–194

61. Zongo D, Kabre BG, Dayeri D, Savadogo B, Poda JN (2012) Étude comparative de la transmission de la schistosomiase (formes urinaire et intestinale) dans dix sites du Burkina Faso. Médecine et Sante´ Tropicales 22:323–329

10

UNCERTAINTY MODELING AND VISUALIZATION FOR TSUNAMI HAZARD AND RISK MAPPING: A CASE STUDY FOR THE 2011 TOHOKU EARTHQUAKE

Katsuichiro Goda1 , Jie Song1

[1] Department of Civil Engineering, Queen's School of Engineering, University of Bristol, Queen's Building, University Walk, Bristol BS8 1TR, UK

ABSTRACT

This study presents a rigorous computational framework for visualizing uncertainty of tsunami hazard and risk assessment. The methodology consists of three modules: (i) earthquake source characterization and stochastic simulation of slip distribution, (ii) tsunami propagation and inundation, and (iii) tsunami damage assessment and loss estimation. It takes into account numerous stochastic tsunami scenarios to evaluate the uncertainty propagation of earthquake source characteristics in probabilistic tsunami risk analysis. An extensive Monte Carlo tsunami inundation simulation is implemented for the 2011 Tohoku tsunami (focusing upon on Rikuzentakata along the Tohoku coast of Japan) using 726 stochastic slip models derived from eleven inverted source models. By integrating the tsunami hazard results with empirical tsunami fragility functions, probabilistic tsunami risk analysis and loss estimation are carried out; outputs from the analyses are displayed using various visualization methods. The developed framework is comprehensive, and can provide valuable insights in promoting proactive tsunami risk management and in improving emergency response capability.

KEYWORDS

Tsunami, Stochastic source modeling, Hazard/risk maps, Uncertainty, Visualization

1. INTRODUCTION

An extremely large subduction earthquake triggers massive tsunamis, which can destroy coastal cities and towns completely. Recent catastrophic tsunamis include the 2004 Indian Ocean tsunami (Borrero 2005; Murata et al. 2010) and the 2011 Great East Japan (Tohoku) tsunami (Fraser et al. 2013; Liu et al. 2013). To mitigate and control potential tsunami risk exposures effectively, physical protection and preparedness against anticipated tsunami events are of critical importance (FEMA 2008). One of the essential pieces of information for enhancing tsunami resilience of coastal communities is accurate tsunami hazard assessment; based on such hazard estimates, effective risk reduction countermeasures (e.g. breakwaters and seawalls) and evacuation strategies can be deployed. Current state-of-the-practice for preparing local/regional tsunami hazard maps is mainly concerned with tsunami hazard parameters that correspond to a single scenario on a single fault. It fails to assess uncertainty related to hazard predictions. Consequently, users of scenario-based tsunami hazard maps are unable to appreciate potential risks in different situations, inhibiting risk communications among tsunami experts, crisis managers, and local residents. For instance, during the 2011 Tohoku tsunami, more than 65 % of all fatalities in Kamaishi, Iwate Prefecture, Japan, were caused outside the areas that were designated as major inundation zones in public tsunami hazard maps presented prior to 2005. The actual tsunami due to the 2011 Tohoku event was beyond historical events/scenarios that were taken into account in preparing the 2005 hazard map. Clearly, a set of tsunami inundation hazard maps, corresponding to different tsunami scenarios and consequences, is useful for tsunami hazard preparedness and evacuation planning in coastal cities and towns. Use of stochastic hazard maps that take into account the main sources of uncertainty related to the tsunami hazard assessment promotes the informed decisions regarding tsunami risk reduction actions by communicating the uncertainty of hazard predictions and by understanding the consequences in different situations.

Recent advances in tsunami hazard and risk mapping entail the quantification of uncertainty associated with tsunami modeling. Major sources of uncertainty in predicting tsunami wave heights and inundation extents can be attributed to source characteristics of future tsunamigenic earthquakes, such as location, magnitude, geometry, and slip distribution (Geist 2002; McCloskey et al. 2008; Løvholt et al. 2012; Fraser et al. 2014; Goda et al. 2014; Wiebe and Cox 2014; Fukutani et al. 2015; Mueller et al. 2015). New methods for developing stochastic earthquake source models, which are based on spectral analysis of slip heterogeneity of an inverted source model and spectral synthesis of slip random fields (Mai and Beroza 2002; Lavallée et al. 2006), facilitate the generation of possible scenarios having different earthquake slips and fault geometry. The approaches also allow the inclusion of epistemic uncertainty related to stochastic source modeling by considering multiple reference source models (Goda et al. 2014). By conducting Monte Carlo tsunami simulation based on numerous synthesized source models, stochastic inundation depth maps can be evaluated (Goda et al. 2015; Mueller et al. 2015). Furthermore, by integrating tsunami hazard estimates with tsunami fragility models (e.g. Koshimura et al. 2009; Reese et al. 2011; Suppasri et al. 2013; Tarbotton et al.2015), stochastic tsunami risk maps (e.g. showing tsunami damage probability) as well as tsunami risk/loss curves (e.g. exceedance probability curve of tsunami loss for a building portfolio) at local/regional scales can be developed. Quantitative estimates of potential tsunami risk and loss are valuable for risk managers and emergency officers in making decisions for tsunami risk management. An advantage of such stochastic hazard/risk maps and curves is that the main sources of uncertainty related to the tsunami source characteristics are taken into account.

The physical processes of tsunami generation/propagation/inundation have crucial influence on how hazard analysis is formulated and conducted, and what mitigation measures can be adopted to protect our society and assets. Different stakeholders require different tsunami hazard/risk information related to potential effects on coastal communities and infrastructure over a variety of temporal and spatial scales. Understanding the meaning of hazard estimates with regard to epistemic uncertainty of hazard modeling processes is an essential part of probabilistic hazard analysis and thus visualization of the outcomes of an uncertainty assessment has become increasingly important. Novel source modeling combined with Monte Carlo tsunami simulation offers new ways of presenting and visualizing tsunami hazard analysis results, which are relevant for risk management actions. Various proposals for presentation and communication of hazard results have been discussed in the literature. The properties of key variables of interest have a significant impact on the selection of visualization methods. The variables may be multi-dimensional, and may be dependent spatially and temporally (e.g. tsunami wave profiles at two locations). The probabilistic characteristics of the variables can differ significantly (e.g. Gaussian variable versus non-Gaussian variable with heavy right tail). In these cases, simple metrics, such as mean and standard deviation (statistical moments), are not sufficient to convey underlying uncertainty of the hazard estimates, and some higher-order statistics as well as probability distribution may be presented graphically (Potter et al. 2010). For representing uncertainty of spatially varying quantities (e.g. tsunami inundation map), visual parameter cues, such as color, texture, transparency, clarity, and size, can be varied in a cartographic graphical system (MacEachren et al. 2005; Pang 2008; Kunz et al. 2011). Typically, these results are shown for average trends and possible ranges (e.g. mean plus/minus one/two standard deviations), and both forms of information can be displayed on the same map (e.g. using color contour and scattered dots/isolines; Kunz et al. 2011; Pappenberger et al. 2013). It is also important to recognize that the visualization of the analysis results itself (e.g. interpolation) creates uncertainty and potentially gives a false impression of accuracy of the hazard results (Pang 2008).

This study presents a rigorous approach for visualizing uncertain tsunami hazard/risk analysis outputs that are obtained from extensive Monte Carlo tsunami simulations by taking into account a wide range of earthquake source characteristics. It aims at demonstrating how rich information of Monte Carlo tsunami simulations can be summarized concisely and how it can be used for tsunami risk mitigation decisions effectively. The investigation is focused upon the 2011 Tohoku tsunami as a case study from retrospective viewpoints. The 2011 Tohoku event offers unique opportunities to carry out detailed tsunami hazard/risk assessments, because various data and models for earthquake source properties, inundation/run-up measurements, and tsunami damage records are available. The quality and amount of available information from the 2011 Tohoku earthquake are unprecedented in comparison with other previous events. For instance, reliable tsunami inundation and run-up survey results (more than 5000 locations) are available from the Tohoku Tsunami Joint Survey (TTJS) team (Mori et al. 2011), whereas tsunami damage data as well as building data for the Tohoku region (more than 250,000 buildings) are available from the Ministry of Land, Infrastructure, and Transportation (MLIT 2014). In representing the possible range of tsunami scenarios/source models in the Tohoku region that may be applicable to tsunami hazard mapping purposes, various inversion models are gathered from the literature (Ammon et al. 2011; Fujii et al. 2011; Hayes 2011; Iinuma et al. 2011, 2012; Shao et al. 2011; Yamazaki et al. 2011; Gusman et al. 2012; Satake et al. 2013), and numerous stochastic source models are then generated based on these reference source models (Goda et al. 2014). In the following, a computational framework of tsunami hazard and risk assessments for coastal cities and towns is presented based on Monte Carlo tsunami inundation simulations (Goda et al. 2015) and empirical tsunami fragility models (Suppasri et al. 2013). Various visualization methods of tsunami hazard/risk analysis are then illustrated through a case study focusing upon Rikuzentakata, a coastal city devastated by the 2011 Tohoku tsunami. A novel aspect of this study is that quantitative tsunami hazard and risk assessments are carried out by accounting for stochastic earthquake source properties and are presented with various visualization methods for improved tsunami risk mitigation actions and risk communications.

2. STOCHASTIC TSUNAMI HAZARD AND RISK ANALYSIS

A generic equation for probabilistic tsunami risk assessment can be expressed as:

$$\nu(L \geq l) = \iiint_{\Omega_{DS,IM,EQS}} P(L \geq l|ds) f_{DS|IM}(ds|im) f_{IM|EQS} \times (im|eqs) f_{EQS}(eqs) |\mathrm{d}ds||\mathrm{d}im||\mathrm{d}\lambda(eqs)| \tag{1}$$

where $\nu(L \geq l)$ is the annual exceedance probability that the tsunami loss L of a structure exceeds a certain loss threshold l, $\lambda(eqs)$ is the annual occurrence rate of earthquake scenarios (EQS) represented by multiple physical parameters (e.g. magnitude, location, geometry, and slip distribution), $P(L \geq l|ds)$ is the tsunami loss function in terms of damage state variable (DS), $f_{DS|IM}$ is the tsunami fragility function in terms of intensity measure (IM), $f_{IM|EQS}$ is the probability density function of IM

given *EQS*, f_{EQS} is the probability density function of *EQS*, and $\Omega_{DS,IM,EQS}$ is the joint domain of integration for *DS*, *IM*, and *EQS*. Note that *DS* is often defined in a discrete manner; in such cases, integration for *DS* in Eq. (1) can be replaced by summation. A typical *IM* is the inundation depth, which is often used as an input parameter for tsunami fragility modeling (i.e. $f_{DS|IM}$). $f_{IM|EQS}$ is evaluated through numerical evaluations of governing equations for tsunami waves and inundation/run-up (e.g. solving the nonlinear shallow water equations for given initial boundary conditions). The uncertainty associated with variable earthquake source characteristics is captured by f_{EQS} . When earthquake scenarios are defined for a single source region (i.e. scenario-based), the interpretation of Eq. (1) becomes conditional. Such conditional assessment may be considered for situations where dominant earthquake source regions are identified through historical tsunami records or probabilistic tsunami hazard analysis (e.g. Geist and Parsons 2006; Horspool et al. 2014), but their occurrence probabilities and potential earthquake sizes cannot be estimated reliably (Kagan and Jackson 2013). An advantage of the conditional evaluations is that the nonlinear physical processes of tsunami wave propagation and inundation as well as the uncertainty of detailed source characteristics can be fully incorporated in the hazard computation (Goda et al. 2014, 2015).

This section succinctly summarizes an analytical procedure for scenario-based tsunami hazard and risk assessments using stochastic source models that are derived from multiple reference models, focusing upon the 2011 Tohoku earthquake. The methodology consists of three major components, i.e. earthquake source characterization and stochastic simulation of slip distribution, tsunami propagation and inundation, and tsunami damage assessment of structures. A computational flow of the tsunami hazard and risk assessments is illustrated in Fig. 1.

2.1 Stochastic source models

An earthquake slip modeling procedure, which is based on spectral analysis of an inversion-based source model and spectral synthesis of random fields, generates earthquake slip distributions with statistical properties equivalent to the inverted source model. The method is based on Mai and Beroza (2002), and has been extended for large mega-thrust subduction earthquakes (Goda et al. 2014). A brief summary of the stochastic method is given below; full details of the method can be found in Goda et al. (2014) and are not repeated here.

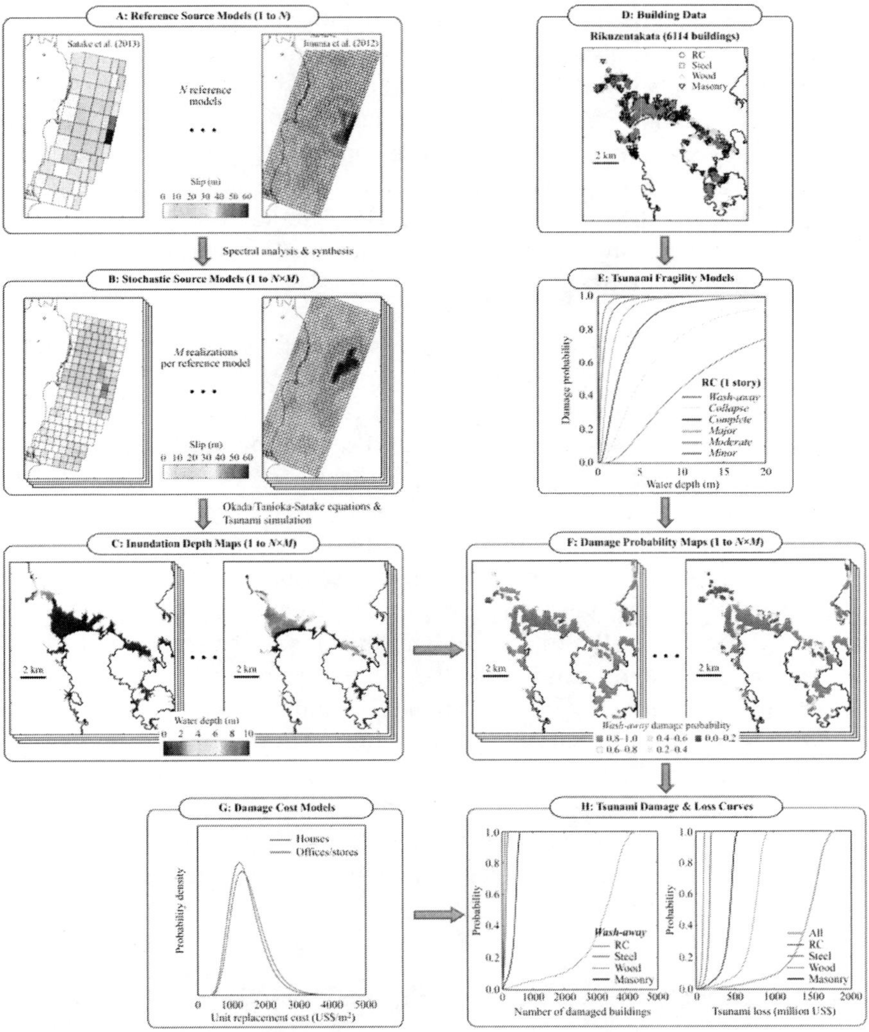

Figure 1. Probabilistic tsunami loss estimation framework

The stochastic source models are dependent on the characteristics of the reference source model that is derived from inversion analysis. Because data and methods used for constraining the slip distribution over a fault plane differ significantly, details of the slip distribution and fault geometry differ significantly among the inversion models (e.g. moment magnitude, fault length and width, fault location, strike, dip, and asperities [i.e. large slip patches]; Panel A in Fig. 1). This reflects the complexity and uncertainty of the rupture process of mega-thrust subduction earthquakes. Therefore, to explore a range of possible slip distributions for a given scenario (i.e. epistemic uncertainty), multiple reference models should be considered for generating stochastic source models. In total, eleven source models that are developed using teleseismic/tsunami/geodetic data are considered (Ammon et al. 2011; Fujii et al. 2011; Hayes 2011; Iinuma et al. 2011, 2012; Shao et al. 2011; Yamazaki et al. 2011; Gusman et al. 2012; Satake et al. 2013), and are used as reference to further produce stochastic source models with variable characteristics.

The analytical procedure of generating stochastic source models from a given reference model is as follows. Firstly, key characteristics of the reference source model are obtained as slip statistics (mean and standard deviation) and asperity zone (i.e. a set of sub-faults that have slip values greater than a specified threshold value; typically, the threshold is set to two to three times the average slip). In addition, right tail characteristics of the slip distribution are modeled via Box-Cox transformation. The Box-Cox analysis is necessary because for the 2011 Tohoku earthquake, very large slip values (e.g. exceeding 40 m) are obtained for a small number of sub-faults and thus distribution of earthquake slip significantly deviates from normal distribution with the same slip statistics (Goda et al. 2014). Secondly, after modifications to the original slip distribution (e.g. interpolation/resampling using finer grid sizes), fast Fourier transform (FFT) of the slip distribution is carried out to obtain the two-dimensional (2D) normalized power spectrum. The normalized power spectra in the down-dip and along-strike directions are then fitted by the power spectrum of a von Kármán auto-correlation function:

$$P(k) \propto \frac{A_x A_z}{(1+k^2)^{H+1}}, \tag{2}$$

where k is the wavenumber, $k = (A_z^2 k_z^2 + A_x^2 k_x^2)^{0.5}$, A_z and A_x are the correlation lengths for the down-dip and along-strike directions, respectively, and H is the Hurst number. A_x and A_z control the absolute level of the power spectrum in the low wavenumber range (i.e. $k << 1$) and capture the anisotropic spectral features of the slip distribution. H determines the slope of the power spectral decay in the high wavenumber range, and is theoretically constrained to fall between 0 and 1.

Thirdly, multiple realizations of slip distributions with desired stochastic properties are generated using a Fourier integral method. The amplitude spectrum of the target slip distribution is specified by the theoretical power spectrum with the estimated correlation lengths and Hurst number, while the phase spectrum is represented by a random phase matrix. The constructed complex Fourier coefficients are transformed into the spatial domain via 2D inverse FFT. The synthesized slip distribution is converted via Box–Cox transformation to achieve realistic heavy right-tail features of the slip distribution. To resemble the synthesized slip distribution with the original one in terms of location and amplitude of asperities, an asperity zone of the synthesized slip distribution is compared with that of the original slip distribution. An acceptable slip distribution is required to have its maximum slip patch and similar slip concentration within the asperity zone of the original distribution. Finally, the mean and standard deviation of the transformed slip distribution are adjusted to achieve similar statistics of the synthesized slip model with respect to the original slip model.

To generate source models that have different geometrical properties from the reference models, the top-edge depth, strike, and dip are varied over certain ranges with respect to the original models by changing one parameter at a time and by keeping a slip distribution identical to the original model. Geometrical parameters are allowed to vary as follows: the top-edge depth varies in steps of 2.5 km from by −2.5 to +10 km with respect to the reference depth; the strike angle takes on values between −5° and +7.5° in 2.5° increment with respect to the reference strike; the dip angle is changed between −5° and +10.0° in 2.5° increment from the reference dip. The ranges of varied geometrical parameters are determined by examining various inverted source models for the 2011 Tohoku earthquake [see Goda et al. (2014) for more detailed information] and are consistent with the seismotectonic setting in the Tohoku region (i.e. physical characteristics of the subduction zone, such as the location of the Japan Trench and typical dip angles for the subducting slab). When parameters of an original source model are variable, an average value is used to define a reference case for each parameter. Moreover, for each of the eleven reference models, 50 realizations of a target slip distribution are generated using the stochastic synthesis method by keeping their geometrical parameters identical to those of the original models. In total, 726 slip distributions are generated for Monte Carlo tsunami simulation (66 cases per reference model, consisting of a reference case, 15 cases with varied geometrical parameters, and 50 cases with stochastic slip distributions; Panel B in Fig. 1).

2.2 Monte Carlo tsunami simulation

Tsunami modeling is carried out using a well-tested numerical code (Goto et al. 1997) that is capable of generating off-shore tsunami propagation and run-up/inundation by evaluating nonlinear shallow water equations using a leap-frog staggered-grid finite difference scheme. The run-up/inundation calculation is performed by a moving boundary approach, where a dry or wet condition of a computational cell is determined based on total water depth in comparison with its elevation. The computational domains are nested at four resolutions (i.e. 1350, 450, 150, and 50-m domains). Computational cells include those on land, and coastal defense structures are taken into account using overflowing formulae as a sub-grid model. In the tsunami simulation code that is used in this study, the coastal defense structures (e.g. breakwater and levee) are represented as vertical walls at one or two sides of the computational cells. To evaluate the volume of the water that overpasses these walls, Homma's overflowing formulae are employed. It is noteworthy that the destruction of the coastal defense structures during the tsunami simulation is not taken into account; this may result in underestimation of the tsunami inundation in coastal cities and towns.

Bathymetry/elevation data, roughness data, and information of coastal defense structures for the Tohoku region are obtained from the Cabinet Office of the Japanese Government. The ocean bathymetry data are based on the digital database and nautical charts developed by Japan Hydrographic Association and Japan Coastal Guard. The land elevation data are based on the 50-m grid digital elevation model (DEM) developed by the Geospatial Information Authority of Japan. The bottom friction is evaluated using Manning's formula by considering four land use classes: 0.02 $m^{-1/3}$ s for agricultural land, 0.025 $m^{-1/3}$ s for ocean/water, 0.03 $m^{-1/3}$ s for forest vegetation, and 0.04 $m^{-1/3}$ s for urban areas. It is noted that the assigned roughness coefficients are crude representation of actual situations.

Differences in earthquake slip result in different boundary conditions for tsunami propagation and run-up. In tsunami simulation, the initial water surface elevation is evaluated based on formulae by Okada

(1985) and Tanioka and Satake (1996). The latter equation accounts for the effects of horizontal seafloor movements in case of steep seafloor, inducing additional vertical water dislocation. Although the seafloor deformations are obtained for the same event, spatial characteristics of the seafloor displacements vary significantly among the models, leading to various tsunami wave profiles at different locations along the Tohoku coast (Goda et al. 2014). The fault rupture is assumed to occur instantaneously, and numerical tsunami calculation is performed for duration of 2 h with an integration time step of 0.5 s. The tsunami simulation is conducted for the 726 source models. For each case, the maximum inundation depths at all in-land computational cells (50-m grids) are obtained by subtracting the DEM data from the calculated maximum wave heights (Panel C in Fig. 1).

2.3 Tsunami fragility and damage analysis

Structural vulnerability against tsunami loading can be modeled by empirical tsunami fragility curves, which relate tsunami intensity measures (*IM*) to tsunami damage states (*DS*) statistically. The damage states of structures are determined during reconnaissance survey and building inspection. The inundation depth is often adopted as *IM*, because the inundation depth is usually observable (e.g. water marks, post-tsunami interview, and numerical simulation). Mathematically, the tsunami fragility is modeled by the lognormal distribution. The exceedance probability of a damage state ds_i for a given value *im* is expressed as:

$$P(DS \geq ds_i | im) = \Phi\left(\left[\ln(im) - \mu_{\ln IM|DS_i}\right] / \sigma_{\ln IM|DS_i}\right) \quad (3)$$

where Φ is the cumulative distribution function of the standard normal variate, and $\mu_{\ln IM|DS_i}$ and $\sigma_{\ln IM|DS_i}$ are the mean and standard deviation of $\ln IM|DS_i$, respectively. For mutually exclusive damage states that are defined in a discrete manner, the probability of being in ds_i is given by:

$$p(ds_i | im) = P(DS \geq ds_i | im) - P(DS \geq ds_{i+1} | im) \quad (4)$$

Note that ds_{i+1} is severer than ds_i (i.e. $P(DS \geq ds_{i+1} | im) < P(DS \geq ds_i | im)$).

In Japan, the MLIT (2014) implements a uniform classification scheme for tsunami damage survey. Seven discrete damage levels are defined: *no damage, minor damage, moderate damage, major damage, complete damage, collapse*, and *wash-away*. Furthermore, information regarding the structural material types and the number of stories is also provided. The material types are categorized into: reinforced concrete (RC), steel, wood, masonry, and unknown, whereas the number of stories is divided into: 1-story, 2-story, and 3 + -story (Panel D in Fig. 1). Using the extensive MLIT tsunami damage database for the 2011 Tohoku tsunami (more than 250,000 data points), Suppasri et al. (2013) developed tsunami fragility models by distinguishing tsunami damage data according to the structural materials and the number of stories (Panel E in Fig. 1). The refinement for the different material types as well as for the number of stories is desirable, because the tsunami capacities for RC, steel, wood, and masonry buildings differ significantly (Koshimura et al. 2009; Suppasri et al. 2013; Tarbotton et al. 2015).

Using the tsunami fragility curves, probabilities of attaining particular damage states $p(ds|im)$ can be estimated for each building and for each scenario (Panel F in Fig. 1). The statistical analysis can be then carried out to develop stochastic tsunami risk maps. For instance, representative percentiles of tsunami damage probability (e.g. 50th and 90th percentiles) can be displayed on a map to show the relative likelihood of tsunami damage occurrence at different spatial scales. Moreover, calculated values of tsunami damage probability can be used in Monte Carlo sampling to generate realizations of individual damage states for the buildings of interest. This resampling facilitates the development of the cumulative distribution functions of aggregate tsunami risk metrics, such as the number of buildings attaining a specific damage state (Panel H in Fig. 1).

Finally, by incorporating damage cost models for different buildings (Panel G in Fig. 1), the tsunami damage information can be transformed into tsunami loss information for individual buildings as well as building portfolios (Panel H in Fig. 1). The loss ratios, which are typically used to represent damage as a percentage of total construction repair/replacement cost, can be assigned as: 0.0, 0.05, 0.2, 0.4, 0.6, 1.0, and 1.0 for the seven damage states (i.e. from *no damage* to *wash-away*) (MLIT 2014). Using the damage state probability $p(ds)$ and loss ratio $R_L(ds)$, tsunami damage cost for a given tsunami hazard intensity can be calculated as (for discrete cases):

$$L = C_R \sum_{i=1}^{7} p(ds_i) \times R_L(ds_i) \tag{5}$$

where C_R is the replacement cost of a building. An advantage of using loss metrics, instead of damage probability or number of damaged buildings, is that the consequences due to tsunami damage in coastal cities/towns can be aggregated for the entire building portfolio; this facilitates the various financial risk management decisions (Yoshikawa and Goda 2014).

3. CASE STUDY: 2011 TOHOKU TSUNAMI

The main objective of this section is to demonstrate the uncertainty and visualization methods for tsunami hazard and risk assessments, which are introduced in the preceding section, through a real case study in the Tohoku region of Japan. Firstly, a brief summary of the analysis set-up is given. Subsequently, tsunami simulation results obtained from Monte Carlo analyses are visualized such that uncertain outcomes of the tsunami hazard and risk predictions are better displayed and used for further investigations.

3.1 Problem set-up

In this study, a set of 726 source models that are generated by considering the eleven reference models and by varying their source characteristics are adopted as a representative set. Figure 2a–c show the earthquake source models by Satake et al. (2013), Ammon et al. (2011), and Iinuma et al. (2012) (for illustration), which have been derived from source inversion analyses using tsunami, teleseismic, and geodetic data, respectively. Inspection of the three reference models indicates that in addition to differences in fault plane geometry, the location and concentration of major asperities vary significantly. Figure 2d–f present three sample realizations of stochastic slip models based on Satake et al. (2013) (i.e. Fig. 2a). Stochastically generated alternative models display the varied asperity characteristics, and thus a wide range of possible source models for the specified scenario can be analyzed to quantify their effects on tsunami wave heights, coastal inundation, and structural damage and loss to buildings.

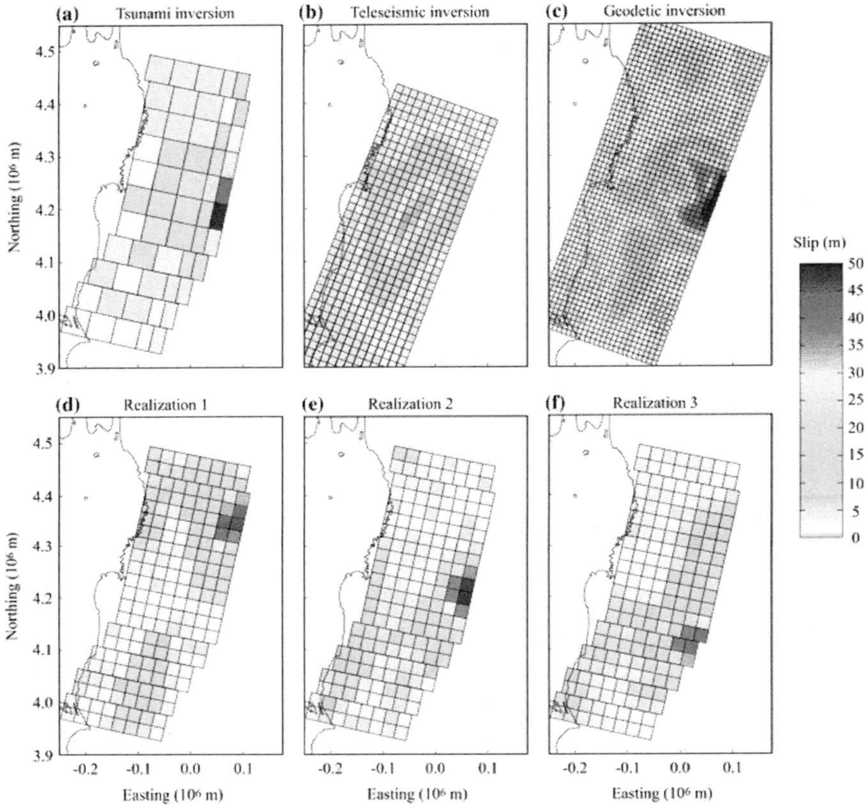

Figure 2. Earthquake source models for the 2011 Tohoku earthquake: **a** Satake et al. (2013), **b** Ammon et al. (2011), **c**Iinuma et al. (2012), and **d–f** three synthesized models based on Satake et al. (2013)

For damage and loss assessments, a pre-2011 building portfolio in Rikuzentakata, Iwate Prefecture, Japan, is considered (Fig. 3). Rikuzentakata is situated on the northern shore of the Hirota Bay which is about 3.5 km wide. It was one of the most heavily impacted areas along the Tohoku ria coast (Fraser et al. 2013; Liu et al. 2013). The city spreads across the northern shore of the Hirota Bay for at least 1.65 km east to west and for around 1 km in the northern (inland) direction, while at its western side it is crossed by the Kesen River. Over the wide zone, the land elevation is very low and the tsunami inundation depth exceeded 12 m in the city center (with run-up reaching 19 m); every timber frame house (predominantly two-story) was totally destroyed. The MLIT database contains 6,114 low-rise buildings (one- or two-story) in Rikuzentakata, which have material type information. The considered building portfolio consists of 143 RC structures, 265 steel structures, 4,963 wood structures, and 743 masonry structures (note: the adopted selection criterion essentially excludes relatively large buildings, such as schools, hospitals, and hotels). The tsunami fragility assessment is carried out using the tsunami fragility curves by Suppasri et al. (2013). Based on the tsunami damage classification scheme and damage-loss ratios suggested by the MLIT, tsunami damage costs for buildings in Rikuzentakata are evaluated. For simplicity (note: the MLIT database does not contain occupancy information for individual buildings), the buildings are classified into residential houses (wood) and commercial stores/offices (RC/steel/masonry). The adopted replacement cost models for houses and stores/offices are the lognormal distribution. The mean and coefficient of variation (CoV) of the unit replacement costs are obtained from the Japanese building cost information handbook published by Construction Research Institute (2011). Using the cost handbook, mean unit cost = 1,600 US$/m^2 and CoV = 0.320 are adopted for wooden houses, whereas mean unit cost = 1,500 US$/m^2 and CoV = 0.318 are adopted for stores/offices (note: 1 US$ = 100 yen). The adopted cost statistics are for the Tohoku region. Moreover, typical floor areas of wooden houses and stores/offices are determined based on the national construction statistics maintained by the MLIT (http://www.mlit.go.jp/toukeijouhou/chojou/stat-e.htm). In this study, floor areas of 130 and 540 m^2 are adopted for wooden houses and stores/offices, respectively, by averaging the construction data for Miyagi and Iwate Prefectures during the period between 2012 and 2014.

3.2 Visualizing uncertain tsunami hazard outcomes

For the Tohoku event, tsunami waveforms recorded at nearshore GPS buoys (water depths between 150 and 200 m) are available (Kawai et al. 2013) and can be used as benchmark in comparing simulated tsunami wave profiles with observed ones. Figure 4 shows a set of 726 simulated tsunami wave profiles at three GPS buoy stations (Fig. 3a), i.e. South Iwate (802G), North Miyagi (803G), and Central Miyagi (801G), in comparison with the observed ones. The simulated wave profiles are shown with grey lines, representing the entire variability of the Monte Carlo tsunami simulations. In addition to the individual results, three percentile curves (i.e. 10th/50th/90th percentiles) are included as summary statistics. It is interesting to observe that the median trend matches well with the observations in magnitude as well as in temporal profile, although the variability of the predicted tsunami waves is significant (which can be more than twice the observed profiles for some cases). Plots, such as Fig. 4, provide rich information regarding the overall performance (consistency) and variability of the tsunami simulations. Moreover, results shown in Fig. 4, when generated for locations at shallower water depths (e.g. 10–20 m), are useful for evaluating the reliability of coastal defense structures (e.g. breakwaters and seawalls) subjected to a wide range of tsunami waves. It is noteworthy that the variability of the tsunami simulations shown in Fig. 4 depends on the locations of the GPS buoy stations, indicating the sensitivity of the tsunami hazard estimates to the source characteristics as well as the topographical effects.

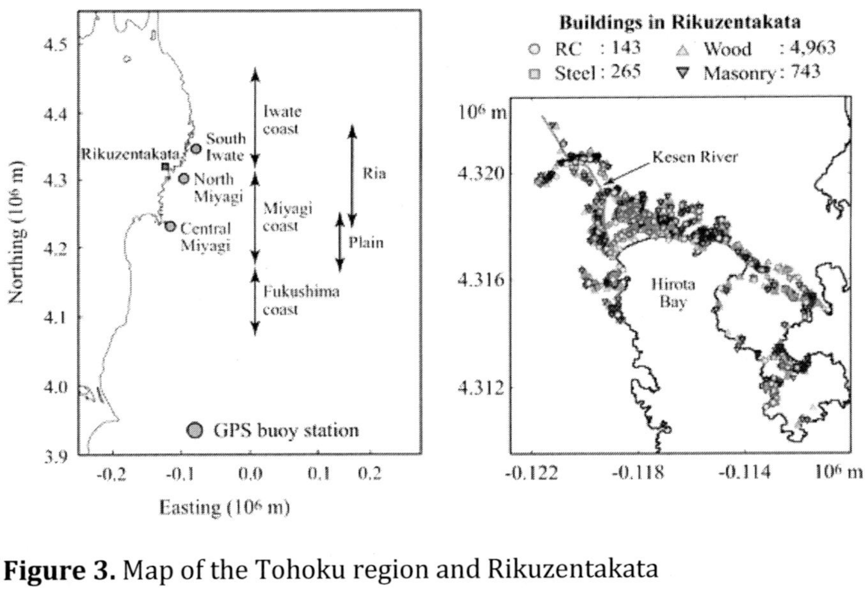

Figure 3. Map of the Tohoku region and Rikuzentakata

Figure 5a shows the variability of the maximum tsunami inundation height along the Tohoku coast; grey lines are the individual simulation results, whereas the black line is the median of the 726 cases. For reference, the tsunami inundation heights from the TTJS database (Mori et al. 2011) are included in the figure. Compared with Fig. 4, Fig. 5a suppresses information related to the temporal variation of tsunami waveforms, but on the other hand presents rich information related to the spatial variability of the maximum inundation height. Individual results from the simulation exhibit significant variability of the inundation height due to stochastic tsunami scenarios. The median simulation results and the TTJS data are generally consistent in the southern and central parts of the Tohoku region (northing between 4.06 × 10^6 and 4.32 × 10^6), while a large discrepancy is noticed in the northern part of the Tohoku region (northing between 4.32 × 10^6 and 4.45 × 10^6). The latter is a known issue for inversion models of the 2011 Tohoku earthquake (Goda et al. 2014). In addition, a hypothetical risk indicator for tsunami inundation is included in Fig. 5a to illustrate how the tsunami hazard assessments can be used for tsunami risk management. Using the simulated tsunami results, for instance, the chance of attaining *cautious/risky/dangerous* risk levels can be evaluated for different locations.

The spatial variability of the maximum tsunami inundation height can be represented by the mean and CoV (Fig. 5b). Generally, the mean is larger in the ria part of the Tohoku coast (North Miyagi to Iwate), compared with other regions (e.g. coastal plains in southern Miyagi and Fukushima). On the other hand, the CoV decreases gradually from the northern Tohoku coast (e.g. Iwate) to the central Tohoku coast (e.g. Miyagi), and then increases towards the southern part of the Tohoku coast (e.g. Fukushima). This is because for the considered tsunami scenarios, large slip asperities are located off the Miyagi coast (i.e. common features of the inverted source models for the 2011 Tohoku tsunami); thus the extent of tsunami hazard variability is greater in the northern and southern Tohoku regions (and also affected by the topographical features). Since current design guidelines of tsunami evacuation buildings specify safety margins independent of coastal regions but buildings in a certain region are subjected to different levels of tsunami hazard variability, coastal infrastructure may not achieve consistent reliability levels across the Tohoku region against rare tsunami events.

Figure 6 shows the variability of the simulated tsunami wave profiles at three locations in Rikuzentakata (i.e. P1, P2, and P3 in the map). Because the three locations are at different elevations, the base levels of the simulated tsunami waves are different at these locations. In Fig. 6, individual results based on 726 scenarios as well as 10th, 50th, and 90th percentile curves are included. The results clearly show that the tsunami inundation height becomes smaller with increasing elevation and that the tsunami wave profiles are highly variable/uncertain, particularly at locations near shoreline (P1). The maximum tsunami inundation height can be greater than the median tsunami inundation height by a factor of 2. It is noteworthy that the tsunami arrival time also varies significantly (10–15 min), implying that a time available for evacuation can differ greatly, depending on tsunami scenarios. The results presented in Fig. 6 are valuable input for analyzing tsunami vulnerability of a building, because tsunami inundation heights can be converted into tsunami forces acting on a building (FEMA 2008). Therefore, reliability of tsunami-proof buildings can be evaluated by considering different tsunami wave loadings.

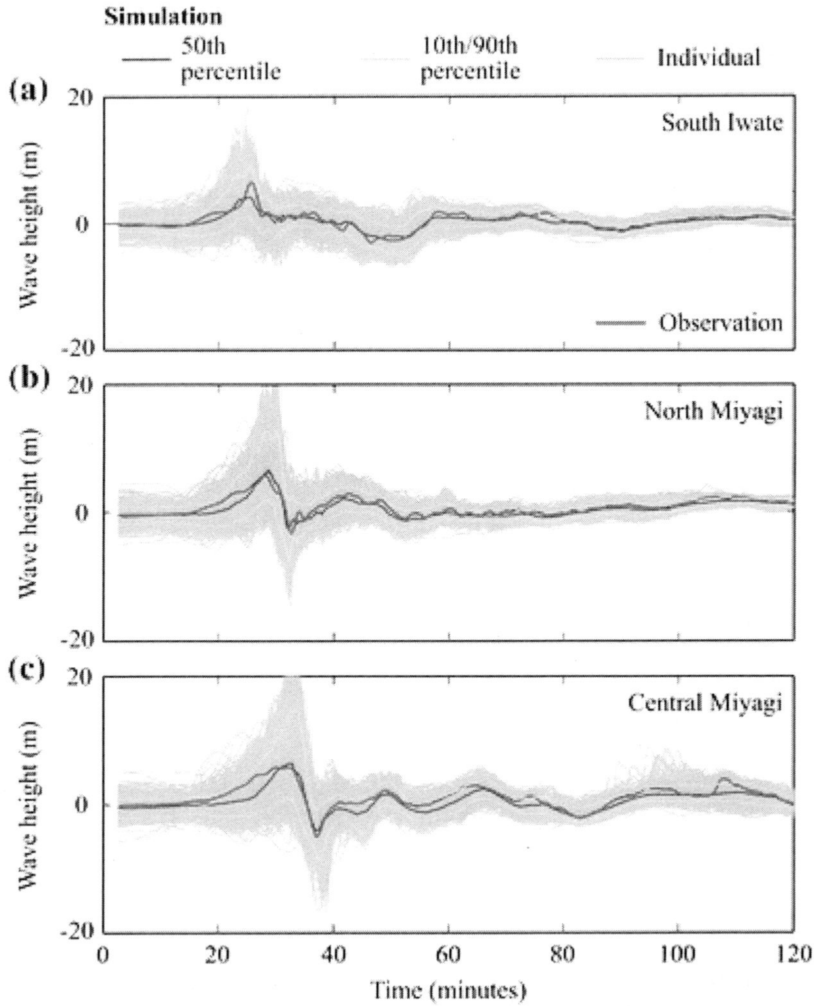

Figure 4. Comparison of the simulated tsunami wave profiles with the observed wave profiles at GPS buoy stations: **a**South Iwate, **b** North Miyagi, and **c** Central Miyagi

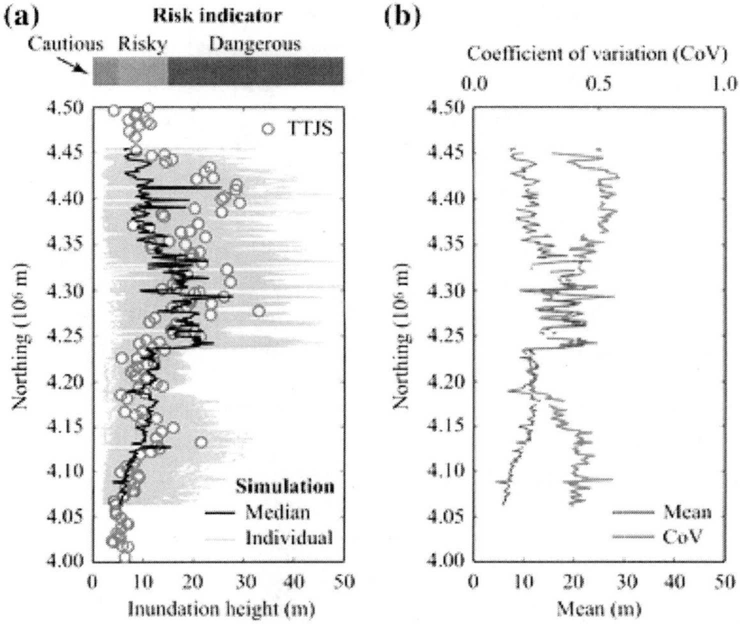

Figure 5. a Variability of the maximum tsunami inundation heights along the coast in comparison with TTJS data, and**b** mean and CoV of the maximum tsunami inundation heights along the coast

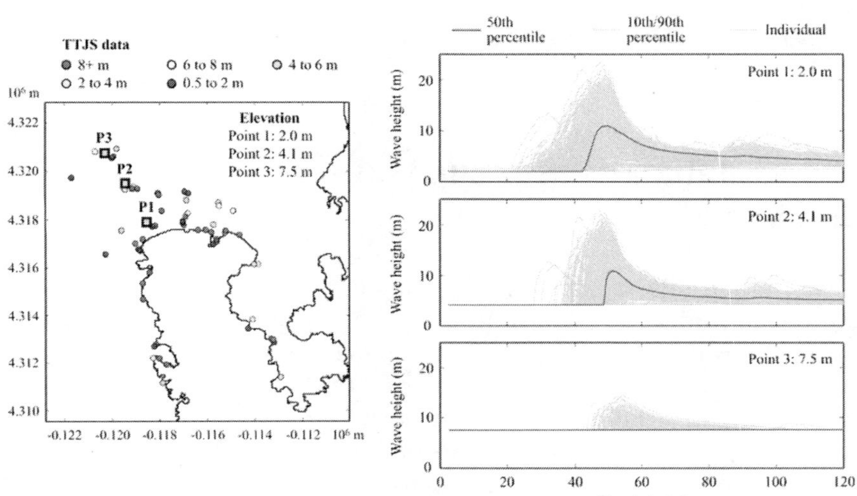

Figure 6. Variability of the simulated tsunami wave profiles at three locations in Rikuzentakata

Moreover, using the Monte Carlo tsunami simulation results for the 726 stochastic slip models, stochastic tsunami hazard maps for an area of interest (i.e. Rikuzentakata) can be developed [see Goda et al. (2014) for stochastic tsunami hazard maps at various locations along the Tohoku coast]. The level of tsunami hazard for a given scenario (i.e. M_w9-class tsunamigenic event in the Tohoku region) can be specified by adopting a relevant tsunami hazard metric. In other words, stochastic tsunami inundation maps show spatial variation of the maximum inundation depth for different situations in terms of a chosen metric, and thus are useful for visualizing the uncertainty of tsunami hazard predictions in the form of cartographic map. An illustration is given in Fig. 7. Figure 7 (middle) displays the probability distribution of inundation area above 1 m depth in Rikuzentakata based on all simulation results (i.e. 726 cases); in this case, the inundation area above 1 m depth is adopted to define the tsunami hazard level by ranking tsunami simulation results (note: different metrics can be adopted). Figure 7 (top) shows inundation depth maps for Rikuzentakata corresponding to three tsunami hazard levels: 50th percentile as a typical case and 10th and 90th percentiles as rare cases. The set of inundation depth maps are useful for communicating with emergency officers and local residents regarding the uncertain tsunami hazard information. For the typical case (50th percentile), the majority of the city center of Rikuzentakata (northern side of the Hirota Bay; see Fig. 3b) is inundated by 8 + m tsunami waves. A similar feature is also observed for the extreme case (90th percentile). In contrast, significantly lower hazard level is indicated in the map for the 10th percentile level, where inundation depth of 8 + m occurs near the shoreline only.

3.3 Visualizing uncertain tsunami damage and loss outcomes

Using applicable tsunami fragility models and damage cost models, inundation depth information from the Monte Carlo tsunami simulations can be transformed into tsunami risk and loss information for a building portfolio (Fig. 1). Figure 7 (bottom) shows the stochastic *wash-away* damage probability maps corresponding to the 10th, 50th, and 90th percentile levels in terms of inundation area above 1 m depth [i.e. Figure 7(middle)]. Similar maps can be developed for different tsunami damage levels. In the maps, *wash-away*damage probabilities for individual buildings in Rikuzentakata (Fig. 3b) are color-coded and displayed. The stochastic damage probability maps are generated by integrating both local tsunami hazard level and physical vulnerability of structures. An advantage of the damage maps over the inundation depth maps [i.e. Fig. 7 (bottom) vs. Fig. 7 (top)] is noticed by comparing the two types of maps for the 10th percentile level. In these maps, a different pattern can be observed in the city center of Rikuzentakata near the shoreline; in the bottom-left map, several red markers are noticeable in the middle of the inundated areas, whereas in the top-left map, such local features are not present. These red markers in the damage probability map correspond to a group of wooden houses in the city center which may be washed out by 4–8 m tsunami waves. Consideration of tsunami vulnerability of structures is useful for developing tsunami emergency response plans for local communities.

By implementing Monte Carlo sampling of the calculated damage probabilities for individual buildings, the cumulative distribution functions of the number of buildings in the *wash-away* damage state can be evaluated for the building portfolio in Rikuzentakata. For instance, when the estimated *wash-away* damage probability of a RC building is 0.2 (estimated using the corresponding tsunami fragility curve and inundation depth for a given tsunami scenario), a uniform random number between 0 and 1 can be generated; if the generated random number is less than 0.2, the RC building is considered as washed out. The procedure can be applied to all buildings in Rikuzentakata, and can be used to develop the cumulative distribution functions of the number of damaged buildings. This transformation of the tsunami risk metric aggregates the tsunami impact to the coastal city/town into the number counts of damaged buildings. Such aggregated tsunami risk metrics are useful for risk management purposes. Figure 8a, b show the cumulative distribution functions of the number of *wash-away* buildings in terms of structural type (RC/steel/wood/masonry) and reference source model (for stochastic slip generation), respectively. Figure 8a shows that the number of *wash-away*buildings is dominated by wooden structures because of the large number of this construction type in Rikuzentakata (see Fig. 3) and their tsunami vulnerability compared with other types of structures. By taking into account the total number of buildings for each structural type, the tsunami damage curves shown in Fig. 8a can be presented in terms of damage ratio. It can be found that about 70 % of wood structures are likely to be washed out at the 50th percentile level, this damage ratio is greater than those for other structural types, indicating greater tsunami vulnerability of wooden construction relative to other construction types (Suppasri et al. 2013).

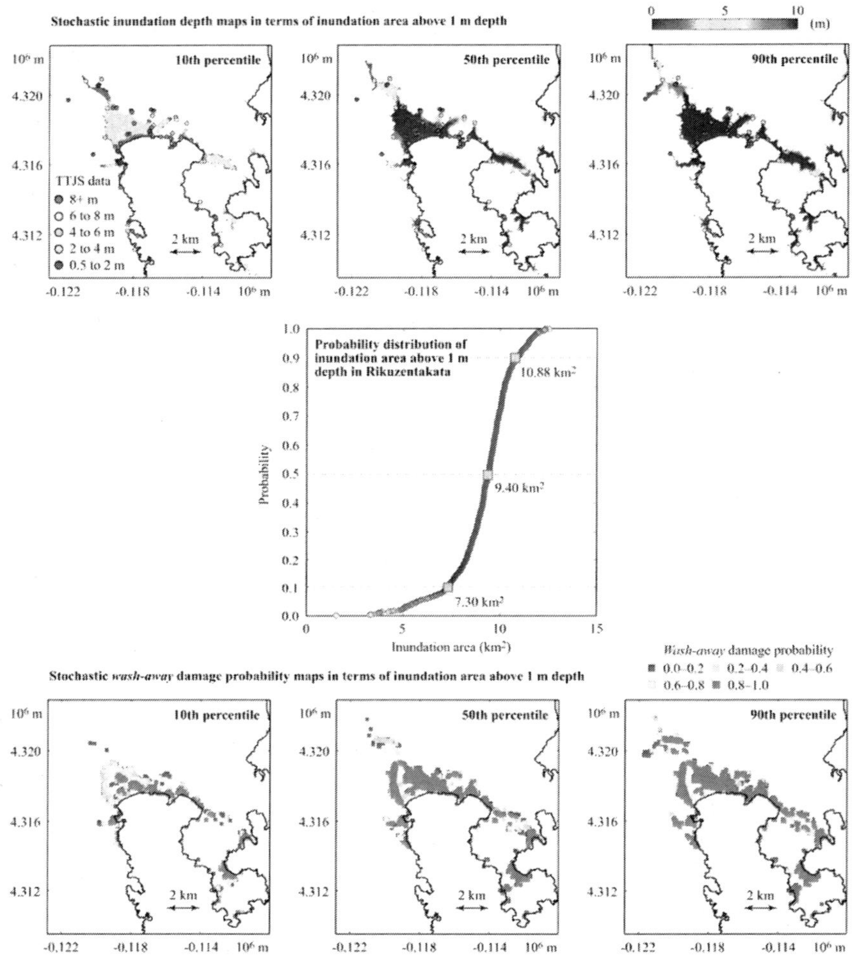

Figure 7. Cumulative distribution function of inundation area above 1 m depth for Rikuzentakata (*middle*), stochastic inundation depth maps at 10th, 50th, and 90th percentile levels (*top*), and stochastic *wash-away* damage probability maps at 10th, 50th, and 90th percentile levels (*bottom*)

Figure 8b reorganizes the same tsunami risk analysis results in terms of reference inversion model for stochastic slip generation (note: the results do not distinguish different structural types). The results facilitate the appreciation of variability of tsunami damage curves due to different tsunami source characteristics. Although damage curves for different source models become more or less similar at high probability levels, wider variation of the damage curves is observed at low probability levels. This is because in worst-case situations, most of the source models place large slips directly off Rikuzentakata, while in the opposite situations, migration of large asperities towards north or south of Rikuzentakata depends on the characteristics of the reference source models. For instance, the difference between the minimum and maximum values of the number of *wash-away* buildings at the 50th percentile level is approximately 1000 with respect to the total number of 6114 buildings. Such results facilitate the quantification of epistemic uncertainty of the tsunami risk due to different source models.

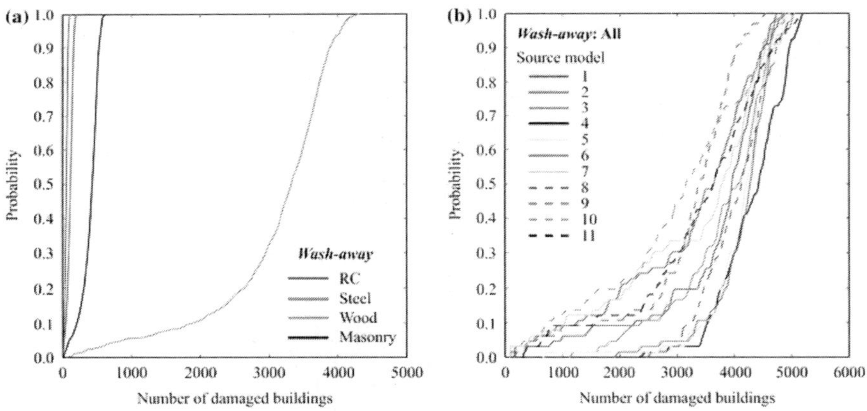

Figure 8. Cumulative distribution functions of the number of buildings in the *wash-away* damage state for: **a** RC, steel, wood, and masonry structures by considering 726 source models, and **b** all structures by distinguishing eleven reference source models

The tsunami damage curves shown in Fig. 8 can be further transformed into the tsunami loss curves by using the damage cost models. Figure 9a, b show the cumulative distribution functions of tsunami loss in terms of structural type and reference source model, respectively. An important difference between Figs. 8and 9 is that in Fig. 9, tsunami loss for all buildings with different damage states is included in the tsunami loss curve by weighting their relative impact based on the tsunami damage cost. This is useful for assessing overall tsunami loss potential for the building portfolio (note: subdivision into different cases/categories is feasible). For instance, Fig. 9a shows that the predicted tsunami loss at the 50th percentile level for RC, steel, masonry and wood structures is approximately 80, 170, 440, and 750 million US$, respectively (in total, 1,440 million US$). It is noteworthy that the tsunami loss curve for masonry structures becomes closer to that for wooden structures (Fig. 9a), in comparison with the counterparts shown in Fig. 8a. This is because in Fig. 9a, tsunami loss for all masonry buildings is taken into account, while in Fig. 8a, masonry buildings in the *wash-away* damage state only is accounted for. In other words, because tsunami damage distribution for different types of structures can be quite different, aggregate risk metrics, such as monetary loss, may be suitable to represent the overall impact. Finally, Fig. 9b shows the similar tendency as Fig. 8b, indicating that the characteristics of the reference source models have major influence in terms of tsunami loss.

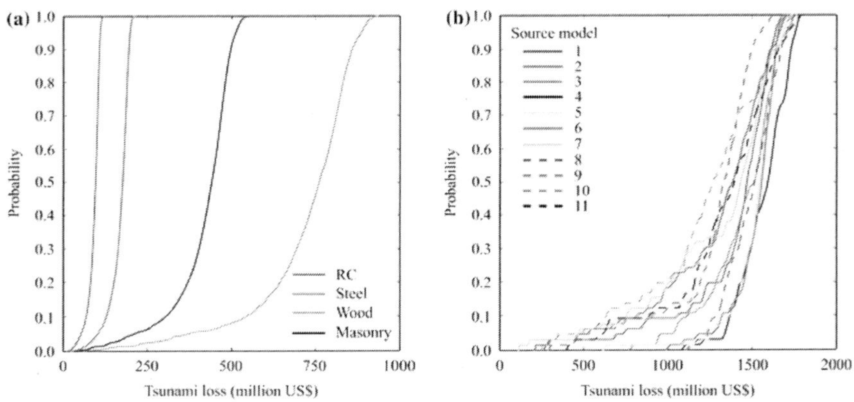

Figure 9. Cumulative distribution functions of tsunami loss for: **a** RC, steel, wood, and masonry structures by considering 726 source models, and **b** all structures by distinguishing eleven reference source models

4. CONCLUSIONS

This paper proposed a computational framework of tsunami hazard and risk assessments for coastal areas by emphasizing the importance of uncertainty modeling and visualization. The improved method for tsunami source characterization generated hundreds of stochastic random-field slip distributions to quantify the uncertainty of tsunami hazard parameters. Subsequently, the uncertainty of tsunami hazard parameters was propagated through tsunami fragility models and damage cost models for an existing building portfolio to evaluate the tsunami damage probability and tsunami loss. Finally, the tsunami hazard and risk assessment results were displayed in various forms by focusing upon temporal and spatial variation of tsunami waves. In particular, stochastic tsunami inundation maps as well as tsunami damage probability maps were useful for succinctly presenting rich information from numerous tsunami simulations. When properly summarized and presented, these tools facilitate the risk communication with emergency officers and residents in a coastal community. In addition, tsunami damage curves in terms of the number of damaged buildings and tsunami loss curves in terms of monetary loss were promoted as effective tools for tsunami risk management. The developed methodology and tools were demonstrated through a case study for the 2011 Tohoku tsunami. Importantly, visualization of variable tsunami hazard and risk assessment results not only conveys a whole picture of tsunami impact to a coastal region but also facilitates deeper understanding of specific aspects of the tsunami hazard and risk assessments. This leads to more proactive and informed decisions for tsunami risk reduction.

ACKNOWLEDGMENTS

The bathymetry and elevation data for the Tohoku region were provided by the Cabinet Office of the Japanese Government. The run-up and inundation survey data were obtained from the 2011 Tohoku Earthquake Tsunami Joint Survey Group (http://www.coastal.jp/tsunami2011/). The tsunami damage data for the 2011 Tohoku earthquake were obtained from the Ministry of Land, Infrastructure, and Transportation (http://www.mlit.go.jp/toshi/toshi-hukkou-arkaibu.html). The national statistics of the construction data in Japan are available from the Ministry of Land, Infrastructure, and Transportation (http://www.mlit.go.jp/toukeijouhou/chojou/stat-e.htm). This work was supported by the Natural Environment Research Council through the Consortium on Risk in the Environment: Diagnostics, Integration, Benchmarking, Learning and Elicitation (CREDIBLE; NE/J017450/1).

REFERENCES

1. Ammon CJ, Lay T, Kanamori H, Cleveland M (2011) A rupture model of the 2011 off the Pacific coast of Tohoku earthquake. Earth Planets Space 63:693–696

2. Borrero J (2005) Field survey of northern Sumatra and Banda Aceh, Indonesia after the tsunami and earthquake of 26 December 2004. Seismol Res Lett 76:309–318

3. Construction Research Institute (2011) Japan building cost information 2011. Tokyo, Japan

4. Federal Emergency Management Agency (FEMA) (2008) Guidelines for design of structures for vertical evacuation from tsunamis. FEMA P646, Washington DC

5. Fraser S, Pomonis A, Raby A, Goda K, Chian SC, Macabuag J, Offord M, Saito K, Sammonds P (2013) Tsunami damage to coastal defences and buildings in the March 11th 2011 M_w9.0 Great East Japan earthquake and tsunami. Bull Earthq Eng 11:205–239

6. Fraser S, Power WL, Wang X, Wallace LM, Mueller C, Johnston DM (2014) Tsunami inundation in Napier, New Zealand, due to local earthquake sources. Nat Hazards 70:415–445

7. Fujii Y, Satake K, Sakai S, Shinohara S, Kanazawa T (2011) Tsunami source of the 2011 off the Pacific coast of Tohoku earthquake. Earth Planets Space 63:815–820

8. Fukutani Y, Suppasri A, Imamura F (2015) Stochastic analysis and uncertainty assessment of tsunami wave height using a random source parameter model that targets a Tohoku-type earthquake fault. Stoch Environ Res Risk Assess.

9. Geist EL (2002) Complex earthquake rupture and local tsunamis. J Geophys Res Solid Earth.

10. Geist EL, Parsons T (2006) Probabilistic analysis of tsunami hazards. Nat Hazards 37:277–314

11. Goda K, Mai PM, Yasuda T, Mori N (2014) Sensitivity of tsunami wave profiles and inundation simulations to earthquake slip and

fault geometry for the 2011 Tohoku earthquake. Earth Planets Space 66:1–20.

12. Goda K, Yasuda T, Mori N, Mai PM (2015) Variability of tsunami inundation footprints considering stochastic scenarios based on a single rupture model: application to the 2011 Tohoku earthquake. J Geophys Res Oceans 120:4552–4575

13. Goto C, Ogawa Y, Shuto N, Imamura F (1997) Numerical method of tsunami simulation with the leap-frog scheme. IOC Manual, UNESCO, No. 35, Paris, France

14. Gusman AR, Tanioka Y, Sakai S, Tsushima H (2012) Source model of the great 2011 Tohoku earthquake estimated from tsunami waveforms and crustal deformation data. Earth Planet Sci Lett 341–344:234–242

15. Hayes GP (2011) Rapid source characterization of the 2011 M_w 9.0 off the Pacific coast of Tohoku earthquake. Earth Planets Space 63:529–634

16. Horspool N, Pranantyo I, Griffin J, Latief H, Natawidjaja DH, Kongko W, Cipta A, Bustaman B, Anugrah SD, Thio HK (2014) A probabilistic tsunami hazard assessment for Indonesia. Nat Hazards Earth Syst Sci 14:3105–3122

17. Iinuma T, Ohzono M, Ohta Y, Miura S (2011) Coseismic slip distribution of the 2011 off the Pacific coast of Tohoku earthquake (M9.0) estimated based on GPS data—was the asperity in Miyagi-oki ruptured? Earth Planets Space 63:643–648

18. Iinuma T, Hino R, Kido M, Inazu D, Osada Y, Ito Y, Ohzono M, Tsushima H, Suzuki S, Fujimoto H, Miura S (2012) Coseismic slip distribution of the 2011 off the Pacific coast of Tohoku earthquake (M9.0) refined by means of seafloor geodetic data. J Geophys Res Solid Earth.

19. Kagan Y, Jackson DD (2013) Tohoku earthquake: a surprise? Bull Seismol Soc Am 103:1181–1194

20. Kawai H, Satoh M, Kawaguchi K, Seki K (2013) Characteristics of the 2011 Tohoku tsunami waveform acquired around Japan by NOWPHAS equipment. Coast Eng J 55:1350008.

21. Koshimura S, Oie T, Yanagisawa H, Imamura F (2009) Developing fragility functions for tsunami damage estimation using numerical model and post-tsunami data from Banda Aceh, Indonesia. Coast Eng J 51:243–273

22. Kunz M, Gret-Regamey A, Hurni L (2011) Visualization of uncertainty in natural hazards assessments using an interactive cartographic information system. Nat Hazards 59:1735–1751

23. Lavallée D, Liu P, Archuleta RJ (2006) Stochastic model of heterogeneity in earthquake slip spatial distributions. Geophys J Int 165:622–640

24. Liu H, Shimozono T, Takagawa T, Okayasu A, Fritz HM, Sato S, Tajima Y (2013) The 11 March 2011 Tohoku tsunami survey in Rikuzentakata and comparison with historical events. Pure Appl Geophys 170:1033–1046

25. Løvholt F, Pedersen G, Bazin S, Kuhn D, Bredesen RE, Harbitz C (2012) Stochastic analysis of tsunami runup due to heterogeneous coseismic slip and dispersion. J Geophys Res Oceans.

26. MacEachren AM, Robinson A, Hopper S, Gardner S, Murray R, Gahegan M, Hetzler E (2005) Visualizing geospatial information uncertainty: what we know and what we need to know. Cartogr Geogr Inf Sci 32:139–160

27. Mai PM, Beroza GC (2002) A spatial random field model to characterize complexity in earthquake slip. J Geophys Res Solid Earth.

28. McCloskey J, Antonioli A, Piatanesi A, Sieh K, Steacy S, Nalbant S, Cocco M, Giunchi C, Huang JD, Dunlop P (2008) Tsunami threat in the Indian Ocean from a future megathrust earthquake west of Sumatra. Earth Planet Sci Lett 265:61–81

29. Ministry of land infrastructure and transportation (MLIT) (2014) Survey of tsunami damage condition. http://www.mlit.go.jp/toshi/toshi-hukkou-arkaibu.html. Accessed 1 July 2014

30. Mori N, Takahashi T, Yasuda T, Yanagisawa H (2011) Survey of 2011 Tohoku earthquake tsunami inundation and run-up. Geophys Res Lett.

31. Mueller C, Power WL, Fraser S, Wang X (2015) Effects of rupture complexity on local tsunami inundation: implications for probabilistic tsunami hazard assessment by example. J Geophys Res Solid Earth 120:488–502.

32. Murata S, Imamura F, Katoh K, Kawata Y, Takahashi S, Takayama T (2010) Tsunami: to survive from tsunami. World Scientific Publishing, New Jersey

33. Okada Y (1985) Surface deformation due to shear and tensile faults in a half-space. Bull Seismol Soc Am 75:1135–1154

34. Pang A (2008) Visualizing uncertainty in natural hazards. Risk Gov Soc 14:261–294

35. Pappenberger F, Stephens E, Thielen J, Slamon P, Demeritt D, van Andel SJ, Wetterhall F, Alfieri L (2013) Visualizing probabilistic flood forecast information: expert preferences and perceptions of best practice in uncertainty communication. Hydrol Process 27:132–146

36. Potter K, Kniss J, Riesenfeld R, Johnson CR (2010) Visualizing summary statistics and uncertainty. Comput Graph Forum 29:823–832

37. Reese S, Bradley BA, Bind J, Smart G, Power WL, Sturman J (2011) Empirical building fragilities from observed damage in the 2009 South Pacific tsunami. Earth Sci Rev 107:156–173

38. Satake K, Fujii Y, Harada T, Namegaya Y (2013) Time and space distribution of coseismic slip of the 2011 Tohoku earthquake as inferred from tsunami waveform data. Bull Seismol Soc Am 103:1473–1492

39. Shao G, Li X, Ji C, Maeda T (2011) Focal mechanism and slip history of the 2011 M_w 9.1 off the Pacific coast of Tohoku earthquake, constrained with teleseismic body and surface waves. Earth Planets Space 63:559–564

40. Suppasri A, Mas E, Charvet I, Gunasekera R, Imai K, Fukutani Y, Abe Y, Imamura F (2013) Building damage characteristics based on surveyed data and fragility curves of the 2011 Great East Japan tsunami. Nat Hazards 66:319–341

41. Tanioka Y, Satake K (1996) Tsunami generation by horizontal displacement of ocean bottom. Geophys Res Lett 23:861–864

42. Tarbotton C, Dall'Osso F, Dominey-Howes D, Goff J (2015) The use of empirical vulnerability functions to assess the response of buildings to tsunami impact: comparative review and summary of best practice. Earth Sci Rev 142:120–134

43. Wiebe DM, Cox DT (2014) Application of fragility curves to estimate building damage and economic loss at a community scale: a case study of Seaside, Oregon. Nat Hazards 71:2043–2061

44. Yamazaki Y, Lay T, Cheung KF, Yue H, Kanamori H (2011) Modeling near-field tsunami observations to improve finite-fault slip models for the 11 March 2011 Tohoku earthquake. Geophys Res Lett.

45. Yoshikawa H, Goda K (2014) Financial seismic risk analysis of building portfolios. Nat Hazards Rev 15:112–120

INDEX